ACID RAIN
AND
DRY DEPOSITION

L. W. Canter

Environmental and Ground Water Institute
University of Oklahoma
Norman, Oklahoma

Library of Congress Cataloging-in-Publication Data

Canter, Larry W.
    Acid rain and dry deposition.

    Includes bibliographies and index.
    1. Acid rain. 2. Acid rain—Bibliography. 3. Acid
deposition. I. Title.
TD196.A25C364   1986   363.7'386   86-7186
ISBN 0-87371-016-9

LEWIS PUBLISHERS, INC.
121 South Main Street, P.O. Drawer 519, Chelsea, Michigan 48118

PRINTED IN THE UNITED STATES OF AMERICA

LARRY W. CANTER, P.E., is the Sun Company Professor of Ground Water Hydrology, and Director, Environmental and Ground Water Institute, at the University of Oklahoma, Norman, Oklahoma. Dr. Canter received his PhD in Environmental Health Engineering from the University of Texas in 1967, MS in Sanitary Engineering from the University of Illinois in 1962, and BE in Civil Engineering from Vanderbilt University in 1961. Before joining the faculty of the University of Oklahoma in 1969, he was on the faculty at Tulane University and was a sanitary engineer in the U.S. Public Health Service. He served as Director of the School of Civil Engineering and Environmental Science at the University of Oklahoma from 1971 to 1979.

Dr. Canter has published several books and has written chapters in other books; he is also the author or co-author of numerous papers and research reports. His research interests include environmental impact assessment and ground water pollution control. In 1982 he received the Outstanding Faculty Achievement in Research Award from the College of Engineering, and in 1983 the Regents' Award for Superior Accomplishment in Research.

Dr. Canter currently serves on the U.S. Army Corps of Engineers Environmental Advisory Board. He has conducted research, presented short courses, or served as advisor to institutions in Mexico, Panama, Colombia, Venezuela, Peru, Scotland, The Netherlands, France, Germany, Italy, Greece, Turkey, Kuwait, Thailand, and the People's Republic of China.

iii

CARNEGIE-MELLON UNIVERSITY
PITTSBURGH, PENNSYLVANIA 15213

PREFACE

Acid rain has become an environmental concern of global importance within the last decade. As additional monitoring data are collected, industrialized nations and industrial areas within developing nations are recognizing the ubiquitous nature of this concern. Due to the movement of air masses over great distances, long-range transport and subsequent deposition of precipitation acidity and dry atmospheric constituents are causing transboundary or transnational pollution to be included on the world's environmental agenda.

The most obvious effects of acid rain are associated with the loss of living components of aquatic ecosystems. This may result from precursor effects of acid rain and dry deposition on soils and various components of terrestrial ecosystems. Much remains to be learned about the longer-term, more subtle effects, such as those which can potentially occur through natural recharge systems for ground water. Many technical and policy decisions are associated with the delineation and implementation of control strategies for acid rain and dry deposition.

A scientific/technical approach should be used in developing an understanding of the causes, consequences, and corrective actions necessary to deal with acid rain and dry deposition. Due to the relative newness of this environmental concern, a large number of technical reports, papers, and conference proceedings are being published annually. This book contains a summary of such published information available in the mid-1980's.

This book is organized into six chapters and eight appendices. Following an introductory chapter, Chapter 2 provides a summary of general information on the acid rain problem. Chapter 3 relates to the atmospheric formation, modeling, and long-range transport of acid rain constituents. Chapter 4 describes the multiple effects of acid rain on both terrestrial and aquatic ecosystems. Strategies for controlling, or at least minimizing, the acid rain problem are discussed in Chapter 5. Dry deposition of atmospheric pollutants is addressed in Chapter 6, with emphasis given to precipitation metals and organics. Abstracts of the 540 included references are divided into eight appendices as follows: general information on acid rain, atmospheric reactions, atmospheric models, long-range transport of air pollutants, effects of acid precipitation, control strategies, dry deposition, and precipitation metals and organics.

The author wishes to express his appreciation to several individuals instrumental in the development of this book. First, Debby Fairchild, Senior Environmental Scientist, Environmental and Ground Water Institute, University of Oklahoma, conducted the computer-based literature searches upon which much of the included information was based. Dr. Jackie Shields, Assistant Professor, University of Houston at Clear Lake, Texas, participated in many discussions with the author relative to a book of this nature. Steve Canter provided assistance in reference organization decisions. Finally, the author is extremely grateful to Mrs. Leslie Rard, Technical Manuscript Specialist, Environmental and Ground Water Institute, for her typing and proofing skills, patience, and dedication to the completion of this effort.

The author also gratefully acknowledges the support and encouragement of the College of Engineering at the University of Oklahoma relative to faculty writing endeavors. Most importantly, the author thanks his family for their patience and understanding.

Larry W. Canter

Sun Company Professor of
   Ground Water Hydrology
University of Oklahoma
Norman, Oklahoma

To Donna, Doug, Steve, and Greg

CONTENTS

LIST OF TABLES

ACID RAIN
AND
DRY DEPOSITION

CHAPTER 1

OVERVIEW OF ACID RAIN AND DRY DEPOSITION

Acid rain has become an environmental concern of global importance within the last decade. This concern has largely been derived from reported losses of living components in aquatic ecosystems, particularly fish in lakes. Deterioration of statuary and building materials, along with damage to forest areas, have also been noted. Acid precipitation is defined as rain or snow with pH values of less than 5.6 (Babich and Davis, 1980). This pH value is used since distilled water in equilibrium with carbon dioxide under laboratory conditions has a pH of 5.65. The incidence and severity of acid precipitation falling over the United States, Canada, and most of western Europe has increased significantly within the last 20 to 25 years. Acid precipitation is associated with industrial and automotive emissions of sulfur oxides, nitrogen oxides, and gaseous hydrogen chloride resulting from the combustion of fossil fuels. These pollutants may remain in the atmosphere for several days, during which they may be transported large distances, sometimes exceeding 500 miles, before being deposited on water or land surfaces.

Atmospheric pollutants can be returned to the earth's surface through wet or dry deposition. Wet deposition via precipitation may involve rainout or washout. Rainout refers to the incorporation of atmospheric pollutants into cloud drops that grow in size and then fall to the earth's surface. Washout refers to the process whereby atmospheric pollutants below clouds are swept out by precipitation from above. Dry deposition involves the removal of atmospheric pollutants via gravitational settling or direct contact with the earth's surface components. Metals in the atmosphere are frequently removed by dry deposition.

OBJECTIVE AND SCOPE OF BOOK

The objective of this book is to summarize information on key technical literature relating to the causes and consequences of acid precipitation and dry deposition. In addition, information on mitigative strategies for these problem issues will also be presented. It is recognized that the publication of new information is a continuing process; therefore, this book represents a summary of information available in the mid-1980's. In providing this current summary, it should be noted that detailed aspects of addressing all issues associated with acid rain and dry deposition are not included. Annotated bibliographies in eight appendices provide key information on 540 references. These bibliographies can point the user to more detailed information.

The primary method used to identify the 540 selected references included the conduction of computer-based searches of the literature and systematic review of the procured abstracts. Descriptor words utilized in the computer-based searches included:

acid rain   environmental impacts
acid precipitation   terrestrial environment

1

        wet deposition   ecosystem effects
        dry deposition   aquatic environment
        rainout   effects
        washout   control measures
        environmental effects

    Over 800 references were identified and screened for inclusion in this book.

ORGANIZATION OF BOOK

    This book is organized into six chapters and eight appendices. Following this introductory chapter, the remaining five chapters summarize information on general issues, atmospheric reactions and transport, effects of acid precipitation, control strategies for acid rain, and dry deposition and precipitation metals and organics. Table 1 identifies the eight appendices and number of references associated with each. This information provided the basis for the five substantive chapters. It should be noted that each of the 540 references were placed in only one appendix, thus making a decision necessary on the key feature of each reference. Because of this, references pertinent to a particular issue could be found in other appendices. For example, some references included in Appendix A on general information may also be relevant to the effects of acid precipitation as addressed in Appendix E.

Table 1: Summary of Selected Reference Materials

Category	Number of Selected References	Pertinent Chapter
General information on acid rain (Appendix A)	159	2
Atmospheric reactions (Appendix B)	30	3
Atmospheric models (Appendix C)	42	3
Long-range transport of air pollutants (Appendix D)	19	3
Effects of acid precipitation (Appendix E)	181	4
Control strategies (Appendix F)	28	5
Dry deposition (Appendix G)	13	6

Table 1: (continued)

Category	Number of Selected References	Pertinent Chapter
Precipitation metals and organics (Appendix H)	68	6

SELECTED REFERENCE

Babich, H. and Davis, D.L., "Acid Precipitation--Part 1--Causes and Consequences", Environment, Vol. 22, No. 4, May 1980, pp. 6-13, 40-41.

CHAPTER 2

GENERAL ISSUES OF ACID RAIN

There are a number of background or general issues associated with acid rain concerns. Examples include studies of the historical occurrences of acid rain in Europe and the United States, the primary causes of acid rain formation, and geographical occurrences of acid rain. In addition, there are a number of emerging issues and concerns, including monitoring, transboundary and local management of air pollution, and research needs. This chapter summarizes each of these issues following a background section highlighting a summary of the abstracts contained in Appendix A.

BACKGROUND INFORMATION

This chapter is based on information from the 159 references whose abstracts are contained in Appendix A. Table 2 provides a brief comment on each of these references in the order in which they appear in Appendix A. Not every reference listed in Table 2 will be specifically addressed in this chapter.

Table 2: References in Appendix A Containing General Information on Acid Rain

Author(s) (Year)	Comments
Ahern and Baird (1983)	Study of acid precipitation in southeast Wyoming.
Alexander (1978)	Environmental costs from coal production and consumption.
Ambler (1978)	Effects of coal-fired power plant on agriculture in Montana.
Anderson and Landsberg (1978)	Study of precipitation pH variations per millimeter of precipitation.
Assam and Corrada (1976)	Bibliography of 1,470 references on atmospheric precipitation chemistry.
Babich and Davis (1980)	Description of causes of acid precipitation and its consequences for terrestrial and aquatic ecosystems.
Balson, Boyd and North (1982)	Framework for the analysis of decisions on acid deposition affecting the electric utility industry.

Table 2: (continued)

Author(s) (Year)	Comments
Beamish and Van Loon (1977)	Precipitation loading of acid and heavy metals in a Canadian lake.
Bilonick and Nichols (1983)	Temporal variations in acid precipitation in New York.
Bowersox and DePena (1980)	Analysis of precipitation chemistry in central Pennsylvania.
Brocksen (1980)	Description of Electric Power Research Institute programs on acid precipitation.
Brosset (1975)	Description of acid particulate air pollutants in Sweden.
Brosset, Andreasson and Ferm (1975)	Nature and origin of acid particulate air pollutants in Sweden.
Brosset and Ferm (1976)	Measurement of man-made airborne acidity.
Brown (1980)	Bibliography of 154 references on precipitation washout.
Budiansky (1981)	General issues and problems associated with understanding acid rain.
Burns, Galloway and Hendrey (1981)	Acidification of surface waters in upland watersheds in New Hampshire and North Carolina.
Carter (1979)	Acid rain-related problems from SO_2 emissions from existing power plants.
Chamberlain et al. (1981)	Overview of causes and transport of acid precipitation in the United States.
Church (1978)	Chemistry of acid rain in Delaware.
Coghill and Likens (1974)	Acid precipitation in New York and other northeastern states.
Cowling (1982)	Chronological development of acid precipitation concerns from both scientific and public issue perspectives.
Cowling (1983)	General description of causes and consequences of acid rain and atmospheric deposition.

Table 2: (continued)

Author(s) (Year)	Comments
Curtis (1981)	Sources of the precursors of acid rain in North America.
Davies (1979)	Precipitation scavenging of SO_2 and SO_4 in a rural and urban area in the United Kingdom.
Dawson (1978)	Ionic composition of rain in Arizona.
Deland (1980)	Description of pH of 214 lakes in the Adirondack region of New York.
DeWalle et al. (1983)	Nature and extent of acid snowpack chemistry in Pennsylvania.
Dillon et al. (1978)	Acid precipitation and lake water pH in south central Ontario in Canada.
Drabloes and Tollan (1980)	Abstracts of 134 papers presented at an international conference in Norway on the ecological impact of acid precipitation.
Duncan (1981)	Acid precipitation in the Cascade Mountains in Washington.
Eaton, Likens and Bormann (1980)	Deposition and reactions of sulfur in a northern hardwood forest ecosystem.
Ellis, Verfaillie and Kummerow (1983)	Nutrients in acid rain and dry deposition in southern California.
Ember (1979)	International cooperation on transboundary air pollution.
Ember (1980)	Discussions of 30 states relative to controlling the acid rain problem.
Evans (1981)	Scientific information basic to developing an air quality standard to protect terrestrial vegetation from acidic precipitation.
Evans et al. (1981)	Describes relationships between precipitation acidity and receptor effects.
Feeley and Liljestrand (1983)	Sources of acid rain chemical composition in Texas.

Table 2: (continued)

Author(s) (Year)	Comments
Forland and Gjessing (1975)	Washout-rainout and dry deposition in western Norway.
Frey (1980)	Bibliography of 138 references on acid precipitation.
Fuzzi, Orsi and Mariotti (1983)	Fog water pH in the Po Valley in Italy.
Galloway (1978)	Temporal and spatial trends of sulfur in wet deposition in the eastern United States.
Galloway and Cowling (1978)	Proposed precipitation chemistry network.
Galloway and Likens (1981)	Study of nitric acid in acid rain in New Hampshire.
Galloway, Likens and Edgerton (1976)	Contribution of sulfuric and nitric acid to acid precipitation in the northeastern United States.
Galloway and Whelpdale (1980)	Atmospheric sulfur budgets for eastern North America.
Galvin et al. (1979)	Sources and transport of sulfate in New York.
Gannon (1978)	Description of general acid rain concerns.
GCA Corporation (1981)	Summary of acid rain problems and research needs.
General Accounting Office (1981)	Debate over immediate control needs versus conduction of additional research on acid rain.
Gibson (1978)	Description of acid rain research needs.
Gibson (1980)	Data management system for atmospheric chemistry monitoring network of the National Atmospheric Deposition Program.
Glover et al. (1980)	Influence of sulfur dioxide concentrations on the pH of rainfall.

Table 2: (continued)

Author(s) (Year)	Comments
Glustrom and Stolzenberg (1982)	Issues paper on the management of acid rain in Wisconsin.
Grennfelt, Bengston and Skaerby (1978)	Atmospheric input of acidifying substances to a forest ecosystem in Sweden.
Gunnerson and Willard (1979)	Proceedings of papers on a conference session on acid rain sponsored by the American Society of Civil Engineers.
Haines (1983)	Inputs and outputs of SO_4-S for seven terrestrial ecosystems in North and South America.
Hales and Slinn (1975)	Atmospheric cleansing and recovery processes.
Handa, Kumar and Goel (1982)	Chemical composition of rainwater in Lucknow, India.
Hendry and Brezonik (1980)	Chemistry of rainfall and dry deposition in Gainesville, Florida.
Hibbard (1982)	Benefits from, costs required, and research needed to reduce acid precipitation.
Hileman (1982)	Summaries of papers presented at an American Chemical Society symposium on acid precipitation.
Hileman (1983a)	Summary of a 1982 European conference on acidification of the environment.
Hileman (1983b)	Summary of key issues and needs related to developing solutions to the acid rain problem.
Hill et al. (1977)	Measurement of biogenic sulfur emission fluxes.
Hinds (1983)	Nature and extent of acid deposition in western North America.
Hitchcock (1975)	Biogenic contributions to atmospheric sulfate levels in the eastern part of the United States.

Table 2: (continued)

Author(s) (Year)	Comments
Hoban (1982)	Interstate aspects of acid rain management in the United States.
Hornbeck (1981)	Delineation of environmental concerns from acid rain.
Interagency Task Force on Acid Precipitation (1982a)	First annual report on the national acid precipitation assessment plan.
Interagency Task Force on Acid Precipitation (1982b)	Description of the national acid precipitation assessment plan of the U.S. Department of Energy.
Interagency Task Force on Acid Precipitation (1983)	Annual report focusing on policy questions related to the acid precipitation problem.
Johannes, Altwicker and Clesceri (1981)	Atmospheric inputs to three lake watersheds in the Adirondack Park region of New York.
Jones and Mirra (1979)	Costs and benefits of acid rain reduction in terms of energy and transportation needs.
Khan (1980)	Rainfall acidity in Bangkok, Thailand.
Kish (1981)	Review of historical information on acid precipitation.
Knudson (1981)	Sulfate and nitrate contributions to acidic deposition.
Kramer (1975)	Contributions of acid precipitation to metal and nutrient loadings in southern Lake Michigan.
Lakhani and Miller (1980)	Rainwater contributions of sodium and potassium to a forest in Scotland.
Lee et al. (1978)	Characterization of acid precipitation by acid-base potentiometric titration.
Leonard, Goldman and Likens (1981)	Chemical analyses of precipitation at Davis and Lake Tahoe, California.
Li and Landsbergn (1975)	Rainwater pH near a power plant in southern Maryland.

Table 2: (continued)

Author(s) (Year)	Comments
Likens (1976)	Description of causes of acid precipitation and its effect on terrestrial ecosystems.
Likens et al. (1979)	Historical information on acid rain in Europe and the United States.
Likens and Butler (1981)	Historical information on acidification of precipitation in North America.
Likens, Edgerton and Galloway (1983)	Organic carbon in wet precipitation in New Hampshire and New York.
Liljestrand and Morgan (1978)	Chemical composition of rainfall in Pasadena, California.
Madsen (1981)	Acid rain at the Kennedy Space Center in Florida.
Manzig and Mulvaney (1982)	Basis for legal regulation of acid rain in Canada.
Martin (1978)	Acidity and sulfur in rainwater in rural areas of central England and Wales.
Martin (1984)	Sulfate and nitrate variations between strong- and weakly-acid rainwater.
McColl (1980)	Wet and dry atmospheric precipitation surveys in northern California.
McColl (1981)	Metals in atmospheric precipitation in a California forest.
McColl and Bush (1978)	Chemical composition of precipitation in the San Francisco Bay area, California.
McNaughton (1981)	Sulfate and nitrate contributions to rainfall pH in the northeastern United States.
McQuaker, Kluckner and Sandberg (1983)	Interferences with pH and acidity measurements in rainfall.
McWilliams and Musante (1982)	Description of Acid Rain Information System data base.
Melo (1981)	Acid rain monitoring network operated by Ontario Hydro in Canada.

Table 2: (continued)

Author(s) (Year)	Comments
Miller (1975)	Rationale for collection of precipitation chemistry data.
Miller (1979)	Suggestions for coordinated acid rain monitoring effort in the United States.
Mohnen (1982)	Weekly versus event basis sampling of acid precipitation.
Moody (1978)	Potential acid rain impacts of Canadian power plant.
Morgan and Liljestrand (1980)	Chemical characteristics of acid precipitation in the Los Angeles area, California.
Nagamoto et al. (1983)	Chemical composition of rain and snow clouds in eastern Colorado.
National Research Council (1981)	Needed research on the ecological consequences of fossil fuel combustion.
National Technical Information Service (1981)	Bibliography of 380 references on acid rain and snow and acidification of soils and fresh and salt water bodies.
National Technical Information Service (1982a)	Bibliography of 289 references on the causes, effects, and control of wet and dry acid precipitation.
National Technical Information Service (1982b)	Bibliography of 156 references on various aspects of acid precipitation.
National Technical Information Service (1982c)	Bibliography of 115 references on the measurement and analysis of acid precipitation effects.
National Technical Information Service (1982d)	Bibliography of 169 references on the sources, ecological effects, and economic effects of acid precipitation.
National Technical Information Service (1982e)	Bibliography of 110 references on the political and legal aspects of acid rain pollution.
National Technical Information Service (1983)	Bibliography of 213 references on the removal of pollutants, radioactive isotopes, and dust by rain and snow.

Table 2: (continued)

Author(s) (Year)	Comments
Newman, Likens and Bormann (1975)	Combustion of fossil fuels as a cause of acidity in rainwater.
Oden and Thorsten (1980)	Mass balance of atmospheric fallout of sulfur in Sweden.
Okland (1980)	Historical information on acidification in 50 Norwegian lakes.
O'Sullivan (1976)	Transboundary transport of acid precipitation to Norway.
Patrick, Binetti and Halterman (1981)	Factors causing increases in acid emissions from fossil fuel burning.
Pellett, Bustin and Harriss (1984)	Variability of acid precipitation in Hampton, Virginia.
Perhac (1981)	Acid rain resulting from coal burning by utility companies.
Popp et al. (1980)	Chemical characteristics of precipitation in south-central New Mexico.
Poundstone (1980)	Upwind emissions as a cause of acid rain.
Pratt (1983)	Need for long-term monitoring data on acid precipitation and effects.
Pratt, Coscio and Krupa (1984)	Chemical characteristics of rainfall in Minnesota and west-central Wisconsin.
Rippon (1980)	Field studies on soil effects of acid rain in the United Kingdom.
Robertson, Dolzine and Graham (1980)	Intensity-weighted sequential sampling of precipitation to determine variations in chemical composition.
Robinson and Ghane (1983)	Acidity of rainwater in Baton Rouge, Louisiana.
Rodhe, Soderlund and Ekstedt (1980)	Wet and dry deposition of airborne pollutants in the Baltic Sea and in nearby Sweden.
Roffman (1980)	Survey of pH levels of precipitation in Pittsburgh, Pennsylvania.

Table 2: (continued)

Author(s) (Year)	Comments
Rosencranz (1980)	Technical and policy issues related to transboundary air pollution.
Rosencranz and Wetstone (1980)	Discussion of international effects of acid rain.
Rosenfeld (1978)	General information on causes and effects of acid rain.
Rosenquist (1978)	Potential cause of river acidification in Norway.
Seip (1980)	Snowmelting as a cause of acidification of soil and natural waters in Norway.
Seymour et al. (1978)	Measurement of the acid components of rainwater in five rainfall events.
Seymour and Stout (1983)	Changes in the chemical composition of rainfall during a single rainfall event.
Sharp (1980)	Summary of an acid rain conference held in Norway.
Shinn and Lynn (1979)	Natural and anthropogenic sources of sulfur in the environment.
Spaite (1980)	Local oil burning and automotive sources contributing to acid rain.
Suleymanov and Listengarten (1966)	Chemical composition of precipitation in Russia.
Szabo, Esposito and Spaite (1982)	Discussion of a broad range of acid rain issues.
Tanaka, Darzi and Winchester (1980)	Chemical composition of rainfall in rural north Florida.
Tennessee Valley Authority (1980)	Description of acid rain research program of the Tennessee Valley Authority.
Trumbule and Tedeschi (1983)	Description of the broad ranges of acid rain literature.
Turk (1983)	Historical trends of the acidification of lakes and streams in the northeastern United States.

Table 2: (continued)

Author(s) (Year)	Comments
Tyree (1981)	Analytical methods for determining the acidity of rainwater.
Ulrich (1980)	Procedure for estimating anthropogenic and natural sources of hydrogen ions in terrestrial ecosystems.
U.S. Environmental Protection Agency (1978)	Environmental effects and costs of air pollutant emissions from coal combustion.
U.S. Environmental Protection Agency (1980a)	Proceedings of an acid rain conference held in Virginia.
U.S. Environmental Protection Agency (1980b)	Background information on the causes and consequences of acid rain.
Vermeulen (1978)	Historical information on acid rain in the Netherlands.
Wagner and Steele (1983)	Nutrients and acid in wet and dry fallout in Fayetteville, Arkansas.
Wetstone (1980)	Delineation of potential solutions to the acid rain problem.
Wetstone (1984)	Institutional framework needed for addressing transboundary pollution between the United States and Canada.
Whelpdale (1972)	Pollution of large bodies of water by airborne pollutants.
Wilson, Mohnen and Kadlecek (1980)	Wet deposition of airborne pollutants in the northeastern United States.
Wisniewski (1982)	Acid precipitation concerns in dews, frosts, and fogs.
Wisniewski and Keitz (1983)	Acid rain deposition patterns throughout the continental United States.
Wisniewski and Kinsman (1982)	Review of acid rain monitoring activities and studies in North America.
Wright and Dovland (1978)	Four regional surveys of the chemistry of snowpacks in Norway.

There has been a tremendous growth in acid rain literature within recent years, including bibliographies, data bases, journals, newsletters, books and monographs, congressional and Federal agency documents, industry-sponsored research, and state agency reports (Trumbule and Tedeschi, 1983).   For example, a total of 20 of the 159 references in Appendix A are bibliographies or proceedings of technical conferences on acid rain; these 20 are listed in Table 3.   Computerized bibliographic retrieval systems are available through the National Technical Information Service of the U.S. Department of Commerce, and a specialized Acid Rain Information System (ARIS).   The subject areas covered by ARIS include: (1) the mechanisms of formation of acid rain; (2) its transport phenomena; (3) its effects on materials; (4) its effects on plants; (5) the health effects of acid rain; and (6) monitoring and analysis of acid rain. Data in ARIS comes from several government and commercial data base producers, and these include EDB DOE Energy Database, Environmental Science Index, Air Pollution Abstracts, National Technical Information Service, and articles of regional interests from various newspapers (McWilliams and Musante, 1982).   An extensive specialized bibliography of 1,470 references on atmospheric precipitation chemistry is an example of a noncomputer based reference document (Assam and Corrada, 1976).   To serve as an example of a specific bibliographic report, Wisniewski and Kinsman (1982) summarized acid rain monitoring associated with 71 studies conducted or on-going in North America. Tables are presented that describe the name or title of the study, the organization or agency that funds each study, the chemical parameters monitored, the geographic extent and location of the study, the time period of operation, the types of samples used, where samples are analyzed, and a contact for further information.

Table 3: Bibliographies and Conference Proceedings in Appendix A

Author(s) (Year)	Comments
Assam and Corrada (1976)	Bibliography of 1,470 references on atmospheric precipitation chemistry.
Brown (1980)	Bibliography of 154 references on precipitation washout.
Drabloes and Tollan (1980)	Abstracts of 134 papers presented at an international conference in Norway on the ecological impact of acid precipitation.
Frey (1980)	Bibliography of 138 references on acid precipitation.
GCA Corporation (1981)	Summary of acid rain problems and research needs.
Gunnerson and Willard (1979)	Proceedings of papers on a conference session on acid rain sponsored by the American Society of Civil Engineers.

Table 3: (continued)

Author(s) (Year)	Comments
Hileman (1982)	Summaries of papers presented at an American Chemical Society symposium on acid precipitation.
Hileman (1983a)	Summary of a 1982 European conference on acidification of the environment.
McWilliams and Musante (1982)	Description of Acid Rain Information System data base.
National Technical Information Service (1981)	Bibliography of 380 references on acid rain and snow and acidification of soils and fresh and salt water bodies.
National Technical Information Service (1982a)	Bibliography of 289 references on the causes, effects, and control of wet and dry acid precipitation.
National Technical Information Service (1982b)	Bibliography of 156 references on various aspects of acid precipitation.
National Technical Information Service (1982c)	Bibliography of 115 references on the measurement and analysis of acid precipitation effects.
National Technical Information Service (1982d)	Bibliography of 169 references on the sources, ecological effects, and economic effects of acid precipitation.
National Technical Information Service (1982e)	Bibliography of 110 references on the political and legal aspects of acid rain pollution.
National Technical Information Service (1983)	Bibliography of 213 references on the removal of pollutants, radioactive isotopes, and dust by rain and snow.
Sharp (1980)	Summary of an acid rain conference held in Norway.
Trumbule and Tedeschi (1983)	Description of the broad ranges of acid rain literature.
U.S. Environmental Protection Agency (1980a)	Proceedings of an acid rain conference held in Virginia.

Table 3: (continued)

Author(s) (Year)	Comments
Wisniewski and Kinsman (1982)	Review of acid rain monitoring activities and studies in North America.

Many national and international conferences have addressed acid rain. Examples of proceedings include those from conferences in the United States in 1978 (Gunnerson and Willard, 1979), 1980 (U.S. Environmental Protection Agency, 1980a), and 1982 (Hileman, 1982); and from conferences in Norway in 1980 (Drabloes and Tollan, 1980; and Sharp, 1980), and Sweden in 1982 (Hileman, 1983a). The 1982 conference in Stockholm, Sweden, consisted of three international meetings. The first was on the ecological effects of acid deposition, and the second was on methods to control sulfur and nitrogen oxide emissions. The third conference was a ministerial meeting of government representatives. The ministers' conference adopted a consensus statement asserting that further concrete action is needed to control transboundary air pollution and called for concerted international programs for the reduction of sulfur emissions.

HISTORICAL CONCERNS AND TRENDS

Acid rain is one aspect of the general concerns associated with man's influence on the climate of the earth. Many features of the acid rain phenomenon were first discovered in the mid-1800s in and around the city of Manchester, England (Cowling, 1982). Concern about acid precipitation and its ecological effects in North America developed first in Canada and later in the United States, initially centering on the effects of sulfur dioxide exposure and associated acid precipitation and heavy metal deposition in the vicinity of metal smelting and sintering operations, especially near Sudbury, Ontario.

Likens et al. (1979) have reviewed the history of data gathering concerning acid rain in Europe and the United States. Analysis of precipitation occurring prior to the industrial revolution, and preserved in glaciers or continental ice sheets, has revealed a pH of greater than 5.0 (Likens and Butler, 1981; and Kish, 1981). Remote areas such as Greenland and Antarctica have a pH of about 5.5 for precipitation currently falling. However, rain and snow falling over certain regions of the world are currently 5 to 30-40 times more acid than the lowest value expected for unpolluted atmospheres. This phenomenon is observed to be widespread in nonurban as well as urban areas throughout most of eastern North America and western Europe. Annual pH values average about 4.0 to 4.5 in parts of the northeast United States and southeast Canada and northwest Europe. Individual storms have been recorded at pH 3.0 (Likens and Butler, 1981).

Turk (1983) noted that acidification of lakes and streams in the northeastern United States has occurred in a time frame compatible with the hypothesis that acidification of precipitation was the cause. The acidification of surface waters appears to have occurred before the mid- to late 1960s. In southeastern Canada, the best-documented cases of acidified lakes point to localized sources of acidic emissions as the cause. In the southeastern United States, most data on acidification of surface waters are ambiguous, and in the west, most of the data reflect local conditions. However, recent analysis of a national network of remote stream sampling stations indicates that, since the mid- to late 1960s, sulfate concentrations have increased in the southeast and the west, with a concurrent decrease in alkalinity.

CAUSES OF ACID RAIN

Acid rain is typically associated with industrial and automotive emissions of sulfur oxides, nitrogen oxides, and gaseous hydrogen chloride resulting from the combustion of fossil fuels (Babich and Davis, 1980). Information on the causes of acid rain is contained in 28 references in Appendix A; these 28 are listed in Table 4.

Table 4: References in Appendix A Dealing with the Causes of Acid Rain

Author(s) (Year)	Comments
Babich and Davis (1980)	Description of causes of acid precipitation and its consequences for terrestrial and aquatic ecosystems.
Carter (1979)	Acid rain-related problems from SO_2 emissions from existing power plants.
Chamberlain et al. (1981)	Overview of causes and transport of acid precipitation in the United States.
Cowling (1983)	General description of causes and consequences of acid rain and atmospheric deposition.
Curtis (1981)	Sources of the precursors of acid rain in North America.
Galloway, Likens and Edgerton (1976)	Contribution of sulfuric and nitric acid to acid precipitation in the northeastern United States.
Galloway and Whelpdale (1980)	Atmospheric sulfur budgets for eastern North America.
Glover et al. (1980)	Influence of sulfur dioxide concentrations on the pH of rainfall.
Haines (1983)	Inputs and outputs of SO_4-S for seven terrestrial ecosystems in North and South America.

Table 4: (continued)

Author(s) (Year)	Comments
Hales and Slinn (1975)	Atmospheric cleansing and recovery processes.
Hill et al. (1977)	Measurement of biogenic sulfur emission fluxes.
Hitchcock (1975)	Biogenic contributions to atmospheric sulfate levels in the eastern part of the United States.
Knudson (1981)	Sulfate and nitrate contributions to acidic deposition.
Li and Landsbergn (1975)	Rainwater pH near a power plant in southern Maryland.
Likens (1976)	Description of causes of acid precipitation and its effect on terrestrial ecosystems.
Martin (1984)	Sulfate and nitrate variations between strong and weakly acid rainwater.
McNaughton (1981)	Sulfate and nitrate contributions to rainfall pH in the northeastern United States.
Newman, Likens and Bormann (1975)	Combustion of fossil fuels as a cause of acidity in rainwater.
Patrick, Binetti and Halterman (1981)	Factors causing increases in acid emissions from fossil fuel burning.
Perhac (1981)	Acid rain resulting from coal burning by utility companies.
Poundstone (1980)	Upwind emissions as a cause of acid rain.
Rosenfeld (1978)	General information on causes and effects of acid rain.
Spaite (1980)	Local oil burning and automotive sources contributing to acid rain.
Ulrich (1980)	Procedure for estimating anthropogenic and natural sources of hydrogen ions in terrestrial ecosystems.

Table 4: (continued)

Author(s) (Year)	Comments
U.S. Environmental Protection Agency (1980b)	Background information on the causes and consequences of acid rain.

Many studies have indicated that atmospheric sulfur oxides and nitrogen oxides are the major causes of acid precipitation (rain and snow). For example, detailed chemical analyses have revealed that acid precipitation in the northeastern United States is caused by two strong mineral acids: sulfuric and nitric. There is a large array of other proton sources in precipitation, including weak and bronsted acids. Although these other acids contribute to the total acidity of precipitation, their influence on the free acidity of acid precipitation is minimal (Galloway, Likens and Edgerton, 1976).

Several studies have been made to develop specific relationships between sulfates, nitrates, and rainfall pH. For example, precipitation chemistry data for the northeastern United States were analyzed to test relationships between anthropogenic sulfate and nitrate ion contributions and rainfall pH. The data set incorporated all observations, with complete measurements of all three ion concentrations being collected from September, 1976, through December, 1979. A relationship was found between sulfates and nitrates and rainfall acidity expressed in terms of pH, and this relationship can be applied to deposition predictions to predict pH from regional air pollutant transport models (McNaughton, 1981). In addition, Martin (1984) indicated that sulfate and nitrate in rainwater vary systematically with the acidities of the samples from certain rural sites.

The relative contributions of sulfates and nitrates as causes of acid rain have been estimated. Knudson (1981) indicated that sulfate accounts for some 60 to 70 percent of the precipitation acidity in the northeastern United States. The contribution due to nitrates in this region increased from about 22 percent in the mid-1950s to about 30 percent in 1973. However, in other regions deposition acidity may be attributable to a much different combination of chemical species, and may even be dominated by an acidic constituent other than sulfate or nitrate (Knudson, 1981). Carter (1979) suggested that sulfur dioxide and the fine sulfate mist into which SO_2 is readily transformed are the principal percursors of acid rain.

As noted earlier, airborne constituents other than sulfates and nitrates can contribute, albeit small, to the formation of acid rain. For example, wet-only precipitation collected at Hubbard Brook (HB), New Hampshire, a heavily forested site, and Ithaca, New York, from June, 1976, to May, 1977, was analyzed for several types of organic carbon (Likens, Edgerton and Galloway, 1983). Particulate plus dissolved macromolecular (greater than 1000 molecular weight) organic matter accounted for 51 percent and 63 percent of the total organic carbon at HB and Ithaca, respectively. The remainder of the carbon, the fraction containing compounds of less than 1000 molecular weight, consisted largely of carboxylic acids, aldehydes, carbohydrates, tannin/lignin, primary amines, and phenols. Sources of the organic materials were airborne soil dust and plant material. The organic acids present in the rain at HB contributed only 2 percent of the total acidity.

Atmospheric constituents which may contribute to acid precipitation can occur from anthropogenic and/or natural (biogenic) sources. Shinn and Lynn (1979) have reviewed the natural and anthropogenic input sources of sulfur. They noted that although the natural input of sulfur exceeds that of anthropogenic sources on a global scale, the environment cannot adapt nor absorb sudden, high-rate anthropogenic loadings of sulfur compounds. A technique for the measurement of sulfur emission fluxes has been described by Hill et al. (1977). In addition, Ulrich (1980) presented a general scheme for estimating anthropogenic and natural sources of H ions in terrestrial ecosystems. An excellent and extensive series of chemical equations is presented.

In northeastern North America, anthropogenic emissions of sulfur dioxide and nitrogen oxides were found to exceed those from natural sources, having increased significantly since 1940. In Canada and Ontario, industrial processes and transportation are by far the major sources of SO_2 and NO_x, respectively. However, electric utilities are major emitters of both these acid gases in the United States (Curtis, 1981). Coal-burning power plants are identified as the major pollutant source category (Poundstone, 1980). Although the total amount of coal used in the United States has not significantly increased since 1920, changes in use patterns and pollution technology have altered the nature and quantity of stack emissions (Patrick, Binetti and Halterman, 1981). Factors contributing to more acid emissions from fossil fuel burning are: (1) increase in the proportion of coal consumed by utilities; (2) gradual increase in use of fly ash precipitators, which remove the neutralizing effect of ash upon sulfur and nitrogen oxides; (3) the fact that sulfur scrubbers are only required on coal-burning plants built since 1975; (4) increase in the height of stacks, which introduces emissions high into the atmosphere where they can be readily transported for long distances; (5) use of hotter combustion flames and resultant increase in nitrogen oxides emitted since 1975; and (6) increase in use of petroleum in utilities and transportation.

Acid rain has often been assumed to be a problem resulting from the long-range transport of pollutants from large fossil fuel combustion sources, namely coal-fired utilities. However, acid rain can be measured within the local environs of major sources. For example, the distribution of pH of rain from heavy summer showers was measured in a network of rain gauges 0.5-5.0 km from a 710 MW power plant in southern Maryland. Values ranged from 3.0-5.7, with modal values from 3.6-4.0. A notable wind dependence of the acidic washout from the plume was found. In calm conditions during a shower, the values dropped off concentrically from the source (Li and Landsbergn, 1975).

Spaite (1980) suggested that local oil burning and automotive sources may be major contributors to the occurrence of acid rain in the Adirondacks, Florida, and California. Oil-fired boilers, especially the smaller commercial, industrial, and residential units, produce at least 3 to 10 times as much primary sulfate per unit of sulfur content as coal-fired units. Moreover, oil-fired units emit comparatively large quantities of catalytic compounds capable of rapidly converting still more sulfur oxide to sulfate in the atmosphere. Thus, in areas where large quantities of oil are burned, the direct impact from locally generated sulfates may equal or even exceed that produced by imported sulfates derived from distant coal-burning sources. Fuel consumption data show that large quantities of oil are being consumed in areas experiencing acid rain. Forty percent of the residual and 36 percent of the distillate oil burned in the United States is consumed in the eight-state area surrounding the Adirondacks. California is the next largest oil-consuming area and Florida is third. Nitric acid is responsible for about 30 percent of rainfall acidity in the northeast and Florida, and for about 30 to 75 percent of the rainfall acidity in California.

GEOGRAPHICAL OCCURRENCES OF ACID RAIN

Observations on the geographical occurrences of acid rain are typically associated with northern European countries, the northeastern United States, and southeastern Canada. Acid rain has been observed in many other locations in the United States and throughout the world. From a geographical perspective, acid rain is a wide-spread problem. This section highlights many locations wherein acid rain has been measured; however, it is not intended to be all-inclusive in that many studies were either not identified or not classified into this portion of Chapter 2.

United States and Canada

Table 5 lists 37 references with abstracts in Appendix A which describe acid rain occurrences in the conterminous United States. A total of 20 states are specifically listed, with their sectional distribution as follows: northeast (Delaware, New Hampshire, New York, and Pennsylvania); southeast (Florida, Louisiana, North Carolina, and Virginia); midwest (Michigan, Minnesota, and Wisconsin); southwest (Arizona, Arkansas, New Mexico, and Texas); west (California, Colorado, Montana, and Wyoming); and northwest (Washington).

Table 5: References in Appendix A Describing Acid Rain Occurrences in the United States

Author(s) (Year)	Comments
Ahern and Baird (1983)	Study of acid precipitation in southeast Wyoming.
Ambler (1978)	Effects of coal-fired power plant on agriculture in Montana.
Bilonick and Nichols (1983)	Temporal variations in acid precipitation in New York.
Bowersox and DePena (1980)	Analysis of precipitation chemistry in central Pennsylvania.
Burns, Galloway and Hendrey (1981)	Acidification of surface waters in upland watersheds in New Hampshire and North Carolina.
Church (1978)	Chemistry of acid rain in Delaware.

Table 5: (continued)

Author(s) (Year)	Comments
Coghill and Likens (1974)	Acid precipitation in New York and other northeastern states.
Dawson (1978)	Ionic composition of rain in Arizona.
Deland (1980)	Description of pH of 214 lakes in the Adirondack region of New York.
DeWalle et al. (1983)	Nature and extent of acid snowpack chemistry in Pennsylvania.
Duncan (1981)	Acid precipitation in the Cascade Mountains in Washington.
Ellis, Verfaillie and Kummerow (1983)	Nutrients in acid rain and dry deposition in southern California.
Feeley and Liljestrand (1983)	Sources of acid rain chemical composition in Texas.
Galloway (1978)	Temporal and spatial trends of sulfur in wet deposition in the eastern United States.
Galloway and Likens (1981)	Study of nitric acid in acid rain in New Hampshire.
Galvin et al. (1979)	Sources and transport of sulfate in New York.
Hendry and Brezonik (1980)	Chemistry of rainfall and dry deposition in Gainesville, Florida.
Hinds (1983)	Nature and extent of acid deposition in western North America.
Johannes, Altwicker and Clesceri (1981)	Atmospheric inputs to three lake watersheds in the Adirondack Park region of New York.
Kramer (1975)	Contributions of acid precipitation to metal and nutrient loadings in southern Lake Michigan.
Leonard, Goldman and Likens (1981)	Chemical analyses of precipitation at Davis and Lake Tahoe, California.
Liljestrand and Morgan (1978)	Chemical composition of rainfall in Pasadena, California.

Table 5: (continued)

Author(s) (Year)	Comments
Madsen (1981)	Acid rain at the Kennedy Space Center in Florida.
McColl (1980)	Wet and dry atmospheric precipitation surveys in northern California.
McColl (1981)	Metals in atmospheric precipitation in a California forest.
McColl and Bush (1978)	Chemical composition of precipitation in the San Francisco Bay area, California.
Morgan and Liljestrand (1980)	Chemical characteristics of acid precipitation in the Los Angeles area, California.
Nagamoto et al. (1983)	Chemical composition of rain and snow clouds in eastern Colorado.
Pellett, Bustin and Harriss (1984)	Variability of acid precipitation in Hampton, Virginia.
Popp et al. (1980)	Chemical characteristics of precipitation in south-central New Mexico.
Pratt, Coscio and Krupa (1984)	Chemical characteristics of rainfall in Minnesota and west-central Wisconsin.
Robinson and Ghane (1983)	Acidity of rain water in Baton Rouge, Louisiana.
Roffman (1980)	Survey of pH levels of precipitation in Pittsburgh, Pennsylvania.
Tanaka, Darzi and Winchester (1980)	Chemical composition of rainfall in rural north Florida.
Wagner and Steele (1983)	Nutrients and acid in wet and dry fallout in Fayetteville, Arkansas.
Wilson, Mohnen and Kadlecek (1980)	Wet deposition of airborne pollutants in the northeastern United States.
Wisniewski and Keitz (1983)	Acid rain deposition patterns throughout the continental United States.

Wisniewski and Keitz (1983) reported on both pH and H(+) deposition in precipitation for the continental United States and southern Canada during the late 1970's.   They used data from approximately 100 stations in nine precipitation chemistry networks.   In spite of a wide variety of collection methods and sampling intervals, they found a remarkable uniformity in the average pH among the various stations.   Precipitation remains most acidic in the heart of the northeastern section of the United States.   The next highest level of precipitation acidity occurs in the surrounding area, which includes the Ohio Valley and southern Ontario regions.   Evidence of precipitation acidity is now appearing in Florida, the Colorado Rockies and in population centers along the west coast, where pH values at or below 5 are now recorded.   Numerous studies of Canadian acid rain have been made, examples include Beamish and Van Loon (1977), Dillon et al. (1978), and Moody (1978).

Europe

Historically, acid rain was first recognized as a problem in Europe, hence there have been many studies of this issue.   Table 6 lists 17 references with abstracts in Appendix A which describe acid rain occurrences in Europe. Countries mentioned include Norway (Forland and Gjessing, 1975; Okland, 1980; O'Sullivan, 1976; Rosenquist, 1978; Seip, 1980; and Wright and Dovland, 1978); Sweden (Brosset, 1975; Brosset, Andreasson and Ferm, 1975; Grennfelt, Bengston and Skaerby, 1978; Oden and Thorsten, 1980; and Rohde, Soderlund and Ekstedt, 1980); the United Kingdom (Davies, 1979; Lakhani and Miller, 1980; Martin, 1978; and Rippon, 1980); the Netherlands (Vermeulen, 1978); and Italy (Fuzzi, Orsi and Mariotti, 1983).

Table 6: References in Appendix A Describing Acid Rain Occurrences in Europe

Author(s) (Year)	Comments
Brosset (1975)	Description of acid particulate air pollutants in Sweden.
Brosset, Andreasson and Ferm (1975)	Nature and origin of acid particulate air pollutants in Sweden.
Davies (1979)	Precipitation scavenging of SO_2 and SO_4 in a rural and urban area in the United Kingdom.
Forland and Gjessing (1975)	Washout-rainout and dry deposition in western Norway.
Fuzzi, Orsi and Mariotti (1983)	Fog water pH in the Po Valley in Italy.
Grennfelt, Bengston and Skaerby (1978)	Atmospheric input of acidifying substances to a forest ecosystem in Sweden.
Lakhani and Miller (1980)	Rainwater contributions of sodium and potassium to a forest in Scotland.

Table 6: (continued)

Author(s) (Year)	Comments
Martin (1978)	Acidity and sulfur in rainwater in rural areas of central England and Wales.
Oden and Thorsten (1980)	Mass balance of atmospheric fallout of sulfur in Sweden.
Okland (1980)	Historical information on acidification in 50 Norwegian lakes.
O'Sullivan (1976)	Transboundary transport of acid precipitation to Norway.
Rippon (1980)	Field studies on soil effects of acid rain in the United Kingdom.
Rodhe, Soderlund and Ekstedt (1980)	Wet and dry deposition of airborne pollutants in the Baltic Sea and in nearby Sweden.
Rosenquist (1978)	Potential cause of river acidification in Norway.
Seip (1980)	Snowmelting as a cause of acidification of soil and natural waters in Norway.
Vermeulen (1978)	Historical information on acid rain in the Netherlands.
Wright and Dovland (1978)	Four regional surveys of the chemistry of snowpacks in Norway.

EMERGING ISSUES AND CONCERNS

Acid rain monitoring, issues related to local and transboundary acid rain problems and management, and needed research will be addressed in this portion of Chapter 2. These issues are not all-inclusive; however, they do represent examples of major areas of concern in effectively addressing acid rain problems.

Acid Rain Monitoring

In order to effectively identify and manage acid rain problems, extensive monitoring information is necessary. While monitoring has increased in recent years, there are still many geographical areas for which monitoring has been sparse. For example, Wisniewski and Keitz (1983) noted that many areas in the western United States have limited acid rain data. Table 7 lists 15 references with abstracts in Appendix A which address a variety of monitoring issues.

Table 7: References in Appendix A Describing Acid Rain Monitoring

Author(s) (Year)	Comments
Anderson and Landsberg (1978)	Study of precipitation pH variations per millimeter of precipitation.
Brosset and Ferm (1976)	Measurement of man-made airborne acidity.
Budiansky (1981)	General issues and problems associated with understanding acid rain.
Galloway and Cowling (1978)	Proposed precipitation chemistry network.
Gibson (1980)	Data management system for atmospheric chemistry monitoring network of the National Atmospheric Deposition Program.
Melo (1981)	Acid rain monitoring network operated by Ontario Hydro in Canada.
Miller (1975)	Rationale for collection of precipitation chemistry data.
Miller (1979)	Suggestions for coordinated acid rain monitoring effort in the United States.
Mohnen (1982)	Weekly versus event basis sampling of acid precipitation.
McQuaker, Kluckner and Sandberg (1983)	Interferences with pH and acidity measurements in rainfall.
Pratt (1983)	Need for long-term monitoring data on acid precipitation and effects.
Robertson, Dolzine and Graham (1980)	Intensity-weighted sequential sampling of precipitation to determine variations in chemical composition.
Seymour and Stout (1983)	Changes in the chemical composition of rainfall during a single rainfall event.
Seymour et al. (1978)	Measurement of the acid components of rain water in five rainfall events.
Tyree (1981)	Analytical methods for determining the acidity of rainwater.

Galloway and Cowling (1978) discussed the need for a network of monitoring stations in the United States and Canada, with the need in part based on the changing chemistry of precipitation and its effects on terrestrial and aquatic ecosystems. A review of early United States monitoring efforts by a number of government and private organizations has been prepared, with the need for coordinated efforts noted (Miller, 1979). Long-term monitoring will be required to accurately determine the causes and consequences of acid rain (Miller, 1979; and Pratt, 1983).

A National Atmospheric Deposition Program atmospheric chemistry monitoring network has been in operation in the 1980's, and when it was organized it became evident that there was a need for a data management system to store and make readily available the extensive data base that would be generated (Gibson, 1980). Such a system has been developed to handle data from United States atmospheric monitoring networks, as well as non-U.S. networks, for example, the Canadian CANSAP program. To serve as an example of a specific monitoring network, Ontario Hydro in Canada initiated a monitoring program in 1975 to establish some facts about their sulfur dioxide emissions and the resultant atmospheric sulfate aerosols (Melo, 1981). The monitoring effort has since been expanded to include precipitation chemistry measurements, in response to concerns about the ecological effects of acid rain. Monitoring results have been presented and discussed in terms of influencing variables such as meteorology and sulphur dioxide source distribution, and compared with information from other networks operating in northeastern North America (Melo, 1981).

One of the most important technical issues in acid rain monitoring is related to the influence of meteorological factors on the time variations of precipitation pH (Budiansky, 1981; and Miller, 1975). Based on a study at three sites in or near College Park, Maryland, Anderson and Landsberg (1978) noted that the precipitation pH, with a mean of 4.7, was not normally distributed when measured per millimeter of precipitation. In many events a strong variation of pH was evident. In the course of the study, pH values ranged from 2.7 to 6.9 units. Similar findings have been reported by Seymour and Stout (1983). During the summer of 1979, 20 different rainfall events were collected and the pH of each sequentially collected sample was measured. The pH of the individual samples varied between 5.8 and 3.6. The most rapid decrease in pH occurred during the early portion of the rainfall event, while only small increases or decreases in the pH occurred during the latter portion of the event. To overcome this phenomenon, Robertson, Dolzine and Graham (1980) have described an intensity-weighted sequential sampler that excludes dry deposition. Their experiments have shown that intensity-weighted sequential sampling is a viable technique for monitoring the rapid changes in precipitation chemistry within a storm.

Anthropogenic airborne acidity can be determined using Gran's titration method (Brosset and Ferm, 1976). Microtitrimetric and coulometric methods have been used to determine the acid components of rain water (Seymour et al. 1978). Three methods which are currently used to determine the acidity of rainwater have been recently studied (Tyree, 1981). The first method relies on pH measurement, the second on titration, and the third assumes that any charge discrepancy in favor of anions is due to the hydronium ion concentration. In this third method only strong acid is measured, and the method is used only for those solutions at or below a pH value of 5.6. The method requires the determination of concentrations of all principal ions in the rainwater sample, usually considered to be chloride, nitrate, sulfate, ammonium, sodium, potassium, calcium, and magnesium. The accuracy and reproducibility of such a method are no better than the properly combined accuracy and reproducibility of each of the principal solute ion values.   The first method, the pH measurement, measures the concentration of strong acids plus some of the weak acids in rainwater. However, the accuracy and reproducibility are not sufficient. Similarly, the accuracy and reproducibility of the titration method are also unacceptable.   The results of two collaborative intercomparison laboratory studies on simulated rainwater show that adequate analytical reproducibility may not be attained.

Local and Transboundary Problems and Management

There are many institutional and policy issues associated with the identification and management of acid rain concerns. Basic to many of these issues is the phenomena of long-range transport of air pollution; this transport may create conditions where the effects or consequences of air rain are experienced far from the sources or precursors of the acid rain. Chapter 5 will provide additional related information in terms of source control measures. Table 8 lists 17 references with abstracts in Appendix A which address a range of issues, including national approaches, state approaches, and transboundary concerns from inter-country transport.

Table 8:   References in Appendix A Describing Local and Transboundary Management Concerns

Author(s) (Year)	Comments
Alexander (1978)	Environmental costs from coal production and consumption.
Balson, Boyd and North (1982)	Framework for the analysis of decisions on acid deposition affecting the electric utility industry.
Ember (1979)	International cooperation on trans-boundary air pollution.
Ember (1980)	Discussions of 30 states relative to controlling the acid rain problem.
Evans (1981)	Scientific information basic to developing an air quality standard to protect terrestrial vegetation from acidic precipitation.
Glustrom and Stolzenberg (1982)	Issues paper on the management of acid rain in Wisconsin.
Hoban (1982)	Interstate aspects of acid rain management in the United States.

Table 8: (continued)

Author(s) (Year)	Comments
Interagency Task Force on Acid Precipitation (1982a)	First annual report on the national acid precipitation assessment plan.
Interagency Task Force on Acid Precipitation (1982b)	Description of the national acid precipitation assessment plan of the U.S. Department of Energy.
Interagency Task Force on Acid Precipitation (1983)	Annual report focusing on policy questions related to the acid precipitation problem.
Jones and Mirra (1979)	Costs and benefits of acid rain reduction in terms of energy and transportation needs.
Manzig and Mulvaney (1982)	Basis for legal regulation of acid rain in Canada.
Rosencranz (1980)	Technical and policy issues related to transboundary air pollution.
Rosencranz and Wetstone (1980)	Discussion of international effects of acid rain.
U.S. Environmental Protection Agency (1978)	Environmental effects and costs of air pollutant emissions from coal combustion.
Wetstone (1980)	Delineation of potential solutions to the acid rain problem.
Wetstone (1984)	Institutional framework needed for addressing transboundary pollution between the United States and Canada.

A national acid rain program was initiated in the United States with the passage of the Acid Precipitation Act of 1980 (Title VII of P.L. 96-294). The Act created the Interagency Task Force on Acid Precipitation to plan and implement the National Acid Precipitation Assessment Program (NAPAP). The Act mandates that the Task Force issue an Annual Report to the President and the Congress on the status of the national program (Interagency Task Force on Acid Precipitation, 1982a). A five-part national plan for addressing acid precipitation has also been developed (Interagency Task Force on Acid Precipitation, 1982b). Part I describes the authorizing legislation and the organization of the Plan itself. Previous federal planning efforts in the current acid rain research programs are discussed. Part II presents a general overview of current understanding of the phenomenon and consequences of acid rain.

Subjects discussed include: evidence of trends in rainfall acidity; possible sources of acid precipitation; atmospheric chemistry and transport; monitoring of acid deposition; the effects of acid precipitation; the relationship of acid precipitation to human health; and control technologies and mitigation of impacts.   Part III identifies nine categories of information needed to enhance the ability to make sound energy, environmental, economic, and resource policy decisions as: (a) natural sources; (b) man-made sources; (c) atmospheric processes; (d) deposition monitoring; (e) aquatic impacts; (f) terrestrial impacts; (g) effects on materials and cultural resources; (h) control technologies; and (i) assessment and policy analysis. Possible indirect effects on human health are addressed in both categories (e) and (f). Part IV presents the research proposed to address these information needs. The management and coordination of the Program are described in Part V.

On occasion, national policies may be developed which have implications for the acid rain issue. One example is related to policies directed toward converting to a coal-based economy. Alexander (1978) noted several environmental costs from major coal production and consumption, including potential environmental damages from acid rain.   The U.S. Environmental Protection Agency (1978) evaluated the ecological and environmental effects of gaseous emissions and aerosols that result from coal combustion. Sulfur oxide emissions and nitrogen oxide emissions were projected to be higher in 1985 and 2000 than they were in 1975.   Since SO_x and NO_x are major contributors to acid precipitation, substantial increases in total acid deposition can be expected in the nation as a whole.

Canada has been developing legislative approaches for dealing with acid precipitation (Manzig and Mulvaney, 1982).   The Canadian Clean Air Act, especially with the 1980 amendments, lays the basis for possible adoption of emission standards for acid rain as an international pollutant.   Ontario's Environmental Protection Act of 1971 may offer a basis for control of emissions resulting in acid rain through regulation of major sources and control orders for medium sized and small sources.

Numerous states within the United States are also concerned with controlling the acid rain problem (Ember, 1980). Potential state and/or national options for controlling acid rain include the following: (1) set tighter tall-stack regulations; (2) improve enforcement and enforcement monitoring; (3) set tighter parameters on atmospheric models to yield more stringent emissions limitations; (4) set an upper limit on pollutants; (5) require universal coal washing at coal-using facilities; (6) establish SO_2 and NO_2 taxes; (6) encourage least emissions rather than least cost dispatching of electric utilities, thereby retiring older, less efficient plants; and (7) encourage favorable tax breaks for facilities that reduce emissions. The technical and policy background for a state program in Wisconsin has been described by Glustrom and Stolzenberg (1982).

Management of inter-country transport of acid precipitation is one of the major current issues of concern in Europe and northeastern North America. Inter-state transport within the United States is also of concern. Future re-authorizations of the Clean Air Act Amendments of 1977 are expected to address inter-state air pollution (Hoban, 1982).

The international effects of acid rain have been discussed by Rosencranz and Wetstone (1980). In 1979, a total of 33 nations signed a pact in Geneva,

Switzerland, to fight transboundary air pollution (Ember, 1979). Various bilateral and internatinoal responses and control strategies have been highlighted by Rosencranz (1980). Over the past several years the Environmental Law Institute in the United States has conducted a series of studies of the institutional framework governing control of international pollution problems. A major conclusion of this research effort is that current national and international legal structures are poorly suited to the effective control of transboundary pollution. Despite the recent increase in the number and severity of international environmental problems, governments continue to make energy policy, pollution control, and land use decisions without explicit consideration of transboundary impacts (Wetstone, 1984).

Management decisions related to acid precipitation require the integration of both technical and policy information. Balson, Boyd and North (1982) have proposed a decision framework intended as a means of summarizing scientific information and uncertainties on the relation between emissions from electric utilities and other sources, acid deposition, and impacts on ecological systems. The methodology for implementing the framework is that of decision analysis, which provides a quantitative means of analyzing decisions under uncertainty. The decisions of interest include reductions in sulfur oxide and other emissions thought to be precursors of acid deposition, mitigation of acid deposition impacts through means such as liming of waterways and soils, and choice of strategies for research. The state of scientific information and the modeling assumptions for the framework are discussed for the three main modules of the framework: emissions and control technologies; long-range transport and chemical conversion in the atmosphere; and ecological impacts. The authors then present two versions of a decision tree model that implements the decision framework. The basic decision tree addresses decisions on emissions control and mitigation in the immediate future and a decade hence, and it includes uncertainties in the long-range transport and ecological impacts. The research emphasis decision tree addresses the effect of research funding on obtaining new information as the basis for future decisions.

Research Needs

Table 9 lists 11 references with abstracts in Appendix A which address basic research needs and programs on acid precipitation. More research is needed on essentially every technical and policy aspect of the acid rain problem: examples include the occurrence of acid precipitation (Hileman, 1983b); measurement of acid precipitation (Lee et al., 1978); effects of acid precipitation (Eaton, Likens and Bormann, 1980; and Evans et al., 1981); transport, transformation, and deposition of acid pollutants (Szabo, Esposito and Spaite, 1982); cost-effective control measures (General Accounting Office, 1981; and Hibbard, 1982); and long-term monitoring and forecasting of future effects of these pollutants, and ecotoxicology (National Research Council, 1981). The research program of the Electric Power Research Institute is described by Brocksen (1980). The Tennessee Valley Authority is presently engaged in active research in rainfall monitoring, atmospheric movement of pollutants, and laboratory and field studies of vegetation, soils, and surface waters to define present conditions and indicate possible future trends. The long-range transport, chemical transformation, and ultimate fate of air pollutants and the effect of acid rain on soil chemistry is also being investigated (Tennessee Valley Authority, 1980).

Table 9: References in Appendix A Addressing Research Needs

Author(s) (Year)	Comments
Brocksen (1980)	Description of Electric Power Research Institute programs on acid precipitation.
Eaton, Likens and Bormann (1980)	Deposition and reactions of sulfur in a northern hardwood forest ecosystem.
Evans et al. (1981)	Describes relationships between precipitation acidity and receptor effects.
General Accounting Office (1981)	Debate over immediate control needs versus conduction of additional research on acid rain.
Gibson (1978)	Description of acid rain research needs.
Hibbard (1982)	Benefits from, costs required, and research needed to reduce acid precipitation.
Hileman (1983b)	Summary of key issues and needs related to developing solutions to the acid rain problem.
Lee et al. (1978)	Characterization of acid precipitation by acid-base potentiometric titration.
National Research Council (1981)	Needed research on the ecological consequences of fossil fuel combustion.
Szabo, Esposito and Spaite (1982)	Discussion of a broad range of acid rain issues.
Tennessee Valley Authority (1980)	Description of acid rain research program of the Tennessee Valley Authority.

SELECTED REFERENCES

Ahern, J. and Baird, C., "Acid Precipitation in Southeastern Wyoming", Sept. 1983, Water Resources Research Institute, University of Wyoming, Laramie, Wyoming.

Alexander, T., "New Fears Surround the Shift to Coal", Fortune, Vol. 98, No. 10, Nov. 1978, pp. 50-56.

Ambler, M., "Power Emissions May Reduce AG Productivity", High Country News, Vol. 10, No. 15, July 28, 1978, pp. 1-4.

Anderson, D.E. and Landsberg, H.E., "Detailed Structure of pH in Hydrometeors", Environmental Science and Technology, Vol. 13, No. 8, Sept. 1978, pp. 992-994.

Assam, W.R.H. and Corrada, L.A., "Bibliography on Precipitation Chemistry", 1976, Utrecht University, Netherlands.

Babich, H. and Davis, D.L., "Acid Precipitation--Part 1--Causes and Consequences", Environment, Vol. 22, No. 4, May 1980, pp. 6-13, 40-41.

Balson, W.E., Boyd, D.W. and North, D.W., "Acid Deposition: Decision Framework--Volume 1--Description of Conceptual Framework and Decision-Tree Models", Aug. 1982, Decision Focus, Inc., Palo Alto, California.

Beamish, R.J. and Van Loon, J.C., "Precipitation Loading of Acid and Heavy Metals to a Small Acid Lake Near Sudbury, Ontario", Fisheries Research Board of Canada Journal, Vol. 34, No. 5, May 1977, pp. 649-659.

Bilonick, R.A. and Nichols, D.G., "Temporal Variations in Acid Precipitation Over New York State--What the 1965-1979 USGS Data Revealed", Atmospheric Environment, Vol. 17, No. 6, 1983, pp. 1063-1072.

Bowersox, V.C. and DePena, R.G., "Analysis of Precipitation Chemistry at a Central Pennsylvania Site", Journal of Geophysical Research, Vol. 85, No. C10, Oct. 1980, pp. 5614-5620.

Brocksen, R.W., "Acid Precipitation: The Complexity of a Perceived Problem", Feb. 1980, Electric Power Research Institute, Palo Alto, California.

Brosset, C., "Acid Particulate Air Pollutants in Sweden", EPA International Environmental Document, Report 30781A-Sweden, Feb. 1975, U.S. Environmental Protection Agency, Washington, D.C.

Brosset, C., Andreasson, K. and Ferm, M., "The Nature and Possible Origin of Acid Particles Observed at the Swedish West Coast", Atmospheric Environment, Vol. 6, No. 7, Sept. 1975, pp. 631-643.

Brosset, C. and Ferm, M., "Man-Made Airborne Activity and Its Determination", 1976, Swedish Water and Air Pollution Research Laboratory, Goteborg, Sweden.

Brown, R.J., "Precipitation Washout--1964-June 1980 (A Bibliography with Abstracts)", June 1980, National Technical Information Service, U.S. Department of Commerce, Springfield, Virginia.

Budiansky, S., "Understanding Acid Rain", Environmental Science and Technology, Vol. 15, No. 6, June 1981, pp. 623-624.

Burns, D.A., Galloway, J.N. and Hendrey, G.R., "Acidification of Surface Waters in Two Areas of the Eastern United States", Water, Air, and Soil Pollution, Vol. 16, No. 3, Oct. 1981, pp. 277-285.

Carter, L.J., "Uncontrolled SO_2 Emissions Bring Acid Rain", Science, Vol. 204, No. 4398, June 15, 1979, pp. 1179-1182.

Chamberlain, J. et al., "Physics and Chemistry of Acid Precipitation", Nov. 1981, SRI International, Arlington, Virginia.

Church, T.M., "Chemistry of Acid Rain at Lewes, Delaware as Part of the Precipitation Chemistry Network of MAP3S", Contract W-7405-ENG-48, 1978, Delaware University-College of Marine Studies, Lewes, Delaware.

Coghill, C.V. and Likens, G.E., "Acid Precipitation in the N.E. United States", Water Resources Research, Vol. 10, No. 6, Dec. 1974, pp. 1133-1137.

Cowling, E.B., "Acid Precipitation in Historical Perspective", Environmental Science and Technology, Vol. 16, No. 2, Feb. 1982, pp. 110A-123A.

Cowling, E.B., "Discovering the Causes, Consequences, and Implications of Acid Rain and Atmospheric Deposition", Tappi Journal, Vol. 66, No. 9, Sept. 1983, pp. 43-46.

Curtis, K.E., "Emissions of Acid Rain Precursors in North America", Research Review Ontario Hydro, May 1981, pp. 5-10.

Davies, T.D., "Dissolved Sulphur Dioxide and the Sulphate in Urban and Rural Precipitation", Atmospheric Environment, Vol. 13, 1979, pp. 1275-1285.

Dawson, G.A., "Ionic Composition of Rain During Sixteen Consecutive Showers", Atmospheric Environment, Vol. 12, No. 10, 1978, pp. 1991-2001.

Deland, M.R., "Acid Rain", Environmental Science and Technology, Vol. 14, No. 6, June 1980, p. 657.

DeWalle, D.R. et al., "Nature and Extent of Acid Snow Packs in Pennsylvania", Feb. 1983, Institute for Research on Land and Water Resources, Pennsylvania State University, University Park, Pennsylvania.

Dillon, P.J. et al., "Acid Precipitation in South-Central Ontario: Recent Observations", Fisheries Research Board of Canada Journal, Vol. 35, No. 6, June 1978, pp. 809-816.

Drabloes, D. and Tollan, A., "Proceedings of International Conference on the Ecological Impact of Acid Precipitation", Oct. 1980, Sandefjord, Norway.

Duncan, L.C., "Acid Precipitation in the Washington Cascades", Sept. 1981, Department of Chemistry, Central Washington University, Ellensburg, Washington.

Eaton, J.S., Likens, G.E. and Bormann, F.H., "Wet and Dry Deposition of Sulfur at Hubbard Brook", Effects of Acid Precipitation on Terrestrial Ecosystems, Plenum Press, New York, New York, 1980, pp. 69-75.

Ellis, B.A., Verfaillie, J.R. and Kummerow, J., "Nutrient Gain from Wet and Dry Atmospheric Deposition and Rainfall Acidity in the Southern California Chaparral", Oecologia, Vol. 60, No. 1, 1983, pp. 118-121.

Ember, L.R., "Acid Rain Focus on International Cooperation", Chemical and Engineering News, Dec. 3, 1979, pp. 15-17.

Ember, L., "States Anguish Over Acid Rain Problem", Chemical and Engineering News, Apr. 21, 1980, pp. 22-23.

Evans, L.S., "Considerations of an Air-Quality Standard to Protect Terrestrial Vegetation from Acidic Precipitation", 1981, Laboratory of Plant Morphogenesis, Manhattan College, Bronx, New York.

Evans, L.S. et al., "Acidic Precipitation: Considerations for an Air Quality Standard", Water, Air, and Soil Pollution, Vol. 16, No. 4, 1981, pp. 469-509.

Feeley, J.A. and Liljestrand, H.M., "Source Contributions to Acid Precipitation in Texas", Atmospheric Environment, Vol. 17, No. 4, 1983, pp. 807-814.

Forland, E.J. and Gjessing, Y.T., "Snow Contamination from Washout-Rainout and Dry Deposition", Atmospheric Environment, Vol. 9, No. 3, Mar. 1975, pp. 339-353.

Frey, J.H., "Acid Precipitation, June, 1976-May, 1980 (Citations from the Energy Data Base)", June 1980, National Technical Information Service, U.S. Department of Commerce, Springfield, Virginia.

Fuzzi, S., Orsi, G. and Mariotti, M., "pH of Fog", Journal of Aerosol Science, Vol. 14. No. 3, 1983, pp. 298-301.

Galloway, J.N., "Sulfur Deposition in the Eastern United States", Air Pollution Control Association Mid-Atlantic States Section Semi-Annual Technical Conference, Philadelphia, Pennsylvania, Apr. 13-14, 1978, pp. 101-115.

Galloway, J.N. and Cowling, E.B., "The Effects of Precipitation on Aquatic and Terrestrial Ecosystems: A Proposed Precipitation Chemistry Network", Air Pollution Control Association Journal, Vol. 28, No. 3, Mar. 1978, pp. 229-236.

Galloway, J.N. and Likens, G.E., "Acid Precipitation: The Importance of Nitric Acid", Atmospheric Environment, Vol. 15, No. 6, 1981, pp. 1081-1085.

Galloway, J.N., Likens, G.E. and Edgerton, E.S., "Acid Precipitation in the Northeastern United States: pH and Acidity", Science, Vol. 194, No. 4266, Nov. 12, 1976, pp. 722-725.

Galloway, J.N. and Whelpdale, D.M., "Atmospheric Sulfur Budget for Eastern North America", Atmospheric Environment, Vol. 14, No. 4, 1980, pp. 409-417.

Galvin, P.J. et al., "Transport of Sulfate to New York State", Environmental Science and Technology, Vol. 12, No. 5, Aug. 1979, pp. 580-584.

Gannon, J., "Acid Rain Fallout: Pollution and Politics", National Parks and Conservation Magazine, Vol. 52, No. 10, Oct. 1978, pp. 17-23.

GCA Corporation, Acid Rain Information Book, May 1981, Bedford, Massachusetts.

General Accounting Office, "The Debate Over Acid Precipitation--Opposing Views--Status of Research", Sept. 1981, Washington, D.C.

Gibson, J.H., "A National Program for Assessing the Problem of Atmospheric Deposition (Acid Rain)--A Report to the Council on Environmental Quality", Report NC-141, Dec. 1978, National Atmospheric Deposition Program, Natural Resources Ecology Laboratory, Colorado State University, Fort Collins, Colorado.

Gibson, J.H., Proceedings of Workshop on Data Management Needs for Atmospheric Deposition, Aug. 1980, Natural Resources Ecology Laboratory, Colorado State University, Fort Collins, Colorado.

Glover, G.M. et al., "Ion Relationships in Acid Precipitation and Stream Chemistry", Effects of Acid Precipitation on Terrestrial Ecosystems, Plenum Press, New York, New York, 1980, pp. 95-109.

Glustrom, L. and Stolzenberg, J., "Acid Rain: A Background Report", July 1982, Wisconsin Legislative Council, Madison, Wisconsin.

Grennfelt, P., Bengston, C. and Skaerby, L., "Estimation of the Atmospheric Input of Acidifying Substances to a Forest Ecosystem", July 1978, Swedish Water and Air Pollution Research Laboratory, Goteborg, Sweden.

Gunnerson, C.G. and Willard, B.E., (editors), Acid Rain, 1979, American Society of Civil Engineers, New York, New York.

Haines, B., "Forest Ecosystem SO_4-S Input-Output Discrepancies and Acid Rain: Are They Related?", Oikos, Vol. 41, No. 1, 1983, pp. 139-143.

Hales, J.M. and Slinn, W.G.N., "Natural Recovery Processes in Polluted Atmospheres", Proceedings of the Second National Conference on Complete Water Reuse: Water's Interface with Energy, Air, and Soils, 1975, American Institute of Chemical Engineers, New York, New York, pp. 338-354.

Handa, B.K., Kumar, A. and Goel, D.K., "Chemical Composition of Rain Water Over Lucknow in 1980", Mausam, Vol. 33, No. 4, 1982, pp. 485-488.

Hendry, C.D. and Brezonik, P.L., "Chemistry of Precipitation at Gainesville, Florida", Environmental Science and Technology, Vol. 14, No. 7, July 1980, pp. 843-849.

Hibbard, W.R., Jr., "Acid Precipitation: A Critique of Present Knowledge and Proposed Action", June 1982, Virginia Center for Coal and Energy Research, Virginia Polytechnic Institute, Blacksburg, Virginia.

Hileman, B., "Acid Deposition", Environmental Science and Technology, Vol. 16, No. 6, June 1982, pp. 323A-327A.

Hileman, B., "1982 Stockholm Conference on Acidification of the Environment", Environmental Science and Technology, Vol. 17, No. 1, Jan. 1983a, pp. 15A-18A.

Hileman, B., "Acid Rain: A Rapidly Shifting Scene", Environmental Science and Technology, Vol. 17, No. 9, 1983b, pp. 401A-405A.

Hill, F.B. et al., "Development of a Technique for Measurement of Biogenic Sulfur Emission Fluxes", Report No. CONF-771102-2, July 1977, Brookhaven National Laboratory, Upton, New York.

Hinds, W.T., "Incursion of Acid Deposition into Western North America", Environmental Conservation, Vol. 10, No. 1, 1983, pp. 53-58.

Hitchcock, D.R., "Biogenic Contributions to Atmospheric Sulfate Levels", Proceedings of the Second National Conference on Complete Water Reuse: Water's Interface with Energy, Air, and Soils, 1975, American Institute of Chemical Engineers, New York, New York, pp. 291-310.

Hoban, T.M., "Acid Rain and the Interstate Provisions of the Clean Air Act", International Journal of Environmental Studies, Vol. 18, No. 2, Feb. 1982, pp. 109-116.

Hornbeck, J.W., "Acid Rain: Facts and Fallacies", Journal of Forestry, Vol. 79, No. 7, July 1981, pp. 438-443.

Interagency Task Force on Acid Precipitation, "Interagency Task Force on Acid Precipitation: First Annual Report to the President and the Congress of the United States", Jan. 1982a, U.S. Department of Energy, Washington, D.C.

Interagency Task Force on Acid Precipitation, "National Acid Precipitation Assessment Plan", June 1982b, U.S. Department of Energy, Washington, D.C.

Interagency Task Force on Acid Precipitation, "National Acid Precipitation Assessment Program Annual Report to the President and Congress (1982)", June 1983, U.S. Department of Energy, Washington, D.C.

Johannes, A.H., Altwicker, E.R. and Clesceri, N.L., "Characterization of Acidic Precipitation in the Adirondack Region", Report No. EPRI EA 1826, May 1981, Electric Power Research Institute, Palo Alto, California.

Jones, K.H. and Mirra, R.R., "Estimating Future Trends in Acid Rainfall", Acid Rain: Proceedings of ASCE National Convention, American Society of Civil Engineers, 1979, pp. 120-131.

Khan, S.M., "The Relationship Between Acid Content of Particulates and Rainfall in Bangkok", Journal of Environmental Science and Health, Vol. A15, No. 6, 1980, pp. 561-572.

Kish, T., "Acid Precipitation: Crucial Questions Still Remain Unanswered", Journal of the Water Pollution Control Federation, Vol. 53, No. 5, May 1981, pp. 518-521.

Knudson, D.A., "Acidic Deposition: Review of Current Knowledge", 1981, Argonne National Laboratory, Argonne, Illinois.

Kramer, J.R., "Effect of Precipitation Chemistry on the Upper Great Lakes Region", Proceedings of the Second National Conference on Complete Water Reuse: Water's Interface with Energy, Air, and Soils, 1975, American Institute of Chemical Engineers, New York, New York, pp. 355-357.

Lakhani, K.H. and Miller, H.G., "Assessing the Contribution of Crown Leaching to the Element Content of Rainwater Beneath Trees", Effects of Acid Rain Precipitation on Terrestrial Ecosystems, 1980, Plenum Press, New York, New York, pp. 161-172.

Lee, Y.H. et al., "The Slope of Grant's Plot: A Use Function in the Examination of Precipitation, the Water-Soluble Part of Airborne Particles and Lake Water", Water, Air, and Soil Pollution, Vol. 10, No. 4, 1978, pp. 457-469.

Leonard, R.L., Goldman, C.R. and Likens, G.E., "Some Measurements of the pH and Chemistry of Precipitation at Davis and Lake Tahoe, California", Water, Air, and Soil Pollution, Vol. 15, No. 2, 1981, pp. 153-167.

Li, T. and Landsbergn, H.E., "Rainwater pH Close to a Major Power Plant", Atmospheric Environment, Vol. 9, No. 1, Jan. 1975, pp. 81-89.

Likens, G.E., "Acid Precipitation", Chemical and Engineering News, Vol. 54, No. 48, Nov. 22, 1976, pp. 29-38.

Likens, G.E. et al., "Acid Rain", Scientific American, Vol. 241, No. 4, Oct. 1979, pp. 43-52.

Likens, G.E. and Butler, T.J., "Recent Acidification of Precipitation in North America", Atmospheric Environment, Vol. 15, No. 7, 1981, pp. 1103-1109.

Likens, G.E., Edgerton, E.S. and Galloway, J.N., "The Composition and Deposition of Organic Carbon in Precipitation", Tellus, Vol. 35B, No. 1, 1983, pp. 16-24.

Liljestrand, H.M. and Morgan, J.J., "Chemical Composition of Acid Precipitation in Pasadena, California", Environmental Science and Technology, Vol. 12, No. 12, Nov. 1978, pp. 1271-1274.

Madsen, B.C., "Acid Rain at Kennedy Space Center, Florida: Recent Observations", Atmospheric Environment, Vol. 15, No. 5, May 1981, pp. 853-862.

Manzig, J.G.W. and Mulvaney, J.N., "Legal Aspects of Acidic Precipitation", International Journal of Environmental Studies, Vol. 18, No. 2, Feb. 1982, pp. 117-127.

Martin, A., "Some Observations of Acidity and Sulfur in Rainwater from Rural Sites in Central England and Wales", Atmospheric Environment, Vol. 12, No. 6-7, 1978, pp. 1481-1487.

Martin, A., "Sulphate and Nitrate Related to Acidity in Rainwater", Water, Air, and Soil Pollution, Vol. 21, No. 1-4, Jan. 1984, pp. 271-277.

McColl, J.G., "A Survey of Acid Precipitation in Northern California", Feb. 1980, Department of Soils and Plant Nutrition, University of California, Berkeley, California.

McColl, J.G., "Trace Elements in the Hydrologic Cycle of a Forest Ecosystem", Plant and Soil, Vol. 62, No. 3, 1981, pp. 337-349.

McColl, J.G. and Bush, D.S., "Precipitation and Throughfall Chemistry in the San Francisco Bay Area", Journal of Environmental Quality, Vol. 7, No. 3, July-Sept. 1978, pp. 352-358.

McNaughton, D.J., "Relationships Between Sulfate and Nitrate Ion Concentrations and Rainfall pH for Use in Modeling Applications", Atmospheric Environment, Vol. 15, No. 6, 1981, pp. 1075-1079.

McQuaker, N.R., Kluckner, P.D. and Sandberg, D.K., "Chemical Analysis of Acid Precipitation: pH and Acidity Determinations", Environmental Science and Technology, Vol. 17, No. 7, July 1983, pp. 431-435.

McWilliams, P. and Musante, L., "ARIS: Acid Rain Information System", Apr. 1982, NASA Industrial Applications Center, University of Pittsburgh, Pittsburgh, Pennsylvania.

Melo, O.T., "Ontario Hydro's Acid-Rain Monitoring Network", Research Review of Ontario Hydro, No. 2, May 1981, pp. 29-38.

Miller, J.M., "The Collection and Chemical Analysis of Precipitation in North America", Proceedings of the Second National Conference on Complete Water Reuse: Water's Interface with Energy, Air, and Soils, 1975, American Institute of Chemical Engineers, New York, New York, pp. 331-337.

Miller, J.M., "The Monitoring of Acid Rain", Acid Rain: Proceedings of the ASCE National Convention, 1979, American Society of Civil Engineers, New York, New York, pp. 111-119.

Mohnen, V.A., "Acid Precipitation: Finding the Source", Mining Congress Journal, Vol. 68, No. 8, Aug. 1982, pp. 42-45.

Moody, J., "Sharing Pollution: Quetico and BWCA", Not Man Apart-Foe, Vol. 8, No. 2, Mid-Jan.-Feb. 1978, pp. 12-14.

Morgan, J.J. and Liljestrand, H.M., "Measurement and Interpretation of Acid Rainfall in the Los Angeles Basin", Feb. 1980, California Institute of Technology, Pasadena, California.

Nagamoto, C.T. et al., "Acid Clouds and Precipitation in Eastern Colorado", Atmospheric Environment, Vol. 17, No. 6, 1983, pp. 1073-1082.

National Research Council, "Atmosphere-Biosphere Interactions: Toward a Better Understanding of the Ecological Consequences of Fossil Fuel Combustion", 1981, Washington, D.C.

National Technical Information Service, "Acid Precipitation--June, 1976-December, 1981 (Citations from the Energy Data Base)", Dec. 1981, U.S. Department of Commerce, Springfield, Virginia.

National Technical Information Service, "Acid Precipitation--1970-December, 1982 (Citations from Pollution Abstracts)", Dec. 1982a, U.S. Department of Commerce, Springfield, Virginia.

National Technical Information Service, "Acid Precipitation--June, 1970-December, 1982 (Citations from the Engineering Index Data Base)", Dec. 1982b, U.S. Department of Commerce, Springfield, Virginia.

National Technical Information Service, "Acid Precipitation--June, 1974-November, 1982 (Citations from the International Aerospace Abstracts Data Base)", Nov. 1982c, U.S. Department of Commerce, Springfield, Virginia.

National Technical Information Service, "Acid Precipitation--1977-1982 (Citations from the Selected Water Resources Abstracts Data Base)", Dec. 1982d, U.S. Department of Commerce, Springfield, Virginia.

National Technical Information Service, "Acid Precipitation: Legal, Political, and Health Aspects--1976-November, 1982 (Citations from the Energy Data Base)", Nov. 1982e, U.S. Department of Commerce, Springfield, Virginia.

National Technical Information Service, "Precipitation Washout--1964-November, 1982 (A Bibliography with Abstracts)", Jan. 1983, U.S. Department of Commerce, Springfield, Virginia.

Newman, L., Likens, G.E. and Bormann, E.H., "Acidity in Rainwater: Has an Explanation Been Presented?", Science, Vol. 188, No. 4191, May 30, 1975, pp. 957-959.

Oden, S. and Thorsten, A., "The Sulfur Budget of Sweden", Effects of Acid Precipitation on Terrestrial Ecosystems, Plenum Press, New York, New York, 1980, pp. 111-122.

Okland, J., "Acidification in 50 Norwegian Lakes", Nordic Hydrology, Vol. 11, No. 4, 1980, pp. 25-32.

O'Sullivan, D.A., "Norway: Victim of Other Nation's Pollution", Chemical and Engineering News, Vol. 54, No. 25, June 14, 1976, pp. 15-17.

Patrick, R., Binetti, V.P. and Halterman, S.G., "Acid Lakes from Natural and Anthropogenic Causes", Science, Vol. 211, No. 4481, Jan. 1981, pp. 446-448.

Pellett, G.L., Bustin, R. and Harriss, R.C., "Sequential Sampling and Variability of Acid Precipitation in Hampton, Virginia", Water, Air, and Soil Pollution, Vol. 21, No. 1-4, Jan. 1984, pp. 33-49.

Perhac, R.M., "Acid Rain--An Overview", Mining Congress Journal, Vol. 67, No. 8, Aug. 1981, pp. 19-21.

Popp, C.J. et al., "Precipitation Analysis in New Mexico", 1980, New Mexico Technical University, Socorro, New Mexico.

Poundstone, W.N., "Let's Get the Facts on Acid Rain", Mining Congress Journal, Vol. 66, No. 7, July 1980, pp. 45-47.

Pratt, G.C., Coscio, M.R. and Krupa, S.V., "Regional Rainfall Chemistry in Minnesota and West Central Wisconsin", Atmospheric Environment, Vol. 18, No. 1, 1984, pp. 173-182.

Pratt, M., "Acid Rain: Filtering Out the Facts", Agricultural Engineering, Vol. 64, No. 5, May 1983, pp. 7-9.

Rippon, J.E., "Studies of Acid Rain on Soils and Catchments", Effects of Acid Precipitation on Terrestrial Ecosystems, Plenum Press, New York, New York, 1980, pp. 499-524.

Robertson, J.K., Dolzine, T.W. and Graham, R.C., "Chemistry of Precipitation from Sequentially Sampled Storms", EPA-600/4-80-004, Jan. 1980, U.S. Environmental Protection Agency, Research Triangle Park, North Carolina.

Robinson, J.W. and Ghane, H., "Continued Studies of Acid Rain and Its Effects on the Baton Rouge Area", Journal of Environmental Science and Health, Vol. A18, No. 2, 1983, pp. 165-174.

Rodhe, H., Soderlund, R. and Ekstedt, J., "Deposition of Airborne Pollutants on the Baltic", Ambio, Vol. 9, No. 3-4, 1980, pp. 168-173.

Roffman, A., "Acid Precipitation in the Pittsburgh, Pennsylvania Area", Journal of Environmental Sciences, Vol. 23, No. 2, Mar. 1980, pp. 33-37.

Rosencranz, A., "The Problem of Transboundary Pollution", Environment, Vol. 22, No. 5, June 1980, pp. 15-20.

Rosencranz, A. and Wetstone, G., "Acid Precipitation: National and International Responses", Environment, Vol. 22, No. 5, June 1980, pp. 6-8.

Rosenfeld, A., "Forecast: Poisonous Rain", Saturday Review, Vol. 5, No. 23, Sept. 2, 1978, pp. 16-19.

Rosenquist, I.T., "Alternative Sources for Acidification of River Water in Norway", Science of the Total Environment, Vol. 10, No. 1, July 1978, pp. 38-49.

Seip, H.M., "Acid Snow-Snowpack Chemistry and Snowmelt", Effects of Acid Precipitation on Terrestrial Ecosystems, Plenum Press, New York, New York, 1980, pp. 77-94.

Seymour, M.D. et al., "Variations in the Acid Content of Rain Water in the Course of a Single Precipitation", Water, Air, and Soil Pollution, Vol. 10, No. 2, Aug. 1978, pp. 147-162.

Seymour, M.D. and Stout, T., "Observations on the Chemical Composition of Rain Using Short Sampling Times During a Single Event", Atmospheric Environment, Vol. 17, No. 8, 1983, pp. 1483-1487.

Sharp, P.G., "Acid Rain", Journal of the Air Pollution Control Association, Vol. 30, No. 6, June 1980, p. 968.

Shinn, J.H. and Lynn, S., "Do Man-Made Sources Affect the Sulfur Cycle of Northeastern States", Environmental Science and Technology, Vol. 13, No. 9, Sept. 1979, pp. 1062-1067.

Spaite, P., "Acid Rain: The Impact of Local Sources", Nov. 1980, PEDCo-Environmental, Inc., Cincinnati, Ohio.

Suleymanov, D.M. and Listengarten, V.A., "Chemical Composition of Precipitation on the Apsheronsk Peninsula", Dokl. Akad. Nauk. Azerb. SSR, Vol. 22, No. 12, 1966, pp. 42-44.

Szabo, M.F., Esposito, M.P. and Spaite, P.W., "Acid Rain: Commenting on Controversial Issues and Observations on the Role of Fuel Burning", DOE/MC/19170-1168, Mar. 1982, U.S. Department of Energy, Washington, D.C.

Tanaka, S., Darzi, M. and Winchester, J.W., "Sulfur and Associated Elements and Acidity in Continental and Marine Rain from North Florida", Journal of Geophysical Research, Vol. 85, No. C8, Aug. 1980, pp. 4519-4526.

Tennessee Valley Authority, "The Acid Connection", Impact, Vol. 3, No. 4, July 1980, pp. 2-10.

Trumbule, R.E. and Tedeschi, M., "Acid Rain Information: Knee Deep and Rising", Science Technology Librarian, Vol. 4, No. 2, Winter 1983, pp. 27-41.

Turk, J.T., "Evaluation of Trends in the Acidity of Precipitation and the Related Acidification of Surface Water in North America", Geological Survey Water Supply Paper No. 2249, 1983, U.S. Geological Survey, Washington, D.C.

Tyree, S.Y., Jr., "Rainwater Acidity Measurement Problems", Atmospheric Environment, Vol. 5, No. 1, Jan. 1981, pp. 57-60.

Ulrich, B., "Production and Consumption of Hydrogen Ions in the Ecosphere", Effects of Acid Rain Precipitation on Terrestrial Ecosystems, Plenum Press, New York, New York, 1980, pp. 255-282.

U.S. Environmental Protection Agency, "Environmental Effects of Increased Coal Utilization: Ecological Effects of Gaseous Emissions from Coal Combustion", EPA-600/7-78-108, June 1978, Washington, D.C.

U.S. Environmental Protection Agency, Proceedings of the Acid Rain Conference: Springfield, Virginia, April 8-9, 1980, Aug. 1980a, Research Triangle Park, North Carolina.

U.S. Environmental Protection Agency, "Acid Rain", July 1980b, Office of Research and Development, Washington, D.C.

Vermeulen, A.J., "Acid Precipitation in the Netherlands", Environmental Science and Technology, Vol. 12, No. 9, Sept. 1978, pp. 1016-1021.

Wagner, G.H. and Steele, K.F., "Nutrients and Acid in the Rain and Dry Fallout at Fayetteville, Arkansas (1980-1982)", Apr. 1983, Arkansas Water Resources Research Center, University of Arkansas, Fayetteville, Arkansas.

Wetstone, G.S., "The Need for a New Regulatory Approach", Environment, Vol. 22, No. 5, June 1980, pp. 9-14, 40.

Wetstone, G.S., "National Recourses for International Pollution: Towards a United States, Canada Solution", Journal Air Pollution Control Association, Vol. 34, No. 2, Feb. 1984, pp. 111-118.

Whelpdale, D.M., "The Contribution Made by Air-Borne Pollutants to the Pollution of Large Bodies of Water", Atmosphere, Vol. 10, No. 1, 1972, pp. 18-22.

Wilson, J., Mohnen, V. and Kadlecek, J., "Wet Deposition in the Northeastern United States", Dec. 1980, U.S. Department of Energy, Washington, D.C.

Wisniewski, J., "The Potential Acidity Associated with Dews, Frosts, and Fogs", Water, Air, and Soil Pollution, Vol. 17, No. 4, 1982, pp. 361-377.

Wisniewski, J. and Keitz, E.L., "Acid Rain Deposition Patterns in the Continental United States", Water, Air, and Soil Pollution, Vol. 19, No. 4, May 1983, pp. 327-339.

Wisniewski, J. and Kinsman, J.D., "An Overview of Acid Rain Monitoring Activities in North America", Bulletin of the American Meteorological Society, Vol. 63, No. 6, June 1982, pp. 598-618.

Wright, R.F. and Dovland, H., "Regional Surveys of the Chemistry of the Snowpack in Norway", Atmospheric Environment, Vol. 12, No. 8, pp. 1755-1759.

CHAPTER 3

ATMOSPHERIC REACTIONS AND TRANSPORT

Acid precipitation is formed in the atmosphere and returns to the earth as a result of chemical and physico-chemical phenomena. The effects of acid precipitation are often experienced at long distances from the locations of sources of emissions of precursor constituents such as sulfur oxides and nitrogen oxides. This chapter provides summary information on atmospheric reactions, models for simulating atmospheric phenomena, and long-range transport of air pollutants. Each of these issues is important in understanding the causes and consequences of acid rain, as well as providing a basis for the development and implementation of effective control and management strategies.

ATMOSPHERIC REACTIONS

Information from 30 references dealing with atmospheric reactions is contained in the abstracts in Appendix B. Table 10 provides a brief comment on each of these references in the order in which they appear in Appendix B. Each reference in Table 10 will not necessarily be addressed in this portion of Chapter 3.

Table 10: References in Appendix B Related to Atmospheric Reactions

Author(s) (Year)	Comments
Bolin, Aspling and Persson (1973)	Atmospheric transfer rates associated with rainout and dry deposition.
Calvert and Stockwell (1983)	Description of multiple chemical reactions in the atmosphere.
Chang (1984)	Coefficients for rain and snow scavenging of HNO_3 vapor in the atmosphere.
Davenport and Peters (1978)	Field studies of changes in particulate concentrations during precipitation.
Dittenhoefer and Dethier (1976)	Influence of meteorological variables in precipitation chemistry.
Dupoux (1973)	Meteorological influences on the behavior of atmospheric aerosols.
Durham, Overton and Aneja (1981)	In-cloud scavenging of HNO_3 and its influence on sulfate production and rainfall acidity.

Table 10: (continued)

Author(s) (Year)	Comments
Engelmann (1971)	Prediction of atmospheric scavenging via pollutant concentration ratios.
Enger and Hogstrom (1979)	Field study of SO_2 to sulfate transformation rates near an oil-fired power plant.
Fowler (1980)	Review of transport, chemistry, and deposition mechanisms for atmospheric sulfur and nitrogen compounds.
Fuquay (1970)	Review of mechanisms of precipitation scavenging in the atmosphere.
Gatz (1975)	Calculation of pollutant aerosol input into Lake Michigan using measured wet and dry deposition rates.
Gatz (1979)	Urban deposition of sulfate and soluble metals in summer rains in St. Louis, Missouri.
Georgii and Beilke (1966)	Laboratory studies of washout of SO_2 by droplets of known size distribution and concentration.
Granat (1972)	Stoichiometric model of pH and chemical constituents in atmospheric precipitation.
Gravenhorst et al. (1980)	Equations for calculating the incorporation of SO_2 into raindrops.
Hales (1977)	Trace studies of convective storm scavenging.
Hales (1979)	Mathematical models for mechanisms of atmospheric recovery.
Harrison and Pio (1983)	Field study of the chemical composition of rainwater and suspended particles in northwest England.
Hegg (1983)	Prediction of sulfate washout based on various mechanisms of occurrence.
Hidy et al. (1983)	Calculation of rainout efficiencies for sulfate and nitrate.

Table 10: (continued)

Author(s) (Year)	Comments
Melo (1981)	Field studies of atmospheric chemical reactions near power plants.
Molenkamp (1974)	Discussion of the mechanisms of nuclei condensation in precipitation scavenging.
Overton, Aneja and Durham (1979)	Physico-chemical model for the accumulation of sulfur species in raindrops.
Overton and Durham (1982)	Physico-chemical model for the mass transfer of SO_2, H_2O_2, O_3, HNO_3, and CO_2 in falling raindrops.
Pack (1979)	Summary of workshop on research needs on the formation of acid precipitation.
Penkett et al. (1979)	Rates of sulfur dioxide oxidation by O_2, O_3, and H_2O_2 across a range of pH and temperature values.
Rogowski et al. (1982)	Laboratory study of the effect of carbon particles on oxidation of SO_2 by air and NO_2.
Semonin et al. (1979)	Field studies of atmospheric pollution scavenging in Illinois.
Sullivan, Wen and Frantisak (1975)	Possible air-water transfer mechanisms for acid sulfur gases or salts.

Oxides of sulfur and nitrogen in the atmosphere may be subject to both chemical and/or physico-chemical reactions as well as both wet and dry deposition mechanisms. Dry deposition will be discussed in Chapter 6. Wet deposition, or precipitation scavenging, can occur via both in-cloud and below-cloud processes (Fuquay, 1970).

Chemical and Physico-Chemical Processes

Many different chemical reactions contribute to the oxidation of SO_2, NO, and NO_2 in the atmosphere. Solution phase and gas phase chemistry are both important to acid rain development, and gas-solid, gas-liquid, liquid-solid as well as simple gaseous molecule interactions are also seemingly important in some circumstances (Calvert and Stockwell, 1983). Penkett et al. (1979) compared the measured rates of sulfur dioxide oxidation by oxygen, ozone, and hydrogen peroxide across a range of pH and temperature to evaluate their relative importance in the formation of sulfate in atmospheric droplets. They found that oxidation will occur readily in cloud or fog droplets. If the pH is above 6, ozone will be primarily responsible. If the pH is below about 5.5, hydrogen peroxide is the favored oxidant. Pollutant combination effects can also occur. For example, Rogowski et al. (1982) conducted a series of experiments using carbon particles (commercial furnace black) as a surrogate for soot particles. Carbon particles were suspended in water, and gas mixtures were bubbled into the suspensions to observe the effect of carbon particles on the oxidation of SO_2 by air and NO_2. Identical gas mixtures were bubbled into a blank containing only pure water. After exposure, each solution was analyzed for pH and sulfate. It was found that NO_2 greatly enhances the oxidation of SO_2 to sulfate in the presence of carbon particles.

Both in-cloud and below-cloud rainout and washout can occur in precipitation scavenging. The influence of atmospheric chemistry as well as meteorological variables is discussed by Fowler (1980) and Gravenhorst et al. (1980); and the mathematical basis for describing these influences is presented by Hales (1979). Georgii and Beilke (1966) conducted detailed laboratory investigations on washout and rainout of SO_2 by droplets of known size distribution and concentration. The results show clearly the effect of drop size, intensity and the chemical composition of rain and fog on the scavenging efficiency. The results of the experiments were used as the basis of a model calculation of the effect of washout and rainout by natural precipitation at a given vertical distribution of SO_2.

A physico-chemical subcloud rain model was used by Durham, Overton and Aneja (1981) to study pollutant chemistry leading to rain acidification. In the model, drops fall through a polluted region containing trace gases, CO_2, O_3, SO_2, HNO_3, NO, and NO_2. The production of acid sulfate and the accumulation of acid nitrate can be calculated by accounting for the mass transfer of SO_2, H_2O_2, O_3, HNO_3, and CO_2. The acidification is postulated to occur through the absorption of free gaseous HNO_3 and the absorption and reaction of SO_2, H_2O_2, and O_3 to yield H_2SO_4 (Overton and Durham, 1982).

The relative contributions of nucleation scavenging, in-solution production, and below-cloud scavenging to the sulfate content of precipitation has been studied by Hegg (1983). Based upon the results of a parameterization scheme which predicts changes in precipitation sulfate concentration by up to a factor of 6 for large variations in key cloud physics parameters, the results suggested in-solution sulfate production can be a substantial contributor to the sulfate content of precipitation.

Using a number of approximations, Chang (1984) derived coefficients for the wet removal of HNO_3 vapor from the atmosphere by rain (in-cloud and below-cloud) and snow. The wet removal coefficients were parameterized in terms of precipitation rate. These coefficients represent the average scavenging from large precipitation bands or frontal systems where there is widespread weakly ascending air motion. Consequently, the derived coefficients would be appropriate for inclusion into regional or mesoscale models which include the wet removal of HNO_3 vapor. These coefficients, when combined with altitude-dependent precipitation rates and vertical profiles of HNO_3 concentrations, are also useful to estimate the flux of HNO_3 to the earth's surface.

Field Studies

Several field studies of atmospheric chemistry and precipitation scavenging have been conducted. Table 11 identifies 7 references from Appendix B which are focused on field measurements. Four studies related to the United States (Dittenhoefer and Dethier, 1976; Gatz, 1975; Gatz, 1979; and Hidy et al., 1983); and one each were in Canada (Melo, 1981), Sweden (Enger and Hogstrom, 1979), and England (Harrison and Pio, 1983).

Table 11: References in Appendix B Describing Field Studies of Atmospheric Chemistry and Precipitation Scavenging

Author(s) (Year)	Comments
Dittenhoefer and Dethier (1976)	Influence of meteorological variables in precipitation chemistry.
Enger and Hogstrom (1979)	Field study of SO_2 to sulfate transformation rates near an oil-fired power plant.
Gatz (1975)	Calculation of pollutant aerosol input into Lake Michigan using measured wet and dry deposition rates.
Gatz (1979)	Urban deposition of sulfate and soluble metals in summer rains in St. Louis, Missouri.
Harrison and Pio (1983)	Field study of the chemical composition of rainwater and suspended particles in northwest England.
Hidy et al. (1983)	Calculation of rainout efficiencies for sulfate and nitrate.
Melo (1981)	Field studies of atmospheric chemical reactions near power plants.

Dittenhoefer and Dethier (1976) studied precipitation chemistry data for western New York through the use of 12- and 24-hour isentropic trajectories. Meteorological variables such as surface wind speed and direction, height of the mixed layer, mean relative humidity below the cloud base, and the past history of precipitation were monitored to assess their importance in controlling rainwater concentrations. It was found that ionic constituents such as sulfate owe much of their presence in rainwater to wet removal from the rain-producing layers (1-5 km), where their concentration is largely influenced by long-range transport.

Hidy et al. (1983) studied ambient aerosol chemistry and precipitation water chemistry from samples taken at rural sites in the northeastern United

States.   Data taken between 1978 and 1979 were selected and analyzed to estimate rainout efficiencies for sulfur oxides and nitrogen oxides. Calculations indicated that the mean rainout efficiency for particulate sulfate based on 500 samples was approximately 0.6 to 0.8. Additional chemical data were used to infer that significant acid production may take place in precipitation by SO_2 absorption and oxidation at some locations, particularly in Ohio and northeastward. In contrast with sulfate, the calculations indicated that nitrate was dominated by scavenging of gases, such as HNO_3, NO_2 or N_2O_5, rather than particles. The rainout efficiencies tended to be smaller in the winter, and were largely independent of the phase of precipitation elements, and storm conditions. There was also a tendency for the rainout efficiencies to decrease with precipitation intensity.

Pollutant aerosol input into Lake Michigan has been calculated by Gatz (1975) using measured wet and dry deposition rates. Deposition estimates suggest that annual wet and dry deposition are approximately equal, that from 3 to 15 percent of the elemental emissions from Chicago and northwestern Indiana enter the Lake, and that the fraction of emissions to be deposited in the Lake increases with particle size. Gatz (1979) also studied the content of sulfate and other materials in the atmosphere and in samples of summer convective rainfall from mesoscale sampling networks near St. Louis. These and other observations in the literature agreed that rainfall deposits locally emitted sulfur at short distances downwind of cities. Nucleation scavenging of atmospheric sulfate appears to be adequate as an explanation of the removal mechanisms.

Ontario Hydro in Canada has studied chemical conversions taking place in power plant plumes (Melo, 1981). Ontario Hydro found that most of the nitrogen oxides are emitted as nitric oxide, about 5 percent of the NO is quickly converted to nitrogen dioxide through reaction with oxygen, and that the rest is converted to NO_2 by ambient ozone in times which are typically 1 to 2 hours. Further conversion of NO_2 to nitric acid proceeds at appreciable rates only after this 1 to 2 hour induction period. The reaction of sulphur dioxide to sulphate aerosol is generally quite slow except on a few occasions when it reaches values of up to 5 percent per hour.

Atmospheric transformation rates near an oil-fired 1000 MW power plant in Sweden have been studied by Enger and Hogstrom (1979). Five out of seven wet deposition tests revealed that 66 percent of the emitted sulfates were deposited within 80-120 km of the source. Sulfur dioxide, sulfate, and an inert tracer were measured in ground-based cross plume traverses and in flight. Two of the dispersion tests indicated that SO_2-to-SO_4 transformation rates vary with relative humidity.

Measurements of the chemical composition of rainwater and suspended particles collected in parallel at a rural site in northwest England have shown that sulfate, nitrate, chloride, sodium, magnesium and potassium exists in similar proportions in both media (Harrison and Pio, 1983). However, rainwater showed a marked enhancement of hydrogen ions, and a corresponding decrease in ammonia ions relative to aerosols. It was concluded that the major contribution to rainwater acidity at this site is due to sulfuric acid incorporated at cloud level. The scavenging of nitric or hydrochloric acids, or incorporation of acid sulfates close to ground level can account for only a minor contribution to acidity in this locality.

ATMOSPHERIC MODELS

Information from 42 references addressing atmospheric modeling is contained in the abstracts in Appendix C. Table 12 provides a brief comment on each of these references in the order in which they appear in Appendix C. Not every reference listed in Table 12 will be addressed in this portion of Chapter 3. Lehmann (1975) provides a review of lower atmospheric modeling of air pollution from both mobile and stationary sources. Included are models concerning local diffusion, climatology, and smog. Horst (1979) reviews and compares the assumptions and predictions of several Gaussian diffusion-deposition models. This portion of Chapter 3 will address basic atmospheric reaction and precipitation scavenging process models, deposition models, integrated air/land/water models, and examples of model applications and user's guides.

Table 12: References in Appendix C Related to Atmospheric Models

Author(s) (Year)	Comments
Adamowicz (1979)	Model for washout of sulfur dioxide, carbon dioxide, and ammonia.
Adamowicz and Hill (1977)	Model for washout of trace atmospheric gases and production of acids in rain.
Barrie (1981)	Presentation of quantitative relationship between the sulfur content of rain, and rainwater pH and temperature.
Bhumralkar et al. (1980)	Application of a trajectory-type regional air pollution model for sulfur transport and deposition in eastern North America.
Bolin and Persson (1974)	Regional dispersion model applied to sulfur deposition in western Europe.
Carmichael, Yang and Lin (1980)	Numerical technique for approximating solutions to the atmospheric advection-diffusion equation.
Chen et al. (1982)	Integrated model including hydrologic, canopy chemistry, snowmelt chemistry, soil chemistry, and stream and lake water quality modules.
Corbett (1981)	Model to account for dry deposition in Gaussian plume models.
Dana et al. (1973)	Field measurements and modeling of the washout of SO_2 and sulfate from power plant plumes.

Table 12: (continued)

Author(s) (Year)	Comments
Dana et al. (1974)	Measurement and modeling of precipitation scavenging of urban pollutants in the St. Louis area.
Davis, Eadie and Powell (1979)	User's guide for long-range transport model involving either no deposition, dry deposition, or wet and dry deposition.
Draxler and Heffter (1981)	Workbook for a regional trajectory model capable of handling up to 70 sources.
Golomb et al. (1983)	Sensitivity analysis of kinetic components of acid rain models.
Hales and Lee (1974)	Precipitation scavenging model for organic contaminants, including diethylamine and ethyl acetoacetate.
Hill and Adamowicz (1976)	Continuous model for washout of sulfur dioxide as a function of several variables.
Hill and Adamowicz (1977)	Model for rain composition and the washout of sulfur dioxide.
Horst (1979)	Comparison of several Gaussian diffusion-deposition models.
Huff et al. (1977)	User's guide for terrestrial ecosystem hydrology model for transport and fate of trace contaminants at a watershed scale.
Junod (1976)	Two conceptual mathematical models of a cooling tower plume.
Lee (1974)	Two numerical models for precipitation scavenging.
Lehman (1975)	Bibliography of references on lower atmospheric modeling of air pollution.
Lewellen and Sheng (1980)	Model for dry deposition of SO_2 and sulfate aerosols.
Liu, Stewart and Henderson (1982)	Regional-scale air quality model for SO_2 and sulfate.
Mayerhofer et al. (1982)	Description of ENAMAP-1A model for long-range air pollution transport.

Table 12: (continued)

Author(s) (Year)	Comments
McNaughton (1979)	Emission source specification in a regional air quality model.
Michael and Raynor (1981)	Difficulties in using trajectory models to calculate source-receptor relationships.
Mills and Reeves (1973)	Multi-source atmospheric transport model for calculating toxic material deposition.
Moore (1980)	Calculation of dry deposition from airborne pollutant plumes.
Moroz (1981)	Models for deposition from large individual and small collective sources.
Patterson, Mankin and Brooks (1973)	Coupled model including atmospheric deposition and hydrologic transport.
Powell et al. (1979)	Model for SO_x that includes trajectories, chemical transformation, and deposition.
Ragland and Wilkening (1983)	Three-dimensional, time-dependent, intermediate-range grid type model for two chemically reactive species.
Ritchie, Bowman and Burnett (1983)	Mesoscale modified Gaussian SO_2 model that incorporates chemical oxidation, dry deposition, and precipitation scavenging.
Ritchie, Brown and Wayland (1978)	Model for effects of washout and runoff on the consequences of atmospheric releases from nuclear reactor accidents.
Schmidt (1982)	Particle dry deposition modeling over lake surfaces.
Schnoor, Carmichael and Van Shepen (1980)	Tiered approach involving a steady state susceptibility model, dynamic regional model, and three-dimensional dynamic event model.
Sehmel (1979)	Dry deposition velocities as a function of particle diameter and gas speciation.
Sehmel and Hodgson (1980)	Model for particle and gas removal rates from the atmosphere by dry deposition.

Table 12: (continued)

Author(s) (Year)	Comments
Shannon (1978)	Description of Advanced Statistical Trajectory Regional Air Pollution (ASTRAP) model.
Shannon (1981)	Features of ASTRAP model.
Tsai and Johnson (1974)	Model for predicting distribution of drift from power plant cooling towers.
Williams (1982)	Model for aerosol particle deposition on natural water surfaces.

Basic Atmospheric Process Models

Table 13 identifies 15 references with abstracts in Appendix C which include basic atmospheric process models. Three categories of models are addressed in Table 13: (1) precipitation scavenging or washout models; (2) regional scale dispersion models; and (3) trajectory and long-range transport models. Lee (1974) described two models for precipitation scavenging; one is for scavenging in stratiform clouds, and the other is for air cleaning by a convective rain-generating system.

Table 13: References in Appendix C Describing Basic Atmospheric Process Models

Author(s) (Year)	Comments
Adamowicz (1979)	Model for washout of sulfur dioxide, carbon dioxide, and ammonia.
Adamowicz and Hill (1977)	Model for washout of trace atmospheric gases and production of acids in rain.
Golomb et al. (1983)	Sensitivity analysis of kinetic components of acid rain models.
Hales and Lee (1974)	Precipitation scavenging model for organic contaminants, including diethylamine and ethyl acetoacetate.
Hill and Adamowicz (1976)	Continuous model for washout of sulfur dioxide as a function of several variables.

Table 13: (continued)

Author(s) (Year)	Comments
Hill and Adamowicz (1977)	Model for rain composition and the washout of sulfur dioxide.
Lee (1974)	Two numerical models for precipitation scavenging.
Lin, Stewart and Henderson (1982)	Regional-scale air quality model for SO_2 and sulfate.
Mayerhofer et al. (1982)	Description of ENAMAP-1A model for long-range air pollution transport.
Powell et al. (1979)	Model for SO_x that includes trajectories, chemical transformation, and deposition.
Ragland and Wilkening (1983)	Three-dimensional, time-dependent, inter-mediate-range grid type model for two chemically reactive species.
Ritchie, Bowman and Burnett (1983)	Mesoscale modified Gaussian SO_2 model that incorporates chemical oxidation, dry deposition, and precipitation scavenging.
Schnoor, Carmichael and van Shepen (1980)	Tiered approach involving a steady state susceptibility model, dynamic regional model, and three-dimensional dynamic event model.
Shannon (1978)	Description of Advanced Statistical Trajectory Regional Air Pollution (ASTRAP) model.
Shannon (1981)	Features of ASTRAP model.

Adamowicz and Hill (1977) developed a model to describe the washout of trace atmospheric gases and the production of acids in rain. The model has been applied to the washout of sulfur dioxide, carbon dioxide and ammonia and incorporates reversible mass transfer of the trace gases, all possible ionic equilibria of the compounds in solution and catalyzed oxidation of the dissolved sulfur species to sulfates. The significance of ammonia and carbon dioxide on the raindrops capacity for sulfur and on sulfate production based solely on bisulfite oxidation has been explored in detail. The influence of raindrop size, rainfall intensity, cloud-base height, the presence of oxidation catalyzing compounds in the atmosphere and the initial composition of the raindrops as they enter the polluted atmospheric layer on the detailed chemical composition of rain at ground level and the time scale for gaseous sulfur dioxide removal have been evaluated. The resulting model has been used to calculate the composition of rain as a function of fall distance and to obtain the time scale for SO_2 removal from the atmosphere (Hill and Adamowicz, 1977).

Regional scale dispersion models which account for atmospheric reactions have been developed (Liu, Stewart and Henderson, 1982; Ragland and Wilkening, 1983; and Ritchie, Bowman and Burnett, 1983). The Liu, Stewart and Henderson (1982) model, which is hybrid in nature, consists of a puff module and a grid module. The puff module computes the evolution of individual puffs, such as the horizontal and vertical standard deviations of the puff spreads and the location of the center of mass, emitted continuously from each major point source. It also determines the location at which the puff will be released to the grid module and the amount of oxidation and deposition along the trajectory. The grid module then follows the transport, diffusion, and chemical reactions of these aged puffs, as well as emissions from a variety of diffuse sources. On the basis of model calculations, atmospheric budgets for SO_2 and sulfate over the modeling region can be estimated.

Ragland and Wilkening (1983) have developed a three-dimensional time-dependent grid type model for two chemically reacting species which undergo atmospheric transport, diffusion and wet and dry deposition over a region of several hundred km. Ritchie, Bowman and Burnett (1983) describe a mesoscale modified Gaussian model that incorporates chemical oxidation of sulfur dioxide, dry deposition and precipitation scavenging for single and multiple sources. The model has been verified in the range of 5-250 km using 1.5 years of monthly deposition data, 1 year of 24-h ambient air SO_2 data, and by comparison to two dispersion models. The calculated deposition rates which were compared to data collected around the Sudbury, Ontario, Canada, smelter complex were found to be within a factor of two to three of measured values depending on the averaging time. The comparison with two existing dispersion models gave results to within a factor of two.

Several models have been developed to describe air mass trajectories and atmospheric reactions (Powell et al., 1979; Shannon, 1978; and Shannon, 1979). Powell et al. (1979) developed a sulfur oxides atmospheric pollution model that calculates trajectories using single-layer historical wind data as well as chemical transformation and deposition following discrete contaminant air masses. A Statistical Trajectory Regional Air Pollution (STRAP) model has been described by Shannon (1978). In order to account for variations in the input parameters, an Advanced Statistical Trajectory Regional Air Pollution (ASTRAP) model has been formulated (Shannon, 1979). The ASTRAP model combines efficient calculation of long-term regional-scale concentrations and fluxes of pollutant sulfur with improved parameterizations of boundary-layer processes. The parameterizations include diurnal and seasonal variations of dry deposition velocities for SO_2 and sulfate, rate of transformation from SO_2 to sulfate, vertical structure of the planetary boundary layer, and emission rates.

The ENAMAP-1 and ENAMAP-1A models address long-range air pollution transport (Mayerhofer et al., 1982). The subsequent section of this chapter will provide additional information on long-range transport concerns. In the ENAMAP-1 model the transformation rate for SO_2 to SO_4 and the deposition rates of SO_2 and SO_4 are all constants; in ENAMAP-1A they are variable in space and time. The transformation rate has been made dependent on the amount of sunshine (i.e., a function of latitude and season) and is about twice as large as the previous rate. In ENAMAP-1A the dry deposition rate has been made dependent on the type of underlying terrain and vegetation, on thermal stability in the boundary layer, and on time of day. Wet deposition is treated as a function of rainfall rate and cloud type (convective, warm process, or Bergeron process).

Deposition Models

Table 14 identifies 8 references, with abstracts in Appendix C, which address deposition models. The depletion of airborne pollutant plumes resulting from dry deposition on ground surfaces should be taken into account when estimating pollutant concentrations in air at distances downwind from their sources. One approach to account for this deposition involves using a reduced release rate instead of the actual release rate for the pollutant in atmospheric dispersion equations (Moore, 1980). The reduced release rate is the actual rate of release multiplied by a depletion fraction. Depletion fractions for use with the Gaussian atmospheric dispersion equation of Pasquill and Gifford have been calculated by numerical integration for downwind distances ranging from 35 to 90,000 m for release heights from 1 to 400 m, and they have been tabulated for convenient interpolation for intermediate values of distance and release height.

Table 14: References in Appendix C Describing Deposition Models

Author(s) (Year)	Comments
Corbett (1981)	Model to account for dry deposition in Gaussian plume models.
Lewellen and Sheng (1980)	Model for dry deposition of SO_2 and sulfate aerosols.
Mills and Reeves (1973)	Multi-source atmospheric transport model for calculating toxic material deposition.
Moore (1980)	Calculation of dry deposition from airborne pollutant plumes.
Moroz (1981)	Models for deposition from large individual and small collective sources.
Sehmel and Hodgson (1980)	Model for particle and gas removal rates from the atmosphere by dry deposition.
Schmidt (1982)	Particle dry deposition modeling over lake surfaces.
Williams (1982)	Model for aerosol particle deposition on natural water surfaces.

A model for dry deposition of SO_2 and sulfate aerosol has been described by Lewellen and Sheng (1980). Sehmel and Hodgson (1980) demonstrated a predictive model for correlating particle and gas removal rates from the atmosphere by dry deposition. They noted that predicted deposition rates can vary over several orders of magnitude and are complex functions of pollutant, air, and surface variables. For example, Schmidt (1982) indicated that dry deposition velocities are a function of particle size as well as windspeed, surface roughness, reference height, stability, and possibly relative humidity. Surface microlayers may affect roughness, but they are unlikely to exert significant control over deposition rates except on a very local level.

Mills and Reeves (1973) developed an Atmospheric Transport Model (ATM) from a Gaussian plume model to calculate the toxic material deposition rate at any point within a watershed, given the location of various air pollution sources. Sources included the point source (stack), area source (city), line source (road or railbed), and windblown source (dry tailings pond).

Integrated Air/Land/Water Models

Several models have been developed which couple atmospheric deposition of pollutants to land surface and/or subsurface transport models. One example is the Unified Transport Model (Patterson, Mankin and Brooks, 1973); this model consists of two major submodels: the Atmospheric Transport Model and the Wisconsin Hydrologic Transport Model. The Atmospheric Transport Model can be used to calculate deposition rates of pollutants (Mills and Reeves, 1973). These rates can be used by the Wisconsin Hydrological Transport Model to determine the subsequent transfer of the pollutants to streams and ground water.

Chen et al. (1982) describe an Integrated Lake-Watershed Acidification Study (ILWAS) model. The ILWAS model includes hydrologic, canopy chemistry, snowmelt chemistry, soil chemistry, and stream and lake water quality modules. For modeling purposes, a drainage basin is divided into terrestrial subcatchments, stream segments, and a lake. Vertically, each subcatchment is further segmented into canopy, snow pack, and soil layers. The lake is also vertically layered. All these physical compartments are coupled to form a network that allows water to be routed through the system to the lake outlet.

Model Applications and User's Guides

The ENAMAP-1 model has been used to calculate monthly, seasonal, and annual distributions of sulfur dioxide and sulfate concentrations and wet and dry depositions over the eastern United States, as well as over the southern portions of the Canadian provinces of Quebec and Ontario (Bhumralkar et al., 1980). This geographical study area was partitioned into 13 different regions and interregional sulfur exchanges calculated. Model calculations were based on emission data that included both the specialized data prepared for the Sulfate Regional Experiment (SURE) and the U.S. Environmental Protection Agency's National Emissions Data System (NEDS). Comparisons were made between calculated and measured SO_2 and SO_4 concentrations. Calculated and measured values show reasonable agreement and indicate that improvements in the calculated values may be achieved by refinements in the modeling of mixing heights and stability.

User's guides are available for three models not heretofore mentioned (Davis, Eadie and Powell, 1979; Draxler and Heffter, 1981; and Huff et al., 1977). A long-range transport model called REGIONAL I is addressed by Davis, Eadie and Powell (1979). REGIONAL I is a computer model set up to run atmospheric assessments on a regional basis. The model has the capability of being run in three modes for a single time period. The three modes are: (1) no deposition, (2) dry deposition, and (3) wet and dry deposition. The guide provides the physical and mathematical basis used in the model for calculating transport, diffusion, and deposition for all three modes. Also, the guide includes a program listing with an explanation of the listings and an example in the form of a short-term assessment for 48 hours.

Draxler and Heffter (1981) have prepared a workbook for use with a version of the Air Resources Laboratories' (ARL) regional trajectory model. This model was developed to permit simultaneous calculation of trajectories from 70 hypothetical sources for very long periods. Five years of data were used to develop a climatology of atmospheric dispersion. Air concentrations can be calculated at receptors coincident with the 70 source locations. Each source and receptor can be treated independently to develop a source/receptor matrix that can be used to produce air concentration patterns over the United States for any combination of sources or to evaluate the impact of different sources on any receptor. Five-year averages of seasonal and annual air concentrations can be calculated for various combinations of wet and dry deposition.

A user's guide is available for the Terrestrial Ecosystem Hydrology Model (TEHM). This model combines mechanistic models for climatic and hydrologic processes with vegetation properties to explicitly simulate interception and throughfall; infiltration; root zone evaporation, transpiration, and drainage; plant and soil water potential; unsaturated and saturated subsurface flow; surface runoff; and open channel flow (Huff et al., 1977). For user convenience, the documentation includes a complete discussion of input formats, example data input sets, output summaries, and a microfiche listing of complete source deck and program output.

LONG-RANGE TRANSPORT OF AIR POLLUTANTS

Information from 19 references dealing with the long-range transport of air pollutants is contained in the abstracts in Appendix D. Table 15 provides a brief comment on each of these references in the order in which they appear in Appendix D. Each listed reference in Table 15 will not necessarily be addressed in this portion of Chapter 3.

Table 15: References in Appendix D Related to Long-Range Transport of Air Pollutants

Author(s) (Year)	Comments
Barnes (1979)	General review of European experience on the long-range transport of air pollution.

Table 15: (continued)

Author(s) (Year)	Comments
Chung (1978)	Influence of long-range transport of air pollutants from the United States on atmospheric sulfate levels in Canada.
Galvin et al. (1979)	Influence of SO_2 emissions in New York on rural atmospheric sulfate levels elsewhere in the state.
Kerr (1979)	Influence on air movement from the United States and Europe on heavy haze in the Arctic region.
Lewis and Ball (1982)	Potential long-range transport of gaseous pollutants and fine particulates.
Martin (1982)	Summary of conference proceedings on the long-range transport of airborne pollutants.
McMillan (1981)	Discussion of meteorological factors affecting long-range transport.
Miller (1979)	Influence on long-range transport on precipitation acidity in Hawaii.
National Technical Information Service (1982)	Bibliography of 69 references on transfrontier transport of air pollution.
OECD (1977)	Study of long-range transport of air pollutants in northwestern Europe.
Ostergaard (1974)	Review of information on long-range transport of air pollutants in northern Europe.
Ottar (1977)	Need for emission reductions as a step in minimizing concerns over long-range transport of air pollutants.
Sandusky, Eadie and Drewes (1979a)	Study of sources and long-range transport of air pollutants in the Pacific Northwest region of the United States.
Sandusky, Eadie and Drewes (1979b)	Study of the long-range transport of sulfur in the western United States.
Scriven and Fisher (1975)	Effect of turbulent diffusion on the long distance transport of SO_2.

Table 15: (continued)

Author(s) (Year)	Comments
Shaw (1979)	Summary of monitoring and research related to the long-range transport of SO_x and NO_x to Canada's eastern coast.
Slinn (1981)	Discussion of reasons for large variabilities in the distances of long-range transport.
Tollan and Hagerhall (1978)	Discussion of general issues associated with the long-range transport of air pollution.
World Meteorological Organization (1979)	Papers presented at a symposium on the long-range transport of air pollutants.

A bibliography of 69 references concerning the long-range transport of air pollution across international barriers is available (National Technical Information Service, 1982). The World Meteorological Organization (1979) has published a proceedings containing 57 papers presented at a symposium on long-range transport of air pollutants. Finally, Martin (1982) summarized 30 papers presented at seven technical sessions in a conference dealing with acid rain and the long-range transport of air pollutants.

Basic Information

Gaseous pollutants such as SO_2, NO_x, and ozone can be transported over distances on the order of 100 km (Lewis and Ball, 1982). Pollutants transported longer distances (several hundred to several thousand kms) tend to be various forms of fine particulates, including sulfates and nitrates formed from emitted SO_2 and NO_x, emitted particles (fly ash, heavy metals, and minerals), and organic products of photochemical processes. Models of sulfur transport indicate that sulfur dioxide is carried a distance of about 250 km on the average, and sulfates about 1000 km, before being deposited, but values vary greatly with atmospheric conditions (Lewis and Ball, 1982).

Long-range transport of air pollutants is of particular concern in the northeastern United States and southeastern Canada (McMillan, 1981), and in all European countries (Ottar, 1977). There are wide ranges in the atmospheric transport of air pollutants (Slinn, 1981), with one of the basic determinants being synoptic-scale meteorological conditions (McMillan, 1981).

Field Studies

Several field studies focused on the long-range transport of air pollutants have been conducted. For example, based on trajectory analysis Galvin et al.

(1979) indicated that high sulfate concentrations observed during summer high-pressure systems at three rural sites in New York State are products of sulfur dioxide emissions to the south and southwest of New York State. The highest concentrations occur in air that first stagnates over the area surrounding the Ohio River Valley and then is advected into New York State as the high-pressure system begins to move eastward.

In discussing atmospheric sulfates in Canada, Chung (1978) indicated that the concentrations were relatively low (less than 5 ug/m^3) in the Western Provinces while high values (less than 15 ug/m^3) were frequently observed in eastern Canada. The natural background level of ambient sulfates appeared to be less than 1.5 ug/m^3 in Canada. Maximum values (greater than 20 ug/m^3) occurred in southern Ontario, however. Generally, values recorded in the Atlantic Provinces were lower than the values observed in Ontario. Air-parcel trajectory analyses of the low-level atmosphere showed that high sulfate levels were often associated with S-SW airflows on the rear side of a warm, moist air mass. The results suggested the long-range transport of airborne sulfur pollutants, mainly from several industrial areas in the United States (Chung, 1978).

Extensive surveys conducted in Sweden and Norway have documented adverse effects of the transport of pollutants through the atmosphere, and research has indicated that the major source of acid precipitation in Scandinavia is the combustion of fossil fuels in the heavily industrialized parts of Europe (Tollan and Hagerhall, 1978), or in the United Kingdom (Ostergaard, 1974). In 1977, the Office of Economic Cooperation and Development (OECD, 1977) completed a study of the relative importance of local and distant sources of sulfur compounds to the air pollution over a region. Special attention was focused on acidity in atmospheric precipitation. The research program covered northwestern Europe and contained three essential elements: a survey of sulfur dioxide emissions all over the region; the measurement of sulfur compounds at a network of ground sampling stations and by aircraft; and the development and testing of mathematical air dispersion models to relate emission data with concentration and deposition data. Results indicated that sulfur compounds do travel long distances in the atmosphere and that the air quality in any one European country is measurably affected by emissions from other European nations.

There may be global implications to the long-range transport of acid rain. For example, heavy haze in the Arctic that can reduce visibility from more than 100 km to less than 10 km may have its origins 10,000 km away in the same polluted air that produces acid rain over the United States and Europe (Kerr, 1979). Analyses of samples collected by the Arctic sampling network tend to support the theory of extremely long-range transport of pollutants into the Arctic.

SELECTED REFERENCES

Adamowicz, R.F., "A Model for the Reversible Washout of Sulfur Dioxide, Ammonia, and Carbon Dioxide from a Polluted Atmosphere and the Production of Sulfates in Raindrops", Atmospheric Environment, Vol. 13, No. 1, 1979, pp. 105-112.

Adamowicz, R.F. and Hill, F.B., "Model for the Reversible Washout of Sulfur Dioxide, Ammonia and Carbon Dioxide from a Polluted Atmosphere and the Production of Sulfates in Raindrops", Report No. CONF-77/102-12, 1977, Brookhaven National Laboratory, Upton, New York.

Barnes, R.A., "The Long Range Transport of Air Pollution--A Review of European Experience", Journal of American Pollution Control Association, Vol. 29, No. 12, Dec. 1979, pp. 1219-1235.

Barrie, L.A., "The Prediction of Rain Acidity and SO_2 Scavenging in Eastern North America", Atmospheric Environment, Vol. 15, No. 1, 1981, pp. 31-41.

Bhumralkar, C.M. et al., "ENAMAP-1 Long-Term Air Pollution Model: Adaptation and Application to Eastern North America", EPA-600/4-80-039, July 1980, U.S. Environmental Protection Agency, Research Triangle Park, North Carolina.

Bolin, B., Aspling, B. and Persson, C., "Residence Time of Atmospheric Pollutants as Dependent on Source Characteristics, Atmospheric Diffusion Processes and Sink Mechanisms", Report No. AC-25, May 1973, Stockholm University-Institute of Meteorology, Stockholm, Sweden.

Bolin, B. and Persson, C., "Regional Dispersion and Deposition of Atmospheric Pollutants with Particular Application to Sulfur Pollution Over Western Europe", Report No. AC-28, May 1974, Stockholm University-Institute of Meteorology, Stockholm, Sweden.

Calvert, J.G. and Stockwell, W.R., "Mechanism and Rates of the Gas Phase Oxidations of Sulfur Dioxide and the Nitrogen Oxides in the Atmosphere", EPA-600/D-83-079, July 1983, U.S. Environmental Protection Agency, Research Triangle Park, North Carolina.

Carmichael, G.R., Yang, D.K. and Lin, C., "Numerical Technique for the Investigation of the Transport and Dry Deposition of Chemically Reactive Pollutants", Atmospheric Environment, Vol. 14, No. 12, 1980, pp. 1433-1438.

Chang, T.Y., "Rain and Snow Scavenging of HNO_3 Vapor in the Atmosphere", Atmospheric Environment, Vol. 18, No. 1, 1984, pp. 191-197.

Chen, C.W. et al., "Acid Rain Model: Hydrologic Module", Journal of the Environmental Engineering Division, American Society of Civil Engineers, Vol. 108, No. EE3, June 1982, pp. 455-472.

Chung, Y.S., "The Distribution of Atmospheric Sulfates in Canada and Its Relationship to Long-Range Transport of Air Pollutants", Atmospheric Environment, Vol. 12, No. 6-7, 1978, pp. 1471-1480.

Corbett, J.O., "Validity of Source-Depletion and Alternative Approximation Methods for a Gaussian Plume Subject to Dry Deposition", Atmospheric Environment, Vol. 15, No. 7, 1981, pp. 1207-1213.

Dana, M.T. et al., "Natural Precipitation Washout of Sulfur Compounds from Plumes", Project No. BNW-389/B46621, June 1973, Battelle Pacific Northwest Laboratories, Richland, Washington.

Dana, M.T. et al., "Precipitation Scavenging of Inorganic Pollutants from Metropolitan Sources", Report No. EPA-650/3-74-005, June 1974, Battelle Pacific Northwest Laboratories, Richland, Washington.

Davenport, M.H. and Peters, L.K., "Field Studies of Atmospheric Particulate Concentration Changes During Precipitation", Atmospheric Environment, Vol. 12, No. 5, 1978, pp. 997-1009.

Davis, W.E., Eadie, W.J. and Powell, D.C., "Users Guide to REGIONAL-1: A Regional Assessment Model", Sept. 1979, U.S. Department of Energy, Washington, D.C.

Dittenhoefer, A.C. and Dethier, B.F., "The Precipitation Chemistry of Western New York States: A Meteorological Interpretation", Report No. GBAI7674, Apr. 1976, Cornell University, Ithaca, New York.

Draxler, R.R. and Heffter, J.L., "Workbook for Estimating the Climatology of Regional-Continental Scale Atmospheric Dispersion and Deposition Over the United States", NOAA-TM-ERL-ARL-96, Feb. 1981, U.S. National Oceanic and Atmospheric Administration, Silver Spring, Maryland.

Dupoux, N., "Variations in Atmospheric Aerosols as a Function of the Properties of the Medium", Report No. NSA2909, July 1973, Centre d'Etudes Nucleaires de Saclay, 91-Gif-sur-Yuette, France.

Durham, J.L., Overton, J.H. and Aneja, V.P., "Influence of Gaseous Nitric Acid on Sulfate Production and Acidity in Rain", Atmospheric Environment, Vol. 15, No. 6, 1981, pp. 1059-1068.

Engelmann, R.J., "Scavenging Prediction Using Ratios of Concentrations in Air and Precipitation", Journal of Applied Meteorology, Vol. 10, June 1971, pp. 493-497.

Enger, L. and Hogstrom, U., "Dispersion and Wet Deposition of Sulfur from a Power Plant Plume", Atmospheric Environment, Vol. 13, No. 6, 1979, pp. 797-811.

Fowler, D., "Wet and Dry Deposition of Sulfur and Nitrogen Compounds from the Atmosphere", Effects of Acid Precipitation on Terrestrial Ecosystems, Plenum Press, New York, New York, 1980, pp. 9-27.

Fuquay, J.J., "Scavenging in Perspective", Report No. CONF-700601-2, July 1970, Pacific Northwest Laboratory, Battelle Northwest, Richland, Washington.

Galvin, P.J. et al., "Transport of Sulfate to New York State", Environmental Science and Technology, Vol. 12, No. 5, Aug. 1979, pp. 580-584.

Gatz, D.F., "Estimates of Wet and Dry Deposition of Chicago and Northwest Indiana Aerosols into Southern Lake Michigan", ERDA 2nd Federal Conference on the Great Lakes, Mar. 25-27, 1975, Argonne National Laboratory, Argonne, Illinois, pp. 277-290.

Gatz, D.F., "Urban Influence on Deposition of Sulfate and Soluble Metals in Summer Rains", 1979, Illinois State Water Survey, Urbana, Illinois.

Georgii, H.W. and Beilke, S., "Atmospheric Aerosol- and Trace-Gas-Washout", Report No. USGRDR6616, Mar. 1966, Institut Fuer Meteorologie Und Geophysik, Frankfurt University, Frankfurt, West Germany.

Golomb, D. et al., "Sensitivity Analysis of the Kinetics of Acid Rain Models", Atmospheric Environment, Vol. 17, No. 3, 1983, pp. 645-653.

Granat, L., "On the Relation Between pH and the Chemical Composition in Atmospheric Precipitation", Tellus, Vol. 24, 1972, pp. 550-560.

Gravenhorst, G. et al., "Sulfur Dioxide Absorbed in Rain Water", Effects of Acid Precipitation on Terrestrial Ecosystems, Plenum Press, New York, New York, 1980, pp. 41-55.

Hales, J.M., "Pollutant Transformation and Removal Measurements at METROMEX", Report No. CONF-770210-8, Jan. 1977, Battelle Pacific Northwest Laboratories, Richland, Washington.

Hales, J.M. and Lee, R.N., "Precipitation Scavenging of Organic Contaminants", July 1975, Battelle Pacific Northwest Laboratories, Richland, Washington.

Hales, J.W., "How the Air Cleans Itself", Proceedings of the SCI Sulfur Symposium, Feb. 1979, Battelle Pacific Northwest Laboratories, Richland, Washington.

Harrison, R.M. and Pio, C.A., "Comparative Study of the Ionic Composition of Rainwater and Atmospheric Aerosols: Implications for the Mechanism of Acidification of Rainwater", Atmospheric Environment, Vol. 17, No. 12, 1983, pp. 2539-2543.

Hegg, D.A., "Sources of Sulfate in Precipitation--Part 1--Parameterization Scheme and Physical Sensitivities", Journal of Geophysical Research, Vol. 88, No. C2, Feb. 1983, pp. 1369-1374.

Hidy, G.M. et al., "Precipitation-Scavenging Chemistry for Sulfate and Nitrate from the SURE and Related Data", EPRI-EA-1914-V.2, Feb. 1983, Environmental Research and Technology, Inc., Westlake Village, California.

Hill, F.B. and Adamowicz, R.F., "Model for Rain Composition and the Washout of Sulfur Dioxide", Contract No. E(30-1)-16, July 1976, Brookhaven National Laboratories, Upton, New York.

Hill, F.B. and Adamowicz, R.F., "A Model for Rain Composition and the Washout of Sulfur Dioxide", Atmospheric Environment, Vol. 11, No. 10, 1977, pp. 917-928.

Horst, T.W., "Review of Gaussian Diffusion-Deposition Models", 1979, Battelle Pacific Northwest Laboratories, Richland, Washington.

Huff, D. et al., "TEHM: A Terrestrial Ecosystem Hydrology Model", EDFB/1BP-76'8, Apr. 1977, Oak Ridge National Laboratory, Oak Ridge, Tennessee.

Junod, A., "Prediction Methods of Cooling Tower Plumes", International Center for Heat and Mass Transfer Conference on Heat Disposal from Power Generation, Aug. 23-28, 1976, Dubrovnik.

Kerr, R.A., "Global Pollution: Is the Arctic Haze Actually Industrial Smog?", Science, Vol. 205, No. 4403, July 20, 1979, pp. 290-293.

Lee, Y., "Numerical Models for Precipitation Scavenging", Report No. C00-1407-52, Apr. 1974, Department of Atmospheric and Oceanic Science, University of Michigan, Ann Arbor, Michigan.

Lehmann, E.J., "Atmospheric Modeling of Air Pollution (A Bibliography with Abstracts)", Report No. GLAT7515, June 1975, National Technical Information Service, U.S. Department of Commerce, Springfield, Virginia.

Lewellen, W.S. and Sheng, Y.P., "Modeling of Dry Deposition of SO_2 and Sulfate Aerosols", July 1980, Aeronautical Research Associates of Princeton, Inc., Princeton, New Jersey.

Lewis, D.H. and Ball, R.H., "Long-Range Transport of Air Pollution", DOE/EP-0037, Jan. 1982, U.S. Department of Energy, Washington, D.C.

Liu, M.K., Stewart, D.A. and Henderson, D., "Mathematical Model for the Analysis of Acid Deposition", Journal of Applied Meteorology, Vol. 21, No. 6, June 1982, pp. 859-873.

Martin, H.L., "Long-Range Transport of Airborne Pollutants and Acid Rain Conference", Water, Air, and Soil Pollution, Vol. 18, No. 1-3, July/Oct. 1982.

Mayerhofer, P.M. et al., "ENAMAP-1A Long-Term SO_2 and Sulfate Air Pollution Model: Refinement of Transformation and Deposition Mechanisms", EPA-600/3-82-063, May 1982, U.S. Environmental Protection Agency, Research Triangle Park, North Carolina.

McMillan, A.C., "Long-Range Transport of Atmospheric Pollutants", Research Review of Ontario Hydro, No. 2, May 1981, pp. 11-20.

McNaughton, D.J., "Emission Source Specification in a Regional Pollutant Transport Model", June 1979, Battelle Pacific Northwest Laboratories, Richland, Washington.

Melo, O.T., "Ontario Hydro Studies of Atmospheric Chemistry", Research Review Ontario Hydro, No. 2, May 1981, pp. 21-28.

Michael, P. and Raynor, G.S., "Modification of Trajectory Models Needed for Pollutant Source-Receptor Analysis", BNL-29924, 1981, Brookhaven National Laboratory, Upton, New York.

Miller, J.M., "Acidity of Hawaiian Precipitation as Evidence of Long-Range Transport of Pollutants", 1979, National Oceanic and Atmospheric Association, Silver Spring, Maryland.

Mills, M.T. and Reeves, M., "A Multi-Source Atmospheric Transport Model for Deposition of Trace Contaminants", ORNL-NSF-EATC-2, 1973, Oak Ridge National Laboratory, Oak Ridge, Tennessee.

Molenkamp, C.R., "Numerical Modeling of Precipitation Scavenging by Convective Clouds", Report No. UCRL-75896, Sept. 1974, California University-Lawrence Livermore Laboratory, Livermore, California.

Moore, R.E., "Calculation of the Depletion of Airborne Pollutant Plumes Through Dry Deposition Processes", Environment International, Vol. 3, No. 1, 1980, pp. 3-10.

Moroz, W.J., "Prediction of Deposition of Acid Precursors", Research Review of Ontario Hydro, No. 2, May 1981, pp. 75-80.

National Technical Information Service, "Acid Precipitation: Transfrontier Transport of Air Pollutants--1976-November, 1982 (Citations from the Energy Data Base)", Nov. 1982, U.S. Department of Commerce, Springfield, Virginia.

OECD, "The OECD Programme on Long Range Transport of Air Pollutants", 1977, Office of Economic Cooperation and Development, Paris, France.

Ostergaard, K., "The Problem of Regional Transportation of Air Pollutants in Northern Europe", Clean Air-Australia, Vol. 8, No. 2, May 1974, pp. 30-33.

Ottar, B., "International Agreement Needed to Reduce Long-Range Transport of Air Pollutants in Europe", Ambio, Vol. 6, No. 5, 1977, pp. 262-270.

Overton, J.H., Aneja, V.P. and Durham, J.L., "Production of Sulfate in Rain and Raindrops in Polluted Atmospheres", Atmospheric Environment, Vol. 13, 1979, pp. 355-367.

Overton, J.H. and Durham, J.L., "Acidification of Rain in the Presence of SO_2, H_2O_2, O_3, and HNO_3", EPA-600/D-82-150, Jan. 1982, U.S. Environmental Protection Agency, Research Triangle Park, North Carolina.

Pack, D.H., editor, Proceedings: Advisory Workshop to Identify Research Needs on the Formation of Acid Precipitation, May 1979, Sigma Research, Inc., Richland, Washington.

Patterson, M.R., Mankin, J.B. and Brooks, P.A., "Overview of Unified Transport Model", Proceedings of First Annual National Science Foundation Trace Contaminant Conference, 1973, National Science Foundation, Washington, D.C., pp. 12-23.

Penkett, S.A. et al., "The Importance of Atmospheric Ozone and Hydrogen Peroxide in Oxidizing Sulfur Dioxide in Clouds and Rainwater", Atmospheric Environment, Vol. 13, No. 1, 1979, pp. 123-128.

Powell, D.C. et al., "Variable Trajectory Model for Regional Assessments of Air Pollution from Sulfur Compounds", Contract EY-76-C-06-1830, Feb. 1979, Battelle Pacific Northwest Laboratories, Richland, Washington.

Ragland, K.W. and Wilkening, K.E., "Intermediate-Range Grid Model for Atmospheric Sulfur Dioxide and Sulfate Concentrations and Depositions", Atmospheric Environment, Vol. 17, No. 5, 1983, pp. 935-947.

Ritchie, I.M., Bowman, J.D. and Burnett, G.B., "Mesoscale Atmospheric Dispersion Model for Predicting Ambient Air Concentration and Deposition Patterns for Single and Multiple Sources", Atmospheric Environment, Vol. 17, No. 7, 1983, pp. 1215-1223.

Ritchie, L.T., Brown, W.D. and Wayland, J.R., "Effects of Rainstorms and Runoff on Consequences of Atmospheric Releases from Nuclear Reactor Accidents", Nuclear Safety, Vol. 19, No. 2, Mar.-Apr. 1978, pp. 220-239.

Rogowski, R.S. et al., "Carbon-Catalyzed Oxidation of SO_2 by NO_2 and Air", Apr. 1982, National Aeronautics and Space Administration, Hampton, Virginia.

Sandusky, W.F., Eadie, W.J. and Drewes, D.R., "Long-Range Transport of Pollutants in the Pacific Northwest", Jan. 1979a, Battellle-Pacific Northwest Laboratories, Richland, Washington.

Sandusky, W.F., Eadie, W.J. and Drewes, D.R., "Long-Range Transport of Sulfur in the Western United States", Contract: EY-76-C-06-1830, Jan. 1979b, Battelle Pacific Northwest Laboratories, Richland, Washington.

Schmidt, J.A., "Models of Particle Dry Deposition to a Lake Surface and Some Effects of Surface Microlayers", Journal of Great Lakes Research, Vol. 8, No. 2, 1982, pp. 271-280.

Schnoor, J.L., Carmichael, G.R. and van Shepen, F.A., "Integrated Approach to Acid Rainfall Assessments", 1980, University of Iowa, Iowa City, Iowa.

Scriven, R.A. and Fisher, B.E.A., "The Long-Range Transport of Airborne Material and Its Removal by Deposition and Washout--II. The Effect of Turbulent Diffusion", Atmospheric Environment, Vol. 9, No. 1, Jan. 1975, pp. 59-69.

Sehmel, G.A., "Model Predictions and a Summary of Dry Deposition Velocity Data", Aug. 1979, Battelle Pacific Northwest Laboratories, Richland, Washington.

Sehmel, G.A. and Hodgson, W.H., "Model for Predicting Dry Deposition of Particles and Gases to Environmental Surfaces", American Institute of Chemical Engineers Symposium Series, Vol. 76, No. 196, 1980, pp. 218-230.

Semonin, R.G. et al., "Study of Atmospheric Pollution Scavenging-- Seventeenth Progress Report", July 1979, Illinois State Water Survey, Urbana, Illinois.

Shannon, J.D., "Advanced Statistical Trajectory Regional Air Pollution Model", 1978, Argonne National Laboratory, Argonne, Illinois.

Shannon, J.D., "Model of Regional Long-Term Average Sulfur Atmospheric Pollution, Surface Removal, and Net Horizontal Flux", Atmospheric Environment, Vol. 15, No. 5, 1981, pp. 689-701.

Shaw, R.W., "Acid Precipitation in Atlantic Canada", Environmental Science and Technology, Vol. 13, No. 4, Apr. 1979, pp. 406-412.

Slinn, W.G., "Estimates for the Long-Range Transport of Air Pollutants", May 1981, Battelle Pacific Northwest Laboratories, Richland, Washington.

Sullivan, J.L., Wen, Y.P. and Frantisak, F., "Smelter Stack Plume Kinetic Studies in Northern Ontario", Proceedings of the Second National Conference on Complete Water Reuse: Water's Interface with Energy, Air, and Soils, 1975, American Institute of Chemical Engineers, New York, New York, pp. 311-314.

Tollan, A. and Hagerhall, B., "Deterioration of Water Quality Due to Long-Range Transport of Air Pollution", Proceedings of the United Nations Water Conference, (mar del Plata, Argentina), 1978, United Nations, New York, New York, pp. 2059-2070.

Tsai, Y.J. and Johnson, D.H., "Cooling Tower Drift Model", Fifth Pittsburgh Conference Proceedings, 1974, Pittsburgh University, Pittsburgh, Pennsylvania, p. 143.

Williams, R.M., "Model for the Dry Deposition of Particles to Natural Water Surfaces", Atmospheric Environment, Vol. 16, No. 8, 1982, pp. 1933-1938.

World Meteorological Organization, "Papers Presented at the WMO Symposium on the Long-Range Transport of Pollutants and Its Relation to General Circulation Including Stratospheric/Tropospheric Exchange Processes", WMO No. 538, 1979, Geneva, Switzerland.

CHAPTER 4

EFFECTS OF ACID PRECIPITATION

The most recognized effects of acid precipitation are associated with the loss of living components of aquatic ecosystems. Specifically, the loss of fish life in many lakes in North America and Europe represent a major environmental concern. Acid precipitation can cause multiple effects on both terrestrial and aquatic ecosystems. Terrestrial ecosystem effects include those on vegetation, crops, and soil. Ground water can potentially be influenced by changes in soil chemistry from direct precipitation and/or dry deposition in recharge areas. Aquatic ecosystem effects encompass changes in surface water chemistry in rivers and lakes, and potential losses of aquatic floral and faunal species. Acid precipitation and dry deposition can also affect materials and even human health. This chapter summarizes each of these categories of effects following a background section featuring the abstracts in Appendix E.

BACKGROUND INFORMATION

This chapter is based on information from 181 references whose abstracts are contained in Appendix E. Table 16 provides a brief comment on each of these references in the order in which they appear in Appendix E. Not every reference listed in Table 16 will be specifically addressed in this chapter.

Table 16: References in Appendix E Addressing the Effects of Acid Precipitation

Author(s) (Year)	Comments
Abrahamsen et al. (1976)	Multiple effects of acid precipitation on terrestrial and aquatic ecosystems in Norway.
Abrahamsen, Houland and Haguar (1980)	Effects of artificial acid rain on decomposer organisms in soil.
Alexander (1980)	Effects of acidity on soil microorganisms and microbial processes.
Allen (1980)	Laboratory studies of the effects of acid rain on fresh water microcrustaceans.
Amthor and Bormann (1983)	Precipitation acidity effects on perennial ryegrass plants.
Andersson, Fagerstrom and Nilsson (1980)	Use of nitrogen dependent tree growth model in studying forest responses to acid deposition.

Table 16: (continued)

Author(s) (Year)	Comments
Arnold, Light and Dymond (1980)	Water quality effects of acid precipitation in Pennsylvania.
Arthur and Wagner (1982)	Proceedings of workshop on response of agricultural soils to acid deposition.
Baath et al. (1980)	Study of soil chemistry and microbial effects from artificial acid rain.
Baath, Lundgren and Soderstrom (1979)	Effects of artificial acid rain on soil microbial activity and biomass.
Babich and Stotzky (1978)	Effects of atmospheric sulfur compounds on soil microorganisms.
Bache (1980a)	Factors influencing soil acidification by acid rain.
Bache (1980b)	Factors influencing soil sensitivity to acidification.
Baker (1981)	Study of Al and hydrogen ion influences on fish survival in New York lakes.
Beamish (1974)	Influence of Zn and pH on white suckers in a lake.
Bell (1981)	Effects of acid precipitation on fish stocks in Norwegian lakes.
Botkin and Aber (1979)	Simulation of effects of acid precipitation on forest productivity and species composition.
Boylen et al. (1983)	Aquatic microbial population effects of acidic deposition in New York lakes.
Bradford, Page and Straughan (1981)	Fish kills in acidified lakes in the California Sierra Nevadas.
Brown and Sadler (1981)	Effects of reduced sulfur inputs on fish populations in Norwegian lakes.
Canfield (1983)	Sensitivity of animal and plant populations in Florida lakes to acidic precipitation.

Table 16: (continued)

Author(s) (Year)	Comments
Chang and Alexander (1983)	Effect of acid precipitation on algal fixation of nitrogen and CO_2 in forest soils.
Chen et al. (1983)	Model for tree canopy influences on acid precipitation.
Ciolkosz and Levine (1983)	Computer simulation model for determining the impact of acid deposition on Pennsylvania soils.
Coffin and Knelson (1976)	Possible human health effects of sulfates and sulfuric acid.
Cole and Johnson (1977)	Influence of acid rainfall on canopy leaching in a Douglas fir ecosystem.
Cole and Stewart (1983)	Acid deposition effects on phosphorus cycling in soils.
Coleman (1983)	Impacts of acid deposition on carbon cycling in agricultural soils.
Conway and Hendrey (1981)	Effects of acid precipitation on lake algae, diatoms, dinoflagellates, macrophytes, and benthic plants.
Cowling and Davey (1981)	General discussion of ecological consequences of acid precipitation.
Cowling and Dochinger (1979)	Summary of information on effects of acid rain on crops and trees.
Crisman and Brezonik (1980)	Biological effects of acid precipitation on 20 lakes in Florida.
Cronan (1981)	Effects of acid precipitation on cation transport in forest soils in New Hampshire.
Dailey and Winslow (1980)	Bibliography of 961 references on the health and environmental effects of acid precipitation.
Davis (1981)	Water quality effects of the atmospheric deposition of energy-related pollutants.

Table 16: (continued)

Author(s) (Year)	Comments
Derrick et al. (1984)	Neutralization of atmospheric acidic species by airborne alkaline soil material in southwest Texas.
Dochinger and Seliga (1975)	General discussion of forest ecosystem effects from acid rain in Norway.
Dochinger and Seliga (1976)	85 papers presented at the international symposium on acid precipitation and the forest ecosystem.
Dynamac Corporation (1982)	Bibliography on effects of air pollution and acid rain on fish, wildlife, and their habitats.
Evans (1981)	Dose-response functions of several crops to simulated acid rain.
Evans et al. (1982)	Effects of acid precipitation on seed yields of field-grown soybeans.
Evans, Conway and Lewin (1980)	Yield responses of field-grown soybeans to simulated acid rain.
Evans, Francis and Raynor (1977)	Effects of acid rain on terrestrial plants.
Evans, Gmur and DaCosta (1968)	Effects of simulated acid rain on six clones of hybrid poplars.
Evans, Gmur and Mancini (1982)	Greenhouse study of effects of simulated acid rain on several crops.
Evans and Hendrey (1979)	Proceedings of workshop on effects of acid precipitation on vegetation, soils, and terrestrial ecosystems.
Evans and Lewin (1979)	Effects of simulated acid rain on seed yields of soybeans and pinto beans.
Faust and McIntosh (1983)	Buffer capacities of fresh water lakes to acid rain.
Ferenbaugh (1975)	General description of effects of acid rain.
Flinn et al. (1983)	Bibliography of over 1,300 references on the effects of acidic deposition on corrosion and deterioration of materials.

Table 16: (continued)

Author(s) (Year)	Comments
Francis, Olson and Bernatsky (1980)	Study of effects of acidity on microbial processes in a forest soil.
Francis, Olson and Bernatsky (1981)	Effects of soil acidity on microbial decomposition of organic matter, transformation of nitrogen, and soil chemical and biological properties.
Fritz (1980)	Effects of low pH on freshwater fish and shellfish.
Frizzola and Baier (1975)	Effects of rain water quality on ground water quality in Long Island, New York.
Galloway et al. (1978)	Acid rain effects on agricultural lands, forests, ranges, parks, surface waters, and aquatic life.
Galloway, Norton and Church (1983)	Model for aquatic ecosystem response to acid deposition.
Gauri (1979)	Effects of acid rain on external structural materials.
Gjessing (1980)	Water quality impacts of acid precipitation in Norway.
Glass (1978)	Potential economic impacts of acid rain.
Glass (1979)	Description of U.S. program on studying the effects of acid rain.
Glass et al. (1982)	Effects of acid precipitation in the Adirondack Mountains of New York.
Glass, Glass and Rennie (1979)	General discussion of effects of acid rain.
Glass, Glass and Rennie (1980)	Effects of acid rain on natural and man-made environments.
Glass, Likens and Dochinger (1978)	General delineation of effects of acid rain.
Gorham and McFee (1980)	Discussion of materials capable of being leached or retained by soils when exposed to acid rain.

Table 16: (continued)

Author(s) (Year)	Comments
Graham and Wood (1981)	Effects of acid type on toxicity to rainbow trout.
Grodzinska et al. (1977)	Effects of acid precipitation on various features of forest ecosystems.
Haines (1981)	Discussion of effects of acid precipitation on various components of aquatic ecosystems.
Haines and Waide (1980)	Tree responses to acid rain in a southern deciduous forest.
Harriman and Morrison (1982)	Invertebrate and fish responses to acid rain in 12 streams in forested and nonforested areas in central Scotland.
Harte, Holdren and Tonnesson (1983)	Potential for micronutrient changes in Sierra Nevada lakes subject to nitrate-dominated acid deposition.
Havas and Hutteneu (1980)	Acid rain effects on coniferous trees in Finland.
Heagle et al. (1983)	Effects of simulated acid rain on field-grown soybeans.
Hendrey et al. (1976)	Changes in freshwater faunal species due to acid precipitation.
Hendrey et al. (1980)	Sensitivity of surface water bodies in the eastern United States to water quality changes.
Hendrey and Kaplan (1982)	Sensitivity of headwater areas in the eastern United States to water quality changes.
Hendrey and Lipfert (1980)	Review of impacts of acid rain on aquatic ecosystems.
Hendrey and Vertucci (1980)	Benthic plant communities in an acidic New York lake.
Henrikson and Kirkhusmo (1982)	Acidification of ground water in Norway areas with low surface water pH.
Hileman (1981)	General information on effects of acid rain in Canada.

Table 16: (continued)

Author(s) (Year)	Comments
Hoffman, Lindberg and Turner (1980)	Role of the forest canopy in acid precipitation quality changes.
Hultberg et al. (1977)	Effects of acid precipitation on salamanders, aquatic ecosystems, and forest soils.
Hutchinson (1980)	Mobilization and leaching of heavy metals from two Canadian soils due to acid precipitation.
Hutchinson et al. (1977)	Effects of acid rainfall and heavy metal particulates on Canadian forest ecosystems.
Irving (1983)	Review of effects of simulated acid rain on crop species.
Jacks and Knutsson (1981)	Acidification of ground water in Sweden due to acid precipitation.
Jacobson (1980)	Role of acid rain in plant phototoxicity.
Johnsen (1980)	Effects of SO_2 on lichens and bryophytes in Denmark.
Johnson (1979)	Neutralization of strong acids from precipitation in the soil zone.
Johnson (1980)	Potential soil leaching changes resulting from H_2SO_4 in acid rainfall.
Johnson (1981a)	Need for quantification of acid rain effects on forest productivity.
Johnson (1981b)	Potential impacts of acid rain on stream water quality in southeastern Alaska.
Johnson (1981c)	Techniques and frameworks for assessing acid rain effects on elemental transport from terrestrial to aquatic ecosystems.
Johnson et al. (1981a)	Acid precipitation as a growth-limiting factor for pine trees in the New Jersey Pinelands.
Johnson et al. (1981b)	Water quality changes in a New Hampshire stream due to acid rain.

Table 16: (continued)

Author(s) (Year)	Comments
Johnson and Richter (1984)	Effects of atmospheric depositions on nutrient leaching patterns in forest ecosystems.
Johnson and Siccama (1983)	Potential mechanisms of acid deposition-induced declines in forest vegetation.
Johnson and Siccama (1984)	Possible role of acid deposition in the decline of the red spruce in the northern Appalachians area.
Johnson, Turner and Kelly (1982)	Effects of acidic atmospheric inputs on forest nutrient cycles.
Kaplan, Thode and Protas (1981)	Soils as a basic determinant of regional sensitivities to acid precipitation.
Killham, Firestone and McColl (1983)	Effects of acid rain on microbial activity in Sierran forest soil in northern California.
Klein, Kreitinger and Alexander (1983)	Influence of simulated acid rain on nitrate formation in forest soils in the Adirondacks region of New York.
Klopatek, Harris and Olson (1979)	Regional ecological overview of the potential effects of acid precipitation on soils.
Kucera (1976)	Effects of SO_2 and acid precipitation in Sweden on various metals.
Lee et al. (1981)	Study of response of major U.S. crops to acid rain.
Lee, Neely and Perrigan (1980)	Study of relative sensitivity of major U.S. crops to sulfuric acid rain.
Lee and Weber (1979)	Effects of simulated acid rain on 11 woody species.
Lee and Weber (1982)	Influence of hardwood canopy and litter layer on acid rain input of chemicals to soil.
Likens et al. (1977)	Study of biogeochemistry of Hubbard Brook in New Hampshire as influenced by acid precipitation.

Table 16: (continued)

Author(s) (Year)	Comments
Likens, Bormann and Eaton (1980)	Stream water chemistry at Hubbard Brook in New Hampshire as influenced by acid precipitation.
Lindberg and Harriss (1981)	Atmospheric deposition influences on chemical changes in and by a forest canopy.
Lindberg, Shriner and Hoffman (1981)	Beneficial and detrimental effects of wet and dry deposition on forest canopies.
Logan, Derby and Duncan (1982)	Study of lake susceptibility to acid precipitation in the central Washington Cascades.
Malmer (1976)	Chemical changes in soil due to acid precipitation in Scandinavia.
Matziris and Nakos (1977)	Simulated acid rain effects on Aleppo pine.
Mayer and Ulrich (1980)	Wet and dry deposition as a source of input of elements into soil.
McColl (1981)	Effects of acid rain on selected California plants and soils.
McKinley and Vestal (1982)	Effects of acidic pH on microbial decomposition of plant litter.
McLaughlin et al. (1983)	Accumulation of trace metals in tree rings.
McLaughlin, West and Blasing (1984)	Dendroecological approaches for evaluating acid precipitation effects on forest trees.
Mitchell, Landers and Brodowski (1981)	Organic and inorganic sulfur fractions in three New York lake sediments.
Mortvedt (1983)	Acid deposition effects on micronutrient cycling and, in turn, crop yields.
National Technical Information Service (1982a)	Bibliography of 180 references on acid precipitation effects on terrestrial ecosystems.

Table 16: (continued)

Author(s) (Year)	Comments
National Technical Information Service (1982b)	Bibliography of 152 references on acid precipitation effects on aquatic ecosystems.
Nelson and Delwiche (1983)	Susceptibility to acidification of 63 Oregon lakes in the Cascades.
Nilssen (1980)	Effects of acid precipitation on faunal species in freshwater lakes in southern Norway.
Nilsson (1980)	Soil leaching changes from acid precipitation.
Norton et al. (1981)	Water quality changes in 94 northern New England lakes from atmospheric inputs of acids and heavy metals.
Nriagu (1978)	Natural buffering ability of 120 Canadian lakes.
Olson (1983)	Soil chemical changes from acid deposition.
Olson, Johnson and Shriner (1982)	Soil chemistry, bedrock geology, terrain characteristics, and land use as determinants in soil susceptibility to acid precipitation.
Overrein, Seip and Tollan (1980)	General effects of acid precipitation in Norway as determined through a research program.
Peterson (1980a)	Effects of acid precipitation on soil podzolization processes.
Peterson (1980b)	Soil orders and classifications according to their response to acid precipitation.
Pfeiffer and Festa (1980)	Water quality features of 849 ponded waters in the Adirondack region of New York.
Pitblado, Keller and Conroy (1980)	Classification of 187 Canadian lakes according to water chemistry data (23 variables).
Pough (1981)	Response of 14 species of amphibians to acid precipitation.

Table 16: (continued)

Author(s) (Year)	Comments
Pucket (1982)	Use of dendroecological analyses to examine acid rain impacts on tree growth in southeastern New York.
Raynal (1980)	Review of sensitivity of forest ecosystem components to acid precipitation.
Raynal, Roman and Eichenlaub (1982)	Response of sugar maple seedling growth to simulated acid rain.
Reuss (1975)	Soil-plant system responses to acid rain.
Richter, Johnson and Todd (1983)	Soil leaching from atmospheric sulfur deposition.
Roberts et al. (1980)	Effects of acid rain on soil microbial and leaching processes in central England.
Rorison (1980)	Constraints of acidic soils on soil chemistry and leaching.
Scherbatskoy and Klein (1983)	Response of yellow birch and white spruce to acidic mists.
Schindler et al. (1980)	Effects of acidification on mobilization of heavy metals and radionuclides from freshwater lake sediments.
Schindler and Turner (1982)	Biological, physical and chemical changes in a small lake subjected to experimental acidification.
Schnitzer (1980)	Reactions of humic and fulvic acids to decreased soil pH.
Schofield (1976)	Summary of information on effects of acid precipitation on freshwater fish.
Schofield (1979)	Water quality changes in surface water due to acid rain input.
Sears and Langmuir (1982)	Soil chemical changes as a function of acid precipitation.
Shinn (1978)	Processes affecting foliar deposition and retention of particles on plants.

Table 16: (continued)

Author(s) (Year)	Comments
Shriner (1978)	Simulated acid rain influences on disease development in plants.
Shriner and Cowling (1978)	Acid rain influences on host-pathogen complexes in plant diseases.
Sigma Research, Inc. (1982)	Proceedings of a workshop on the ecological effects of acid precipitation.
Singh, Abrahamsen and Stuanes (1980)	Effect of simulated acid rain or sulfate mobility in Norwegian forest soils.
Sparks and Curtis (1983)	Effects of acid rain on leaching kinetics in three Delaware soils.
Sposito, Page and Frink (1980)	Computer calculations for effects of acid precipitation on soil leachate quality.
Sprules (1975)	Crustacean zooplankton species and species associations in 47 Canadian lakes.
Stednick and Johnson (1982)	Changes in chemical composition of acid precipitation upon passage through a forest soil.
Strayer and Alexander (1981)	Effects of simulated acid rain on glucose mineralization in forest soils.
Strayer, Chyi-Jiin and Alexander (1981)	Effects of simulated acid rain on nitrification and nitrogen mineralization in forest soils.
Tamm (1976)	Biological effects of acid precipitation on soils and forest vegetation.
Taylor and Symons (1984)	Potential health effects of acid rain as reflected in water supplies in the northeastern United States.
Tetra Tech, Inc. (1981)	Seven papers associated with the Integrated Lake-Watershed Acidification Study in New York.
Torrenueva (1981)	Review of effects of acid precipitation on vegetation and soils.
Tschupp (1974)	Influence of acid precipitation on ground water quality in recharge areas.

Table 16: (continued)

Author(s) (Year)	Comments
Tukey (1980)	Summary of factors that may produce injury to plant species.
Turk and Adams (1983)	Sensitivity of lakes in northwestern Colorado to water quality changes from acid precipitation.
Tveite and Abrahamsen (1980)	Effects of artificial acid rain on growth and nutrient status of Scots pines in southern Norway.
Tyler (1978)	Influence of artificial acid rain on metals leaching from soils in southeastern Sweden.
Ulrich, Mayer and Khanna (1980)	Soil chemical changes due to acid precipitation in a forest in the Federal Republic of Germany.
U.S. Fish and Wildlife Service (1982)	Report on series of Adaptive Environmental Assessment workshops on effects of acid precipitation on aquatic ecosystems.
Vangenechten (1980)	Water quality changes in 53 Belgium surface waters due to acid rain.
Vaughan, Underwood and Ogden (1982)	Responses of diatoms to lake acidification in Nova Scotia.
Wainwright, Supharungsun and Killham (1982)	Solubilization of heavy metals in soils subject to low pH rain.
Webb (1982)	Influence of rainfall chemistry on water chemistry in the Tovdal River in southern Norway.
Wetstone and Foster (1983)	Summary of forest impacts of acid precipitation.
Wiklander (1980)	Review of sensitivity of soils to acid precipitation.
Wood and Bormann (1975)	Leaching changes in foliage caused by artificial acid mists.

Table 16: (continued)

Author(s) (Year)	Comments
Wright and Gjessing (1976)	Discussion of changes in the chemical composition of lakes in southern Scandinavia and eastern North America due to acid precipitation.

Several bibliographies on the effects of acid precipitation are available. For example, Dailey and Winslow (1980) prepared an abstracted bibliography containing 961 citations on the health and environmental effects of acid precipitation. Flinn et al. (1983) issued a bibliography containing more than 1300 article citations and abstracts on the effects of acidic deposition, air pollutants, and biological and meteorological factors on the corrosion and deterioration of materials in the atmosphere. The listing includes citations for the years 1950 to 1982, with selected citations for the years 1880 to 1949. In addition, the National Technical Information Service of the U.S. Department of Commerce has available bibliographies of 180 citations on the effects on terrestrial ecosystems (National Technical Information Service, 1982a), and 152 citations on the effects on aquatic ecosystems (National Technical Information Service, 1982b).

Several technical conferences have been held on the effects of acid precipitation, and the proceedings have been subsequently published.   For example, a 1975 symposium assessed known information about atmospheric chemistry, transport, and precipitation, and the effects of acid precipitation on aquatic ecosystems, forest soils, and forest vegetation.   The proceedings contain 85 papers presented at the symposium (Dochinger and Seliga, 1976). A 1979 workshop determined the levels of current knowledge of the effects of acid precipitation on vegetation, soils, and terrestrial ecosystems; and research needed in these areas to understand the environmental impacts of acid rain (Evans and Hendrey, 1979). In 1981 a workshop was held on the response of agricultural soils to acid deposition (Arthur and Wagner, 1982).   In 1982, participants from the United States, Canada, Norway, Sweden, and the United Kingdom exchanged information about their acid precipitation programs; areas of emphasis, approaches and techniques being used; and planned directions (Sigma Research, Inc., 1982).   Among the key recommendations made by workshop participants were that definition and study of linkages between system inputs and responses should be emphasized, and effects should be quantified.

TERRESTRIAL ECOSYSTEM EFFECTS

Acid precipitation can have multiple effects on terrestrial ecosystems. For example, demonstrated biological effects include necrotic lesions on foliage, nutrient loss from foliar organs, reduced resistance to pathogens, accelerated erosion of waxes on leaf surfaces, reduced rates of decomposition of leaf litter, inhibited formation of terminal buds, increased seedling mortality, and heavy metal accumulation (Cowling and Davey, 1981). Soil and vegetation/crop-related effects include soil acidification, calcium removal, aluminum and manganese solubilization, tree growth reduction, crop quality and quantity reduction, elimination of useful soil microorganisms, and heavy metal elements selectively exchanged for more beneficial mono and divalent cations (Glass, Glass and Rennie, 1979). Soil microbiological processes such as nitrogen fixation, mineralization of forest litter, and nitrification of ammonium compounds can be inhibited, the degree depending on the degree of cultivation and soil buffering capacity (Cowling and Davey, 1981). This portion of Chapter 4 summarizes these effects relative to vegetation, crops, soils, and ground water.

Effects on Vegetation

Table 17 identifies 38 references from Appendix E that address the effects of acid precipitation on vegetation. The information can be considered in three categories: laboratory or experimental studies; field studies; and process models. Laboratory or experimental studies include those by Matziris and Nakos (1977); Scherbatskoy and Klein (1983); Raynal, Roman and Eichenlaub (1982); and Shriner (1978).

Table 17: References in Appendix E Dealing with Effects of Acid Precipitation on Vegetation

Author(s) (Year)	Comments
Andersson, Fagerstrom and Nilsson (1980)	Use of nitrogen dependent tree growth model in studying forest responses to acid deposition.
Botkin and Aber (1979)	Simulation of effects of acid precipitation on forest productivity and species composition.
Chen et al. (1983)	Model for tree canopy influences on acid precipitation.
Cole and Johnson (1977)	Influence of acid rainfall on canopy leaching in a Douglas fir ecosystem.
Dochinger and Seliga (1975)	General discussion of forest ecosystem effects from acid rain in Norway.
Evans, Francis and Raynor (1977)	Effects of acid rain on terrestrial plants.
Evans, Gmur and DaCosta (1968)	Effects of simulated acid rain on six clones of hybrid poplars.
Grodzinska et al. (1977)	Effects of acid precipitation on various features of forest ecosystems.

Table 17: (continued)

Author(s) (Year)	Comments
Haines and Waide (1980)	Tree responses to acid rain in a southern deciduous forest.
Havas and Hutteneu (1980)	Acid rain effects on coniferous trees in Finland.
Hoffman, Lindberg and Turner (1980)	Role of the forest canopy in acid precipitation quality changes.
Hutchinson et al. (1977)	Effects of acid rainfall and heavy metal particulates on Canadian forest ecosystem.
Jacobson (1980)	Role of acid rain in plant phototoxicity.
Johnsen (1980)	Effects of SO_2 on lichens and bryophytes in Denmark.
Johnson and Siccama (1983)	Potential mechanisms of acid deposition-induced declines in forest vegetation.
Johnson and Siccama (1984)	Possible role of acid deposition in the decline of the red spruce in the northern Appalachians area.
Johnson et al. (1981)	Acid precipitation as a growth-limiting factor for pine trees in the New Jersey Pinelands.
Johnson (1981a)	Need for quantification of acid rain effects on forest productivity.
Johnson and Richter (1984)	Effects of atmospheric depositions on nutrient leaching patterns in forest ecosystems.
Johnson, Turner and Kelly (1982)	Effects of acidic atmospheric inputs on forest nutrient cycles.
Lee and Weber (1979)	Effects of simulated acid rain on 11 woody species.
Lee and Weber (1982)	Influence of hardwood canopy and litter layer on acid rain input of chemicals to soil.
Lindberg, Shriner and Hoffman (1981)	Beneficial and detrimental effects of wet and dry deposition on forest canopies.

Table 17: (continued)

Author(s) (Year)	Comments
Lindberg and Harriss (1981)	Atmospheric deposition influences on chemical changes in and by a forest canopy.
Matziris and Nakos (1977)	Simulated acid rain effects on Aleppo pine.
McLaughlin et al. (1983)	Accumulation of trace metals in tree rings.
McLaughlin, West and Blasing (1984)	Dendroecological approaches for evaluating acid precipitation effects on forest trees.
Pucket (1982)	Use of dendroecological analyses to examine acid rain impacts on tree growth in southeastern New York.
Raynal (1980)	Review of sensitivity of forest ecosystem components to acid precipitation.
Raynal, Roman and Eichenlaub (1982)	Response of sugar maple seedling growth to simulated acid rain.
Scherbatskoy and Klein (1983)	Response of yellow birch and white spruce to acidic mists.
Shinn (1978)	Processes affecting foliar deposition and retention of particles on plants.
Shriner (1978)	Simulated acid rain influences on disease development in plants.
Shriner and Cowling (1978)	Acid rain influences on host-pathogen complexes in plant diseases.
Tukey (1980)	Summary of factors that may produce injury to plant species.
Tveite and Abrahamsen (1980)	Effects of artificial acid rain on growth and nutrient status of Scots pines in southern Norway.
Wetstone and Foster (1983)	Summary of forest impacts of acid precipitation.
Wood and Bormann (1975)	Leaching changes in foliage caused by artificial acid mists.

Matziris and Nakos (1977) determined the effect of simulated acid rain on growth and other characteristics of Aleppo pine, and investigated intraspecific variability.   The effect of acid rain on soil properties was also examined. During one growing season, seedlings from 1 yr old half sib families of Aleppo pine were irrigated with simulated acid rains of pH 3.1 and 3.5. A control group was irrigated with pH 5.1 water. Seedlings that were treated with acid rain with a pH of 3.1 reached a mean height of 22.6 cm, which was 8.2 percent shorter than the mean height of the control seedlings. Irrigation with acid rain increased the mortality of the seedlings, negatively influenced the formation of terminal buds, and dissolved and leached considerable amounts of calcium carbonate from the soil.

Scherbatskoy and Klein (1983) exposed seedlings of yellow birch and white spruce to mists consisting of distilled water at pH 5.7, or sulfuric acid in distilled water at pH 4.3 or 2.8. Misting treatments, 4 hours each, were applied once, twice or three times, separated by 72 hours. Decreased leaching of carbohydrate, protein, K, and H2PO4 occurred with increasing number of mistings, indicating that these substances were not readily resupplied to leaching sites. Protein leaching was reduced by misting at pH 2.8. Leaching of amino acids, K, and Ca increased with decreasing mist pH, and leachate pH was higher than the pH of applied mists, suggesting that cation exchange plays a role in foliar leaching by acidic solution. Amino acid leaching from birch increased with an increasing number of mistings at pH 2.8.

Sugar maple seedlings were subjected to laboratory studies involving exposure to simulated acidified canopy throughfall at pH 3.0 and below (Raynal, Roman and Eichenlaub, 1982).   Seedling growth was significantly reduced; seedlings exposed to low pH were susceptible to bacterial infection; and survival of seedlings transplanted to soil declined with increasing acidity of simulated canopy throughfall.   The authors noted that while these studies indicate that sugar maple seedlings are potentially susceptible to direct and indirect effects of acid precipitation, evaluation of the findings in relation to natural conditions is complicated by the capacity of vegetation and the forest floor to buffer the seedling environment from extreme pH changes, the direct effects of acidity on pathogenic microorganisms and the predisposition of seedlings to infection, and the episodic nature and varying acidity of precipitation.

Acid precipitation has the potential for influencing plant diseases by influencing not only the pathogen, but also the host organism, and the host-pathogen complex. Shriner (1978) reported on a study to determine whether simulated rain acidified with sulfuric acid influences disease development in plants. The effects of simulated rain acidified with H2SO4 were studied in five host-parasite systems.   Plants were exposed in greenhouses or fields to simulated rain of pH 3.2 or 6.0 in amounts and intervals common to weather patterns of North Carolina. Simulated acid rain resulted in: an 86 percent inhibition of the number of Telia produced by Cronartium Fusiforme on Willow Oak; a 66 percent inhibition in the reproduction of root-knot nematodes on field-grown kidney beans; a 29 percent decrease in the percentage of leaf area of field-grown kidney beans affected by Uromyces Phaseoli; and either stimulated or inhibited development of Halo blight on kidney beans, depending on the state of the disease cycle in which the treatments were applied.   These results indicated that the acidity of rain is an environmental parameter that should be of concern to plant pathologists and ecologists.

Field studies of the effects of acid precipitation on forests have been conducted in the United States and Europe. Johnson and Siccama (1983) have noted that while available evidence does not show a clear cause and effect relationship between acid deposition and forest decline and dieback in the United States, of particular concern is the possibility that acid deposition could have caused or may eventually cause soil changes detrimental to forest vegetation, either by stripping nutrients from the soil or by mobilizing phytotoxic elements. Such changes would constitute a problem that could be extremely expensive to correct. Johnson and Siccama (1984) have also noted that red spruce in the northern Appalachians have died in abnormally large numbers over the past two decades. While the exact mechanisms causing death have not been determined, the authors pointed out that high-elevation spruce-fir forests of the eastern United States receive particularly high rates of acidic deposition (up to 4 keq of $H+$/ha/yr), vegetation is exposed to highly acidic cloud moisture for up to 2000 hours/yr, and very high levels of trace metals have accumulated.

Reports of decreased growth and increased mortality of forest trees in areas receiving high rates of deposition of atmospheric pollutants have emphasized the need to better understand and quantify both the mechanisms and kinetics of potential changes in forest productivity. The complex chemical nature of combined pollutant exposures, and the fact that these changes may involve both direct effects to vegetation, as well as indirect and possibly beneficial effects mediated by a wide variety of soil processes, make quantification of such effects particularly challenging. One possible approach for quantification involves dendroecological analysis. This analysis combines detailed analysis of long-term growth trends as influenced by tree age, competitive status, climatic variables, and pollutant levels (McLaughlin, West and Blasing, 1984). Puckett (1982) considered whether dendroecological analyses could be used to detect changes in the relationship of tree growth to climate that might have resulted from chronic exposure to components of the acid rain-air pollution complex in southeastern New York based on data from the early 1950's. Tree-ring indices of white pine, pitch pine, and chestnut oak were regressed against orthogonally transformed values of temperature and precipitation in order to derive a response-function relationship. Results of the regression analyses for three time periods suggested that the relationship of tree growth to climate has been altered. Statistical tests of the temperature and precipitation data suggest that this change was nonclimatic. Temporally, the shift in growth response appears to correspond with the suspected increase in acid rain.

Evidence is growing on the possible severity of forest problems in central Europe due to acid precipitation. For example, in West Germany, fully 560,000 hectares of forests have recently been damaged (Wetstone and Foster, 1983). In addition, more than 500,000 ha of forests in Czechoslovakia have been damaged. Some of the damage is brought about as a consequence of contact between acid waters and the plant tissues. Rainfall in the pH range of 3.0 to 4.0 has been shown to cause tissue injury which can ultimately reduce forest growth. However, rainfall at such a low pH level is not common. Additional evidence suggests that indirect effects resulting from acid-induced soil changes are far more important. Acid deposition upsets the natural balance in the soils. Negatively charged sulfate ions, after accumulating in soils, can be washed out by rains, carrying with them nutrients such as calcium and magnesium. Acid conditions also slow the bacterial decomposition needed for the continued production of nutrients. In the long term, pollution-induced nutrient depletion accelerates the natural forest aging process, leading to the eventual exhaustion of the soil's ability to sustain tree growth.

Some modeling efforts have been accomplished relative to acid precipitation effects on forests. For example, Andersson, Fagerstrom and Nilsson (1980) have developed diagramatic models of soil and forest hydrogen ion budgets along with discussions of input-output, root uptake, mineralization, and weathering effects. Chen et al. (1983) have developed a canopy model to calculate throughfall characteristics based on canopy properties, ambient air quality, and precipitation quantity and quality. The processes considered include wet and dry deposition, leaf exudation, nitrification, and oxidation of SO_2 and NO_x. The canopy model requires only standard input data that can be obtained by use of rain gauges, thermometers, and dry and wet collectors. The model has been calibrated with data collected at Woods Lake in the Adirondack Mountains of New York. The model has accurately simulated throughfall volume and the concentration of 15 throughfall chemical constituents. Ammonium ions accumulated on the canopy are nitrified rapidly, resulting in an increase in acidity and nitrate fluxes. The dominant process occurring in the coniferous canopy is dry deposition. Calculations suggest that dry deposition measured by bulk collectors grossly underestimates the total amount of dry deposition to the forested watershed. The enrichment of acidity in coniferous throughfall is derived primarily from the accumulation of acidic air particulates. The dominant process occurring in the deciduous canopy is exudation. This partially neutralizes the acidic deposition.

Effects on Crops

Table 18 identifies 11 references from Appendix E that address the effects of acid precipitation on single or multiple crop species. Some studies identify observable effects, while others indicate the effects may be small and subtle. For example, in a review and analysis of research on this subject, Irving (1983) indicated that the majority of crop species studied in field and controlled environment experiments exhibited no effect on growth nor yield as a result of simulated acidic rain. However, it was also noted that some crops were affected either positively or negatively. Irving (1983) also indicated that more complex experimental designs and analyses may be necessary in order to examine and describe the possible subtle responses of agricultural systems to acidic precipitation.

Table 18: References in Appendix E Dealing with Effects of Acid Precipitation on Crops

Author(s) (Year)	Comments
Amthor and Bormann (1983)	Precipitation acidity effects on perennial ryegrass plants.
Evans (1981)	Dose-response functions of several crops to simulated acid rain.
Evans, Conway and Lewin (1980)	Yield responses of field-grown soybeans to simulated acid rain.

Table 18: (continued)

Author(s) (Year)	Comments
Evans et al. (1982)	Effects of acid precipitation on seed yields of field-grown soybeans.
Evans, Gmur and Mancini (1982)	Greenhouse study of effects of simulated acid rain on several crops.
Evans and Lewin (1979)	Effects of simulated acid rain on seed yields of soybeans and pinto beans.
Heagle et al. (1983)	Effects of simulated acid rain on field-grown soybeans.
Irving (1983)	Review of effects of simulated acid rain on crop species.
Lee et al. (1981)	Study of response of major U.S. crops to acid rain.
Lee, Neely and Perrigan (1980)	Study of relative sensitivity of major U.S. crops to sulfuric acid rain.
Mortvedt (1983)	Acid deposition effects on micronutrient cycling and, in turn, crop yields.

Lee et al. (1981) reported on a study designed to reveal patterns of response of 28 major United States crops to sulfuric acid rain. Potted plants were grown in field chambers and exposed to simulated sulfuric acid rain (pH 3.0, 3.5, or 4.0) or to a control rain (pH 5.6). At harvest, the weights of the marketable portion, total aboveground portion and roots were determined. Of these, marketable yield production was inhibited for five crops (radish, beet, carrot, mustard greens, and broccoli), stimulated for six crops (tomato, green pepper, strawberry, alfalfa, orchardgrass, and timothy), and ambiguously affected for one crop (potato). In addition, stem and leaf production of sweet corn was stimulated.

Evans, Gmur and Mancini (1982) determined the effects of simulated acid rain on radishes, lettuce, wheat, and alfalfa grown under greenhouse conditions. Simulated rainfalls of pH 5.6, 4.6, 4.2, 3.4, 3.0, and 2.6 decreased root yields of radishes by 26, 42, 37, 41, 66, and 73 percent, respectively, compared with controls. Lettuce yields (fresh mass) were 11, 10, and 14 percent less than controls for plants exposed to rainfall of pH 4.0, 3.1, and 2.7, respectively. Yields of wheat exposed to 46 rainfalls with pH as low as 2.7 during anthesis and caryopsis development were unaffected. Alfalfa showed no differences in fresh mass among treatments even after 57 simulated rainfalls of pH 2.7 water over 105 days.

Several studies have been conducted on the effects of acid precipitation on soybeans (Evans and Lewin, 1979; Evans, Conway and Lewin, 1980; and Evans et al., 1982). Evans and Lewin (1979) performed experiments to determine the change in seed yield after exposure to simulated rain of pH 5.7, 3.1, 2.9, 2.7, and 2.5. The simulated rain decreased the dry mass of both stems and leaves. Seed yield also decreased after treatment with rain of pH 2.5. However, an increase in seed yield occurred when plants were exposed to rain of pH 3.1. A larger dry mass per seed was responsible for the larger dry mass of seed per plant.

The response of field grown soybeans has been studied by Evans, Conway and Lewin (1980). Soybeans were seeded to provide six Latin Squares. Five treatments (no rain, simulated rainfalls of pH levels of 4.0, 3.1, 2.7, and 2.3) replicated five times in each Latin Square were used to produce a total of 30 plots per treatment. These results show that additions of small amounts of simulated acid rain to soybeans decreased the number of pods per plant. This decrease in the number of pods per plant produced a small but significant decrease in seed mass. Two experiments were performed to determine changes in seed yields of soybeans grown under standard agronomic practices exposed to simulated acidic rain during the summer of 1981 (Evans et al., 1982). Seed yields of soybeans exposed twice weekly to simulated rainfalls of pH 4.1, 3.3, and 2.7 were decreased 10.7, 16.8, and 22.9 percent, respectively, compared with plants exposed to simulated rainfalls of pH 5.6.

Perhaps one explanation for the generally minimal effects of acid precipitation on field crops results from the fact that changes in soil pH due to acid depositions are minimal in most agricultural soils because of relatively high buffering capacities of these soils. Modern farming practices such as liming and return of crop residues also may override depositional effects. Therefore, micronutrient cycling in most agro-ecosystems should not be significantly affected by acid depositions (Mortvedt, 1983). However, the long-term effects of acid depositions on micronutrient cycling are not known.

Effects on Soils

Acid precipitation can affect soil chemistry, leaching, and microbiological processes. In addition, various types of soils exhibit a range of sensitivities to the effects of acid rain; for example, some soils are more sensitive than others. Table 19 lists 7 references from Appendix E that primarily address changes in soil chemistry that result from acid precipitation. Soils in Scandinavia have experienced a decreased pH, decreased base saturation, and increased leaching from acid precipitation (Malmer, 1976). The factors affecting soil acidification have been discussed by Bache (1980a).

Table 19: References in Appendix E Dealing with Changes in Soil Chemistry

Author(s) (Year)	Comments
Bache (1980a)	Factors influencing soil acidification by acid rain.

Table 19: (continued)

Author(s) (Year)	Comments
Malmer (1976)	Chemical changes in soil due to acid precipitation in Scandinavia.
Mayer and Ulrich (1980)	Wet and dry deposition as a source of input of elements into soil.
Reuss (1975)	Soil-plant system responses to acid rain.
Rorison (1980)	Constraints of acidic soils on soil chemistry and leaching.
Sears and Langmuir (1982)	Soil chemical changes as a function of acid precipitation.
Ulrich, Mayer and Khanna (1980)	Soil chemical changes due to acid precipitation in a forest in the Federal Republic of Germany.

Sears and Langmuir (1982) have reported on the chemistry of soil moisture studied in central Pennsylvania sandy loam soils over a 12 month period. The dolomite bedrock depth was 6-14 m, and depth to the water table was 60-90 m. A total of 146 samples were collected at 1-9 m depths with suction lysimeters. Total dissolved solids in the soil water increased with decreasing precipitation and increasing temperature. This was a result of evapotranspiration, which concentrated the salts in the soil water, reduced soil relative permeabilities, and increased the contact time between minerals and water. The increased acidity of precipitation in warmer months and drier periods also contributed to the increase in total dissolved solids.

A Fagus silvatica forest in the Solling highlands, Federal Republic of Germany, showed increasing Al concentrations in the soil during a study of the effect of acid precipitation on the ecosystem between 1966 and 1979 (Ulrich, Mayer and Khanna, 1980). A noticeable change occurred in 1973. Before this, carbon and nitrogen stores in the forest floor were increasing, but afterwards, they decreased. This indicates a shift from one decomposition process to another as a response to acid precipitation. The pH has remained between 3 and 4, decreasing slightly with time. Aluminum and Fe ion concentrations increased between 1966 and 1973 to 2 mg/l and 3 mg/l of equilibrium soil solution, with the buffer reaction reaching deeper into the soil. Acid precipitation has induced soil internal hydrogen ion production at a rate of 2.9-5.5 kmole hydrogen ions per hectare per year, partly from changes in nitrogen nutrition and partly from nitrogen deficiency. Proton consumption by silicate weathering is estimated at 0.2-2 kmole hydrogen ions per hectare per year, meaning that the buffering capacity of this soil is reaching its capacity. Since this forest has reached a critical stage in less than 20 years of acid rain, further acid precipitation will produce serious results, especially on the ridges and plateaus of the medium highlands.

Various soil types exhibit different sensitivities to the effects of acid precipitation. Table 20 lists 9 references from Appendix E that provide information on these sensitivities. Factors influencing soil sensitivity to acidification include the lime capacity, soil profile buffer capacity, and water-soil reactions (Bache, 1980b). Wiklander (1980) presents a review of the sensitivity of various soils, and Peterson (1980b) identifies soil orders and classifications according to their response to acid precipitation.

Table 20:   References in Appendix E Dealing with Soil Sensitivity to Acid Precipitation

Author(s) (Year)	Comments
Bache (1980b)	Factors influencing soil sensitivity to acidification.
Ciolkosz and Levine (1983)	Computer simulation model for determining the impact of acid deposition on Pennsylvania soils.
Johnson (1979)	Neutralization of strong acids from precipitation in the soil zone.
Kaplan, Thode and Protas (1981)	Soils as a basic determinant of regional sensitivities to acid precipitation.
Klopatek, Harris and Olson (1979)	Regional ecological overview of the potential effects of acid precipitation on soils.
Olson, Johnson and Shriner (1982)	Soil chemistry, bedrock geology, terrain characteristics, and land use as determinants in soil susceptibility to acid precipitation.
Peterson (1980b)	Soil orders and classifications according to their response to acid precipitation.
Torrenueva (1981)	Review of effects of acid precipitation on vegetation and soils.
Wiklander (1980)	Review of sensitivity of soils to acid precipitation.

Several soil classification schemes have been developed as a result of sensitivity studies. For example, Kaplan, Thode and Protas (1981) collected data on rocks, soils, land use, and water quality from 283 counties in New England and the Middle Atlantic States. Four of the nine possible soil classes were found: alfisols, inceptisols, spodosols, and ultisols. Bedrocks were classified as intrusive igneous, metamorphic, consolidated sedimentary, and unconsolidated and weakly consolidated sedimentary. The area of concern contained no extrusive igneous rocks. Cluster analysis was applied to the rock and soil data. Path analysis produced two models, one relating rock types to soil classes, and the second, the effect of rocks and soils on water quality. This showed that the only rock which contributed to water quality was consolidated sedimentary. The presence of spodosols, ultisols, and inceptisols indicated surface waters of lower alkalinity and greater susceptibility to acidification from acid rain. Areas with larger percentages of alfisols have greater resistance to the effects of acid precipitation.

Olson, Johnson and Shriner (1982) described a regional assessment of the potential sensitivity of soils in the eastern United States. Areas in the eastern United States were evaluated for their sensitivity to acid deposition by combining county-level information on soil chemistry, bedrock geology, terrain characteristics, and land-use information. The final analysis covered the eastern 37 states and excluded the 1,013 counties that are predominantly agricultural or urban. Soils were characterized for their potential to undergo acidification from acid deposition. The criteria were moderate pH and low cation exchange capacity. The one soil type meeting these criteria occurs extensively only in 16 counties in Nebraska. Soils characterization maps also show low pH soils (potential for aluminum leaching) and peat soils (potential for acidifying precipitation). Areas are also classified for their potential to reduce the acidity of acid deposition prior to the transfer of acid inputs to aquatic systems. Low soil pH, low soil sulfate adsorption capacity, bedrock with no buffering capacity, and steep terrain are factors associated with low potential to reduce acidity. Eight percent of the 2,660 counties in the east were found to have low potential to reduce acidity with an additional 20 percent having moderate potential. These areas occur in northern Minnesota, Wisconsin, and Michigan; the New England states; parts of New York and Pennsylvania; the Appalachian mountains; and Florida. Maps showing the sensitivity classifications are available (Olson, Johnson and Shriner, 1982).

A computer simulation model was developed to determine the impact of various inputs to acid deposition on Pennsylvania soils (Ciolkosz and Levine, 1983). The model simulates the changes that occur in the solid phase of soils in humid, temperate climates that are undergoing acidification and cation leaching. To obtain predictions about soil properties which closely matched actual field conditions, the state was divided into 10 regions based on physiography, county borders, and proximity to one of the 10 precipitation monitoring stations established by the Pennsylvania Department of Environmental Resources. Information from the Pennsylvania State University Soils Data Base for the major soil associations within each of these 10 regions, and the accompanying precipitation data, were used as input to the model. Simulations creating the "worst case scenario" were designed using data for uncultivated and unlimed soils, and without a forest canopy or litter layers. Soils were classified into sensitivity classes based on the amount of time required for the soil to reach a threshold value of pH less than or equal to 4.0, aluminum concentrations in the soil solution of 1.0 mg/l in the upper horizon, or 0.1 mg/l in the lower horizon. Very sensitive soils reached a threshold value within 30 years, sensitive soils attained this value within 60 years, slightly sensitive soils reached it within 90 years; nonsensitive soils did not attain critical levels within 90 years. Using the model, simulation results showed the Pennsylvania soils were either very sensitive or nonsensitive. Soils at the threshold values at the beginning of the simulation were very sensitive, while those not at the critical levels from the start of the simulation were classified as nonsensitive. Thus, the nonsensitive soils contain sufficient buffer capacity to withstand acid deposition inputs at present rates for at least 90 years. Correlation analysis of these results indicates that base saturation, exchangeable cations, exchangeable acidity, and pH are most closely associated with the sensitivity of Pennsylvania soils to acid deposition. Cation exchange capacity and sulfate adsorption showed very little association with soil sensitivity class predicted by the model.

Two very important effects of acid precipitation on soils are associated with changes in the leaching patterns of soil constituents, and with potential removals and subsequent leaching of chemical constituents found in the precipitation. Table 21 lists 13 references from Appendix E that address leaching. Chemical constituents capable of being leached from soils and retained by soils when exposed to acid rain are summarized by Gorham and McFee (1980).

Table 21: References in Appendix E Dealing with Soil Leaching

Author(s) (Year)	Comments
Cronan (1981)	Effects of acid precipitation on cation transport in forest soils in New Hampshire.
Gorham and McFee (1980)	Discussion of materials capable of being leached or retained by soils when exposed to acid rain.
Hutchinson (1980)	Mobilization and leaching of heavy metals from two Canadian soils due to acid precipitation.
Johnson (1980)	Potential soil leaching changes resulting from H_2SO_4 in acid rainfall.
Nilsson (1980)	Soil leaching changes from acid precipitation.
Peterson (1980a)	Effects of acid precipitation on soil podzolization processes.
Richter, Johnson and Todd (1983)	Soil leaching from atmospheric sulfur deposition.
Singh, Abrahamsen and Stuanes (1980)	Effect of simulated acid rain on sulfate mobility in Norwegian forest soils.
Sparks and Curtis (1983)	Effects of acid rain on leaching kinetics in three Delaware soils.
Sposito, Page and Frink (1980)	Computer calculations for effects of acid precipitation on soil leachate quality.

Table 21: (continued)

Author(s) (Year)	Comments
Stednick and Johnson (1982)	Changes in chemical composition of acid precipitation upon passage through a forest soil.
Tyler (1978)	Influence of artificial acid rain on metals leaching from soils in south-eastern Sweden.
Wainwright, Supharungsun and Killham (1982)	Solubilization of heavy metals in soils subject to low pH rain.

Two studies identifying some of the leaching effects of concern will be cited. Hutchinson (1980) presented data obtained from studies involving soils in the Smoking Hills area of Canada and soils collected near Sudbury, Canada. Mobilization and leaching of heavy metals were observed to increase with increases in acid precipitation; and losses of "base" elements also can occur. Cronan (1981) has described the results of an investigation of the effects of regional acid precipitation on forest soils and watershed biogeochemistry in New England. Key findings included the following: (1) acid precipitation may cause increased aluminum mobilization and leaching from soils to sensitive aquatic systems; (2) acid deposition may shift the historic carbonic acid/organic acid leaching regime in forest soils to one dominated by atmospheric H_2SO_4; (3) acid precipitation may accelerate nutrient cation leaching from forest soils and may pose a particular threat to the potassium resources of northeastern forested ecosystems; and (4) progressive acid dissolution of soils in the laboratory may provide an important tool for predicting the patterns of aluminum leaching from soils exposed to acid deposition.

Since sulfates are typically a major constituent in acid precipitation, Johnson (1980) reviewed the mechanisms of sulfate adsorption in soils, soil properties in relation to sulfate adsorption and susceptibility to leaching by H_2SO_4. Cation leaching from soils as caused by atmospheric inputs of sulfuric acid only occurs if the sulfate is mobile in the soil or can displace other mobile anions. Soils rich in sesquioxides are likely to be resistant to leaching by sulfuric acid. Singh, Abrahamsen and Stuanes (1980) have studied the effect of simulated acid rain on sulfate mobility in iron-podzol and semipodzol forest soils of southern Norway. The study was carried out with lysimeters containing undisturbed soil. The lysimeters were watered with "rain" having pH 5.6 and 4.3. It was found that sulfate mobility was higher in the semipodzol than in the iron-podzol, and it was dependent on their sulfate adsorption capacities which in their turn were dependent on the aluminum contents of these soils. Sulfate losses from applied sulfur radioisotopes increased with increasing volume and decreasing pH of the "rain". The element losses were also higher in the semipodzol reflecting further higher mobility of sulfate in this soil.

The effects of acid rain on the kinetics of elemental release from three major Delaware soil types has been investigated (Sparks and Curtis, 1983). All three soils contained low clay and organic matter contents which would indicate rather low buffering capacities.   The kinetics of elemental release were evaluated by leaching the soils with simulated acid rain at pH levels of 2.5, 3.4, 4.8, and 5.6.   Leachates were collected until an apparent equilibrium in elemental release was obtained.   In all soils and at all acid rain pH levels, quantities of basic elemental release were in the order Ca, K, Mg, and Na.   The amount of heavy metals and Si release were very low.   As the pH of the acid rain decreased, the total quantity of each released element increased.   In all cases, a rapid initial release of each element was followed by slower release rates.   The kinetics of aluminum release were characterized by a similar time trend.   The greatest Al release rates occurred at pH 2.5 and from the soil type with the lowest clay and organic matter contents and thus the poorest buffering capacity.   The Al that was released at pH 2.5 from the soils would probably be either monomeric or polymeric Al; however, some release could have occurred from the clay structures.

Soil microorganisms and microbiological processes can be altered by acid precipitation.   Table 22 lists 18 references from Appendix E that address microorganisms and their processes.   The effects of acid precipitation can be considered in terms of changes in bacterial numbers and activity, alterations in nutrient and mineral cycling, and changes in the decomposition of organic matter.

Table 22:   References in Appendix E Dealing with Soil Microbiological Processes

Author(s) (Year)	Comments
Abrahamsen, Houland and Haguar (1980)	Effects of artificial acid rain on decomposer organisms in soil.
Alexander (1980)	Effects of acidity on soil microorganisms and microbial processes.
Baath et al. (1980)	Study of soil chemistry and microbial effects from artificial acid rain.
Baath, Lundgren and Soderstrom (1979)	Effects of artificial acid rain on soil microbial activity and biomass.
Babich and Stotzky (1978)	Effects of atmospheric sulfur compounds on soil microorganisms.
Chang and Alexander (1983)	Effect of acid precipitation on algal fixation of nitrogen and CO_2 in forest soils.
Cole and Stewart (1983)	Acid deposition effects on phosphorus cycling in soils.
Coleman (1983)	Impacts of acid deposition on carbon cycling in agricultural soils.

Table 22: (continued)

Author(s) (Year)	Comments
Francis, Olson and Bernatsky (1980)	Study of effects of acidity on microbial processes in a forest soil.
Francis, Olson and Bernatsky (1981)	Effects of soil acidity on microbial decomposition of organic matter, transformation of nitrogen, and soil chemical and biological properties.
Killham, Firestone and McColl (1983)	Effects of acid rain on microbial activity in Sierran forest soil in northern California.
Klein, Kreitinger and Alexander (1983)	Influence of simulated acid rain on nitrate formation in forest soils in the Adirondacks region of New York.
McKinley and Vestal (1982)	Effects of acidic pH on microbial decomposition of plant litter.
Olson (1983)	Soil chemical changes from acid deposition.
Roberts et al. (1980)	Effects of acid rain on soil microbial and leaching processes in central England.
Schnitzer (1980)	Reactions of humic and fulvic acids to decreased soil pH.
Strayer and Alexander (1981)	Effects of simulated acid rain on glucose mineralization in forest soils.
Strayer, Chyi-Jiin and Alexander (1981)	Effects of simulated acid rain on nitrification and nitrogen mineralization in forest soils.

Bacterial populations in soil may or may not be altered by acid precipitation. For example, Baath et al. (1980) noted decreases in the numbers of bacteria and fungus in some Swedish experiments involving three block units of Scots pine subjected to artificial acid rain. In contrast, other studies indicated that total fungal length and bacterial numbers did not change significantly in response to artificial acid rain (Baath, Lundgren and Soderstrom, 1979). However, they did note that the bacterial cell size was smaller in most acidified plots. The mean cell size with pH 2 water was 0.10 m^3 compared to 0.13 m^3 in the control and the pH 3 treated plot. The reduced cell size could be due to changes in the bacterial population, or it could reflect a lower bacterial growth rate.

Killham, Firestone and McColl (1983) have reported on a study of soil microbial activity. A Sierran forest soil planted with Ponderosa pine seedlings was exposed to simulated rain with an ionic composition reflecting that found in northern California. The soils were collected in two samples (top 1 cm and 4 to 5 cm) which were assayed separately for respiration and enzyme activities. Changes in microbial activity were most significant in surface soils. Only the pH 2.0 input caused inhibition of both respiration and enzyme activities. The overall microbial response to the pH 3.0 and 4.0 acid regimes was one of stimulation, although the response of individual enzymes was more varied.

Changes in bacterial numbers and activity may be reflected by changes in nutrient and mineral cycling patterns. One reason this is important is that the productivity of agricultural soils requires an active cycling of soil phosphorus pools through plants and microbial populations. Phosphorus cycling rates are thus closely linked to transformations of C, N, S and other mineral nutrients (Cole and Stewart, 1983). Two experimental studies related to nitrogen cycling will be used as examples (Strayer, Chyi-Jiin and Alexander, 1981; and Chang and Alexander, 1983).

Columns containing samples of forest soils were leached with either a continuous application of 100 cm of simulated acid rain (pH 3.2-4.1) at 5 cm/hr or an intermittent 1.5-hr application of 1.2 cm of simulated acid rain twice weekly for 19 weeks (Strayer, Chyi-Jiin and Alexander, 1981). The upper 1.0 to 1.5 cm portions of soil from treated columns were used to determine the changes in inorganic N levels in the soil. Nitrification of added ammonium was inhibited following continuous exposure of soil to simulated acid rain of pH 3.2-4.1. The extent of the inhibition was directly related to the acidity of the simulated rain solutions. The production of inorganic N in the absence of added ammonium was either stimulated or unaffected following continuous treatment of soils with pH 3.2 simulated acid rain.

Indigenous algae growing in soil samples from the Panther, Woods, and Sagamore Lake watersheds of the Adirondacks regions of New York were exposed to acid precipitation of pH 3.5 or 5.6 at rates of 50, 100, 200, and 300 cm (Chang and Alexander, 1983). Rates of nitrogen fixation as measured by acetylene reduction were significantly less for soils watered with pH 3.5 water compared with the pH 5.6 water for all rainfall intensities, and for light and dark. As rain volume increased, acetylene reduction decreased for the pH 3.5 water and increased for the pH 5.6 water. The carbon dioxide fixation rate was significantly less for all soils exposed to the more acid rain, and the extent of inhibition became more pronounced with increasing volumes of rain applied.

Acid precipitation can also affect microbial decomposition processes. Illustrations of these effects will be cited from two experimental studies. For example, Strayer and Alexander (1981) exposed samples of forest soils to a continuous application of 100 cm of simulated acid rain (pH 3.2-4.1) at 5 cm/hr, or to intermittent 1-hr applications of 5 cm of simulated acid rain 3 times per week for 7 weeks. The major effects of the simulated acid rain were localized at the top of the soil and included lower pH values and glucose mineralization rates. Glucose mineralization in the test soils (pH values of 4.4-7.1) was inhibited by the continuous exposure to simulated acid rain at pH 3.2 but not at pH 4.1. The extent of inhibition depended on the soil and the initial glucose concentration.

Francis, Olson and Bernatsky (1981) investigated the effects of soil acidity on the microbial decomposition of organic matter. The rates of organic matter decomposition by natural acid soil and by pH-adjusted acid and neutral soils which were preincubated for 14 and 150 days were determined by monitoring CO_2 evolution. In the control (unamended) pH-adjusted acid soil, reductions in CO_2 production of 14 percent by 14-day preincubated samples and of 52 percent by 150-day samples were observed. In the oak-leaf-amended acidified soils, the CO_2 production of 14- and 150-day preincubated samples decreased by about 6 and 37 percent, respectively. These results suggest that further acidification of acid forest soils by acid precipitation may lead to significant reductions in leaf litter decomposition.

Effects on Ground Water

As ground water quality is becoming increasingly important, there is a growing concern relative to the potential effects of acid precipitation on quality constituents. Direct precipitation in recharge areas is of particular concern. Tschupp (1974) noted that most pollutants, both particulates and gases, are soluble in precipitation and are carried into the soil with the precipitation that does not run off. The most pronounced effects are those associated with the increased acidity of precipitation caused by atmospheric pollutants. The increased acidity causes accelerated weathering and chemical reactions as the precipitation passes through the soil and rock in the process of recharging an aquifer. The net effect on the ground water resource is the reduction of water quality because of increased mineralization.

Acid ground water has been detected in some places in southwestern Sweden. In order to shed light on the possible acidification of ground water a study has been performed on ground water supplies for communities in three counties, Kronoberg, Kopparberg and Vaesternorrland (Jacks and Knutsson, 1981). In each county, 20 to 30 ground water supplies were selected for closer study. The county of Kronoberg has a considerably larger number of water supplies with acid ground water than the two other counties. The reason is likely to be twofold, the more acid precipitation in southern Sweden in combination with the more shallow aquifers in the county of Kronoberg. In the counties of Kopparberg and Vaesternorrland no apparent effects of acid precipitation could be traced. In the county of Kronoberg on the other hand, it is obvious that marked changes have taken place. Shallow wells in the county of Kronoberg show a decrease in alkalinity with time, as well as an increase in carbonic acid.

AQUATIC ECOSYSTEM EFFECTS

Acid precipitation can cause many observable, as well as nonobservable, effects on aquatic ecosystems. Included are changes in water chemistry and aquatic faunal and floral species. This portion of Chapter 4 summarizes these three categories of effects.

Effects on Water Quality

Table 23 lists 25 references from Appendix E that address the effects of acid precipitation on surface water quality. Field studies of water chemistry changes have been reported by Norton et al. (1981) and Pfeiffer and Festa (1980). Ninety-four low-humic lakes in northern New England (82 in Maine) located in largely noncalcareous terrain were studied by Norton et al. (1981) to detect possible changes in pH and water quality constituents. Eighty-five percent of the lakes are more acidic than they previously were. Studies of present water chemistry indicate that Fe, Mn, Pb, Zn and Al concentrations tend to be higher with decreasing pH for the set of lakes; and Na, K, Ca, Mg, and alkalinity tend to be higher with increasing pH for the set of lakes.

Table 23:   References in Appendix E Dealing with Effects of Acid Precipitation on Water Quality

Author(s) (Year)	Comments
Arnold, Light and Dymond (1980)	Water quality effects of acid precipitation in Pennsylvania.
Davis (1981)	Water quality effects of the atmospheric deposition of energy-related pollutants.
Faust and McIntosh (1983)	Buffer capacities of freshwater lakes to acid rain.
Galloway, Norton and Church (1983)	Model for aquatic ecosystem response to acid deposition.
Gjessing (1980)	Water quality impacts of acid precipitation in Norway.
Harte, Holdren and Tonnesson (1983)	Potential for micronutrient changes in Sierra Nevada lakes subject to nitrate-dominated acid deposition.
Hendrey et al. (1980)	Sensitivity of surface water bodies in the eastern United States to water quality changes.
Hendrey and Kaplan (1982)	Sensitivity of headwater areas in the eastern United States to water quality changes.
Johnson (1981b)	Potential impacts of acid rain on stream water quality in southeastern Alaska.
Johnson et al. (1981)	Water quality changes in a New Hampshire stream due to acid rain.
Likens et al. (1977)	Study of biogeochemistry of Hubbard Brook in New Hampshire as influenced by acid precipitation.

Table 23: (continued)

Author(s) (Year)	Comments
Likens, Bormann and Eaton (1980)	Stream water chemistry at Hubbard Brook in New Hampshire as influenced by acid precipitation.
Logan, Derby and Duncan (1982)	Study of lake susceptibility to acid precipitation in the central Washington Cascades.
Mitchell, Landers and Brodowski (1981)	Organic and inorganic sulfur fractions in three New York lake sediments.
Nelson and Delwiche (1983)	Susceptibility to acidification of 63 Oregon lakes in the Cascades.
Norton et al. (1981)	Water quality changes in 94 northern New England lakes from atmospheric inputs of acids and heavy metals.
Nriagu (1978)	Natural buffering ability of 120 Canadian lakes.
Pitblado, Keller and Conroy (1980)	Classification of 187 Canadian lakes according to water chemistry data (23 variables).
Pfeiffer and Festa (1980)	Water quality features of 849 ponded waters in the Adirondack region of New York.
Schindler et al. (1980)	Effects of acidification on mobilization of heavy metals and radionuclides from freshwater lake sediments.
Schofield (1979)	Water quality changes in surface water due to acid rain input.
Turk and Adams (1983)	Sensitivity of lakes in northwestern Colorado to water quality changes from acid precipitation.
Vangenechten (1980)	Water quality changes in 53 Belgium surface waters due to acid rain.
Webb (1982)	Influence of rainfall chemistry on water chemistry in the Tovdal River in southern Norway.

Table 23: (continued)

Author(s) (Year)	Comments
Wright and Gjessing (1976)	Discussion of changes in the chemical composition of lakes in southern Scandinavia and eastern North America due to acid precipitation.

The Adirondack region of northeastern New York State contains approximately 2,900 individual lakes and ponds, encompassing approximately 282,000 surface acres. Many of these surface waters have low alkalinities due to a carbonate-poor geology; therefore, they are particularly sensitive to the high acid ion deposition associated with the region's airshed. Since 1975, pH and alkalinity measurements have been made on 849 ponded waters throughout the region to determine the scope of water quality impacts associated with acid ion deposition and to provide a baseline inventory for indexing future measurements (Pfeiffer and Festa, 1980). Twenty-five percent of the waters in the survey registered pH readings below 5.0. Comparisons of historic and post-1974 acidities were made where data points from comparable methodologies exist. Relationships between meter pH, colorimetric pH, alkalinity, conductivity, calcium, lake surface area, lake surface elevation, and geographical location are presented.

One reason for changes in surface water chemistry is associated with the release of metals from stream or lake sediments. For example, Wright and Gjessing (1976) noted that concentrations of aluminum, manganese, and other heavy metals are higher in acid lakes due to enhanced mobilization of these elements in acidified areas. Schindler et al. (1980) sealed large (10 m) diameter enclosures to the sediments in 2.0 to 2.5 meters of water in Lake 223 in Canada. Two enclosures were held at control pH (6.7-6.8), one was lowered to pH 5.7 and one to pH 5.1, using H_2SO_4. Aluminum, Mn, Zn, and Fe were released from lake sediments at pH 5 and 6.

Due to their extant water chemistry and sediment characteristics, some surface waters are most susceptible to changes in water chemistry than others. For example, lakes of the Flat Tops Wilderness Area in northwestern Colorado are sensitive to acidification if precipitation in the area becomes as acidic as that of the northeastern United States (Turk and Adams, 1983). About 370 lakes having a total surface area of about 157 ha are calculated to be sensitive to acidification.

Total alkalinity is frequently used as a measure of the buffer capacity of surface water. However, Faust and McIntosh (1983) have suggested that the Van Slyke definition of buffer capacity, the increment of a strong base or strong acid that causes an incremental change in the pH value of water, is better than total alkalinity for defining a water's resistance to acid rain. This Van Slyke value, designated by beta, shows a peak at pH 6.3 for the bicarbonate-carbonate pair, indicating that the effect of acid rain on the pH and alkalinity of natural waters is not deleterious until this peak is traversed. A beta value of zero indicates a dead water with no capacity to neutralize acid.

Data clearly show that pH and alkalinity alone cannot determine buffer capacity. For example, in New Jersey, Fairview Lake (pH of 5.5 and alkalinity of 10.2 mg/l) has a beta value 11 times that of Clyde Potts Reservoir (pH of 7.3, alkalinity of 8.1 mg/l).

Several surface water sensitivity studies leading to classification schemes have been conducted. For example, Hendrey et al. (1980) analyzed bedrock geology maps of the eastern United States for the relationship between geological materials and surface water pH and alkalinity. Map accuracy was verified by examining current alkalinities and pH's of waters in several test states, including Maine, New Hampshire, New York, Virginia and North Carolina. In regions predicted to be highly sensitive, alkalinities in upstream sites were generally low, less than 200 microequivalents per liter. Many areas of the eastern United States are pinpointed in which some of the surface waters, especially upstream reaches, may be sensitive to acidification.

Acidification susceptibility has been studied for 63 Oregon Cascade lakes (Nelson and Delwiche, 1983). Lake alkalinities averaged 137.6 ueq/l; 25 lakes had alkalinities of less than 50 ueq/l. Calcium was the major cation (36 percent) and bicarbonate was the major anion (82 percent) present. The study determined that Oregon's Cascade lakes do not presently have an acid precipitation problem, but they are extremely susceptible to acidification. In addition, 29 remote lakes in the Alpine Lakes Wilderness Area in the state of Washington were sampled in 1981 (Logan, Derby and Duncan, 1982). The measured alkalinities ranged from 4.0 to 190 ueq/l, and the median was 57 ueq/l. Based on these findings, it was concluded that all 29 lakes, though not at present acid, are susceptible to acidification.

Effects on Faunal Species

Acid precipitation can affect microdecomposers, algae, aquatic macrophytes, zooplankton, benthos, and fish (Hendrey et al., 1976). Table 24 lists 14 references from Appendix E which are related to effects on fish, amphibians, and microcrustaceans. For background information, Fritz (1980) reviews the effects of low pH on freshwater fish and shellfish. The discussion of direct mortality includes information obtained from laboratory experiments and field investigations. Several aspects of indirect mortality are discussed; examples include susceptibility to disease, destruction or modification of osmoregulatory and ionregulatory tissues, endocrine imbalance and genetic damage, changes in predator-prey relationships, habitat degradation, and changes in the availability of toxic substances. In addition, Schofield (1976) has summarized information on the effects of acid precipitation on freshwater fish.

Table 24: References in Appendix E Dealing with Effects of Acid Precipitation on Aquatic Fauna

Author(s) (Year)	Comments
Allen (1980)	Laboratory studies of the effects of acid rain on fresh water microcrustaceans.

Table 24: (continued)

Author(s) (Year)	Comments
Baker (1981)	Study of Al and hydrogen ion influences on fish survival in New York lakes.
Beamish (1974)	Influence of Zn and pH on white suckers in a lake.
Bell (1981)	Effects of acid precipitation on fish stocks in Norwegian lakes.
Bradford, Page and Straughan (1981)	Fish kills in acidified lakes in the California Sierra Nevadas.
Brown and Sadler (1981)	Effects of reduced sulfur inputs on fish populations in Norwegian lakes.
Fritz (1980)	Effects of low pH on freshwater fish and shellfish.
Graham and Wood (1981)	Effects of acid type on toxicity to rainbow trout.
Harriman and Morrison (1982)	Invertebrate and fish responses to acid rain in 12 streams in forested and nonforested areas in central Scotland.
Hendrey et al. (1976)	Changes in freshwater faunal species due to acid precipitation.
Nilssen (1980)	Effects of acid precipitation on faunal species in freshwater lakes in southern Norway.
Pough (1981)	Response of 14 species of amphibians to acid precipitation.
Schofield (1976)	Summary of information on effects of acid precipitation on freshwater fish.
Sprules (1975)	Crustacean zooplankton species and species associations in 47 Canadian lakes.

Many of the over 2,000 lakes in the Adirondack Region of New York are experiencing acidification and declines or loss of fish populations. Baker (1981) found that on the average, aluminum complexed with organic ligand was the dominant aluminum form in the dilute acidified Adirondack surface waters studied. In laboratory bioassays, speciation of aluminum was shown to have a substantial effect on aluminum toxicity to early life history stages of fish. Based upon available field and laboratory data, concentrations of inorganic aluminum and hydrogen ions appeared to be important factors determining fish survival in Adirondack surface waters affected by acidification. For the range of pH and aluminum levels, and fish species and life history stages studied, sensitivity to low pH levels decreased with increasing age while sensitivity to elevated inorganic aluminum levels increased with increasing age (through the post-swim-up fry stage).

Graham and Wood (1981) described an experimental study of the effects of water hardness (14 or 140 mg/l $CaCO_3$), acid (HCl vs. H_2SO_4), and activity level (rest vs. exercise) on acid toxicity to rainbow trout fingerlings at 15°C in 7-day exposures. The 7-day LC50 pH levels were 4.1 to 4.5. Sulfuric acid was less toxic in hard water than in soft water at all pH levels during rest and exercise. Hard water decreased the toxicity of HCl at pH less than 3.5 and increased toxicity at pH above 3.8. Sulfuric acid was less toxic than HCl in hard and soft waters except for soft waters above pH 3.8. Exercise increased H_2SO_4 toxicity in both hard and soft waters except in soft water at pH less than 3.3. Critical swimming speeds were not significantly different between hard and soft water at the same pH in the ranges of 3.5-4.2 and neutral. However, below pH 4.6 in soft water and 4.4 in hard water, critical swimming speeds decreased 4 percent per 0.1 pH units.

Fishery effects from acid precipitation have occurred in Norway (Bell, 1981; and Brown and Sadler, 1981). Bell (1981) noted that the mobilization of aluminum from soils is a major factor in the destruction of fish populations. Brown and Sadler (1981) have indicated that of 471 lakes above 200 meters elevation in southern Norway, 48 percent have no fish, 43 percent are sparsely populated, and 9 percent have good populations. Regression analysis indicated that the pH and excess (nonsea) sulfate concentration were correlated, with a slope of 0.225. Assuming that the excess sulfate is largely derived from the combustion of fossil fuels, a 50 percent reduction in European sulfur emissions would increase the pH by 0.2 units. This would improve the fishery status in lakes with pH above 4.9 (9 percent of all lakes). No improvement would take place in the 67 percent of lakes in the pH range 4.3-4.9, where pH and fish status are independent.

Effects on Floral Species

Aquatic floral species include bacteria, algae, diatoms, and macrophytic plants. Boylen et al. (1982) reported on nine high-altitude oligotrophic Adirondack lakes having water of pH 4.3 to 7.0 which were studied for their total bacterial numbers and possible adaptation of the microbial communities to their environmental pH. The number of heterotrophic bacteria from water samples recoverable on standard plate count agar were low (10 to 1000 per ml) for most of the lakes. Acridine orange direct counts were approximately two orders of magnitude higher than plate counts for each lake. Sediment aerobic heterotrophs recovered on standard plate count agar ranged from 14,000 to 1,300,000 per gm of sediment. Direct epifluorescence counts of bacteria in sediment samples ranged from 3,000,000 to 1.4×10^7 per gm. Low density values were consistent with the oligotrophic nature of all the lakes surveyed. There were no apparent differences in numbers of bacteria originally isolated at pH 5.0 and pH 7.0 between circumneutral lakes (pH greater than 6.0) and acidic lakes (pH less than 5.0). Approximately 1,200 isolates were recultured over a range of pH from 3.0 to 7.0. Regardless of the original isolation pH (pH 5.0 or pH 7.0), less than 10 percent of the isolates grew at pH less than 5.0. Those originally isolated at pH 5.0 also grew at pH 6.0 and 7.0. Those originally isolated at pH 7.0 preferred pH 7.0, with 98 percent able to grow at pH 6.0 and 44 percent able to grow at pH 5.0.

In a study of the ecological effects of acid precipitation on primary producers, Conway and Hendry (1981) indicated that nonacidic, oligotrophic lakes are typically dominated by golden-brown algae, diatoms and green algae. With increasing acidity, the number of species decrease and the species composition changes to dinoflagellates and golden-brown algae, with blue-green algae dominating in some cases. For macrophytic plants, dense stands of Sphagnum and Utricularia found in some acidic lakes may reduce nutrient availability and benthic regeneration.

MATERIALS EFFECTS

Acid precipitation may cause damage to man-made materials such as buildings, metals, paints, and statuary (Glass, Glass and Rennie, 1980). The effects on structures has been illustrated by the example of the deterioration of the exterior of the Field Museum of Natural History in Chicago (Gauri, 1979). The field survey revealed that the marble cornices protected from the direct impact of rain have developed black crust. The unprotected areas, however, are clean due to perpetual dissociation of calcite grains. Fractures in marble blocks were found along dark banks which terminated at joints with polymeric caulking. Atomic absorption spectrophotometry, scanning electron microscopy, and x-ray diffraction of samples from the museum revealed that the black crust consisted of gypsum which had formed due to SO_2 attack on calcite. The gypsum was absent at the naturally cleaned surface.

Data on the corrosion rates of unprotected carbon steel, zinc and galvanized steel, nickel and nickel-plated steel, copper, aluminum, and anti-rust painted steel due to sulfur dioxide and acid precipitation in Sweden have been reported (Kucera, 1976). Corrosion rates are significantly higher in polluted urban atmospheres than in rural atmospheres because of the high concentrations of airborne sulfur pollutants in urbanized areas. Economic damage is significant in the case of galvanized, nickel-plated, and painted steel, and painted wood.

HUMAN HEALTH EFFECTS

Recognition of the potential human health effects of acid precipitation is in its infancy. While human health impairment has been attributed to pollution by sulfur dioxide, data from inhalation studies in animals show that its oxidation products are more irritating (Coffin and Knelson, 1976). Population surveys in which suspended sulfate was a covariant suggest that certain health parameters are associated more strongly with sulfate than with SO_2. Recent work with biological models indicate that sulfates and sulfuric acid act on the lung through the release of histamine, and the degree of release is related to the specific cation present.

Acid precipitation can affect water supplies which in turn can affect their users. Taylor and Symons (1984) reported on the results of the first study concerning the impacts of acid precipitation on drinking water; the results were reported in terms of health effects in humans as measured by U.S. Environmental Protection Agency maximum contaminant levels. The study focused on sampling surface water and ground water supplies in the New England states, but also included other sites in the Northeast and the Appalachians. No adverse effects on human health were demonstrated, although the highly corrosive nature of New England waters may be at least partly attributable to acidic deposition in poorly buffered watersheds and aquifers.

SELECTED REFERENCES

Abrahamsen, G. et al., "Impact of Acid Precipitation on Forest and Freshwater Ecosystems in Norway", Report 6/76, Mar. 1976, Agricultural Research Council of Norway, Oslo, Norway.

Abrahamsen, G., Houland, J. and Haguar, S., "Effects of Artificial Acid Rain and Liming on Soil Organisms and the Decomposition of Organic Matter", Effects of Acid Rain Precipitation on Terrestrial Ecosystems, Plenum Press, New York, New York, 1980, pp. 341-362.

Alexander, M., "Effects of Acidity on Microorganisms and Microbial Processes in Soil", Effects of Acid Rain Precipitation on Terrestrial Ecosystems, Plenum Press, New York, New York, 1980, pp. 363-374.

Allen, J.D., "Laboratory Simulation of Acid Rain Effects on Freshwater Microcrustaceans", OWRT-A-054-MD(1), 1980, Water Resources Research Center, University of Maryland, College Park, Maryland.

Amthor, J.S. and Bormann, F.H., "Productivity of Perennial Ryegrass as a Function of Precipitation Acidity", Environmental Pollution Series A, Vol. 32, No. 2, 1983, pp. 137-145.

Andersson, F., Fagerstrom, T. and Nilsson, S.I., "Forest Ecosystem Responses to Acid Deposition--Hydrogen Ion Budget and Nitrogen/Tree Growth Model Approaches", Effects of Acid Rain Precipitation on Terrestrial Ecosystems, Plenum Press, New York, New York, 1980, pp. 319-334.

Arnold, D.E., Light, R.W. and Dymond, V.J., "Probable Effects of Acid Precipitation on Pennsylvania Waters", EPA-600/3-80-012, Jan. 1980, U.S. Environmental Protection Agency, Washington, D.C.

Arthur, M.F. and Wagner, C.K., "Response of Agricultural Soils to Acid Deposition", July 1982, Battelle Columbus Laboratories, Columbus, Ohio.

Baath, E. et al., "Soil Organisms and Litter Decomposition in a Scots Pine Forest--Effects of Experimental Acidification", Effects of Acid Rain Precipitation on Terrestrial Ecosystems, Plenum Press, New York, New York, 1980, pp. 375-380.

Baath, E., Lundgren, B. and Soderstrom, B., "Effects of Artificial Acid Rain on Microbial Activity and Biomass", Bulletin of Environmental Contamination and Toxicology, Vol. 23, 1979, pp. 737-740.

Babich, H. and Stotzky, G., "Atmospheric Sulfur Compounds and Microbes", Environmental Research, Vol. 15, No. 3, June 1978, pp. 513-552.

Bache, B.W., "The Acidification of Soils", Effects of Acid Rain Precipitation on Terrestrial Ecosystems, Plenum Press, New York, New York, 1980a, pp. 183-202.

Bache, B.W., "The Sensitivity of Soils to Acidification", Effects of Acid Rain Precipitation on Terrestrial Ecosystems, Plenum Press, New York, New York, 1980b, pp. 569-572.

Baker, J.P., "Aluminum Toxicity to Fish as Related to Acid Precipitation and Adirondack Surface Water Quality", Ph.D. Dissertation, Jan. 1981, Cornell University, Ithaca, New York.

Beamish, R.J., "Growth and Survival of White Suckers (Catostomus Commersoni) in an Acidified Lake", Journal of the Fisheries Research Board of Canada, Vol. 31, No. 1, 1974, pp. 49-54.

Bell, J.N., "Acid Precipitation--A New Study from Norway", Nature, Vol. 292, No. 5820, July 1981, pp. 199-200.

Botkin, D.B. and Aber, J.D., "Some Potential Effects of Acid Rain on Forest Ecosystems: Implications of a Computer Simulation", BNL-50889, Apr. 1979, Brookhaven National Laboratory, Upton, New York.

Boylen, C.W. et al., "Microbiological Survey of Adirondack Lakes with Various pH Values", Applied and Environmental Microbiology, Vol. 45, No. 5, May 1983, pp. 1538-1544.

Bradford, G.R., Page, A.L. and Straughan, I.R., "Are Sierra Lakes Becoming Acid", California Agriculture, Vol. 35, No. 5/6, May/June 1981, pp. 6-7.

Brown, D.J. and Sadler, K., "The Chemistry and Fishery Status of Acid Lakes in Norway and Their Relationship to European Sulphur Emissions", Journal of Applied Ecology, Vol. 18, No. 2, Aug. 1981, pp. 433-441.

Canfield, D.E., Jr., "Sensitivity of Florida Lakes to Acidic Precipitation", Water Resources Research, Vol. 19, No. 3, June 1983, pp. 833-839.

Chang, F.H. and Alexander, M., "Effect of Simulated Acid Precipitation on Algal Fixation of Nitrogen and Carbon Dioxide in Forest Soils", Environmental Science and Technology, Vol. 17, No. 1, Jan. 1983, pp. 11-13.

Chen, C.W. et al., "Acid Rain Model: Canopy Module", Journal of Environmental Engineering, Vol. 109, No. 3, June 1983, pp. 585-603.

Ciolkosz, E.J. and Levine, E.R., "Evaluation of Acid Rain Sensitivity of Pennsylvania Soils", OWRT-A-058-PA(1), Sept. 1983, Institute for Research on Land and Water Resources, Pennsylvania State University, University Park, Pennsylvania.

Coffin, D.L. and Knelson, J.H., "Acid Precipitation: Effects of Sulfur Dioxide and Sulfate Aerosol Particles on Human Health", Ambio, Vol. 5, No. 5-6, 1976, pp. 239-243.

Cole, C.V. and Stewart, J.W., "Impact of Acid Deposition and P Cycling", Environmental and Experimental Botany, Vol. 23, No. 3, 1983, pp. 235-241.

Cole, D.W. and Johnson, D.W., "Atmospheric Sulfate Additions and Cation Leaching in a Douglas Fir Ecosystem", Water Resources Research, Vol. 13, No. 2, Apr. 1977, pp. 313-318.

Coleman, D.C., "Impacts of Acid Deposition on Soil Biota and C Cycling", Environmental and Experimental Botany, Vol. 23, No. 2, 1983, pp. 225-233.

Conway, H.L. and Hendrey, G.R., "Ecological Effects of Acid Precipitation on Primary Producers", 1981, Brookhaven National Laboratory, Upton, New York.

Cowling, E.B. and Davey, C.B., "Acid Precipitation: Basic Principles and Ecological Consequences", Pulp and Paper, Vol. 55, No. 8, Aug. 1981, pp. 182-185.

Cowling, E.B. and Dochinger, L.S., "Effects of Acid Rain on Crops and Trees", Acid Rain: Proceedings of ASCE National Convention, 1979, American Society of Civil Engineers, New York, New York, pp. 21-54.

Crisman, T.L. and Brezonik, P.L., "Acid Rain: Threat to Sensitive Aquatic Ecosystems", June 1980, University of Florida, Gainesville, Florida.

Cronan, C.S., "Effects of Acid Precipitation on Cation Transport in New Hampshire Forest Soils", DOE/EV/04498-1, July 1981, U.S. Department of Energy, Washington, D.C.

Dailey, N.S. and Winslow, S.G., "Health and Environmental Effects of Acid Rain--An Abstracted Literature Collection--1966-1979", Mar. 1980, Toxicology Information Response Center, Oak Ridge National Laboratory, Oak Ridge, Tennessee.

Davis, M.J., "Effects of Atmospheric Deposition of Energy-Related Pollutants on Water Quality: A Review and Assessment", ANL/AA-26, May 1981, Argonne National Laboratory, Argonne, Illinois.

Derrick, M.R. et al., "Aerosol and Precipitation Chemistry Relationships at Big Bend National Park", Water, Air, and Soil Pollution, Vol. 21, No. 1-4, Jan. 1984, pp. 171-181.

Dochinger, L.S. and Seliga, T.A., "Acid Precipitation and the Forest Ecosystem", Journal of the Air Pollution Control Association, Vol. 25, No. 18, Nov. 1975, pp. 1103-1106.

Dochinger, L.S. and Seliga, T.A., Proceedings of the International Symposium on Acid Precipitation and the Forest Ecosystem, Aug. 1976, Ohio State University -Atmospheric Sciences Program, Columbus, Ohio.

Dynamac Corporation, "Bibliography on Air Pollution and Acid Rain Effects on Fish, Wildlife, and Their Habitats", Mar. 1982, Rockville, Maryland.

Evans, L.S., "Generation of Dose-Response Relationships to Assess the Effects of Acidity in Precipitation on Growth and Productivity of Vegetation", 1981, Brookhaven National Laboratory, Upton, New York.

Evans, L.S. et al., "Comparison of Experimental Designs to Determine Effects of Acidic Precipitation on Field-Grown Soybeans", BNL-32176, 1982, Brookhaven National Laboratory, Upton, New York.

Evans, L.S., Conway, C.A. and Lewin, K.F., "Yield Responses of Field-Grown Soybeans Exposed to Simulated Acid Rain", Mar. 1980, Brookhaven National Laboratory, Upton, New York.

Evans, L.S., Francis, A.J. and Raynor, G.S., "Acid Rain Research Program--Annual Progress Report--July 1976-September 1977", Contract No. EV-76-C-02-0016, Dec. 1977, Brookhaven National Laboratory, Upton, New York.

Evans, L.S., Gmur, N.F. and DaCosta, F., "Foliar Response of Six Clones of Hybrid Poplar", Phytopathology, Vol. 78, No. 6, June 1968, pp. 847-857.

Evans, L.S., Gmur, N.F. and Mancini, D., "Effects of Simulated Acid Rain on Yields of Raphanus Sativus, Lactuca Sativa, Triticum Aestivum and Medicago Sativa", Environmental and Experimental Botany, Vol. 22, No. 4, 1982, pp. 445-453.

Evans, L.S. and Hendrey, G.R., Proceedings of the International Workshop on the Effects of Acid Precipitation on Vegetation, Soils, and Terrestrial Ecosystems, Brookhaven National Laboratory, June 12-14, 1979, 1979, Brookhaven National Laboratory, Upton, New York.

Evans, L.S. and Lewin, K.F., "Effects of Simulated Acid Rain on Growth and Yield of Soybeans and Pinto Beans", 1979, Laboratory for Plant Morphogenesis, Manhattan College, Bronx, New York.

Faust, S.D. and McIntosh, A., "Buffer Capacities of Fresh Water Lakes Sensitive to Acid Rain Deposition", Journal of Environmental Science and Health, Part A, Vol. 18, No. 1, 1983, pp. 155-161.

Ferenbaugh, R.W., "Acid Rain: Biological Effects and Implications", Environmental Affairs, Vol. 4, No. 4, Fall 1975, pp. 745-759.

Flinn, D.R. et al., "Acidic Deposition and the Corrosion and Deterioration of Materials in the Atmosphere: A Bibliography, 1880-1982", EPA-600/3-83-059, July 1983, U.S. Environmental Protection Agency, Research Triangle Park, North Carolina.

Francis, A.J., Olson, D. and Bernatsky, R., "Effect of Acidity on Microbial Processes in a Forest Soil", Mar. 1980, Brookhaven National Laboratory, Upton, New York.

Francis, A.J., Olson, D. and Bernatsky, R., "Microbial Activity in Acid and Acidified Forest Soils", Mar. 1981, Brookhaven National Laboratory, Upton, New York.

Fritz, E.S., "Potential Impacts of Low pH on Fish and Fish Populations", FWS/OBS-80/40.2, Oct. 1980, U.S. Environmental Protection Agency, Washington, D.C.

Frizzola, J.A. and Baier, J.H., "Contaminants in Rainwater and Their Relation to Water Quality, Part I", Water and Sewage Works, Vol. 122, No. 8, Aug. 1975, pp. 72-76.

Galloway, J.N. et al., "A National Program for Assessing the Problem of Atmospheric Deposition (Acid Rain)", Dec. 1978, North Carolina Agricultural Experiment Station, North Carolina State University, Raleigh, North Carolina.

Galloway, J.N., Norton, S.A. and Church, M.R., "Freshwater Acidification from Atmospheric Deposition of Sulfuric Acid: A Conceptual Model", Environmental Science and Technology, Vol. 17, No. 11, Nov. 1983, pp. 541A-545A.

Gauri, K.L., "Effects of Acid Rain on Structures", Acid Rain: Proceedings of ASCE National Convention, 1979, American Society of Civil Engineers, New York, New York, pp. 70-91.

Gjessing, E., "Effect of Polluted Precipitation on Water Quality--the Situation in Norway", Aqua, No. 7, 1980, pp. 139-140.

Glass, N.R., "Environmental Effects of Increased Coal Utilization: Ecological Effects of Gaseous Emissions from Coal Combustion", EPA-600/7-78-108, June 1978, Corvallis Environmental Research Laboratory, U.S. Environmental Protection Agency, Corvallis, Oregon.

Glass, N.R., "U.S. Federal Program on Effects of Acid Rain", Acid Rain: Proceedings of ASCE National Convention, 1979, American Society of Civil Engineers, New York, New York, pp. 92-110.

Glass, N.R. et al., "Effects of Acid Precipitation", Environmental Science and Technology, Vol. 16, No. 3, Mar. 1982, pp. 162A-169A.

Glass, N.R., Glass, G.E. and Rennie, P.J., "Effects of Acid Precipitation", Environmental Science and Technology, Vol. 13, No. 11, Nov. 1979, pp. 1350-1352.

Glass, N.R., Glass, G.E. and Rennie, P.J., "Effects of Acid Precipitation in North America", Environment International, Vol. 4, No. 5-6, 1980, pp. 443-452.

Glass, N.R., Likens, G.E. and Dochinger, L.S., "The Ecological Effects of Atmospheric Deposition", EPA Energy/Environment 3rd National Conference, June 1-2, 1978, U.S. Environmental Protection Agency, Washington, D.C., pp. 113-120.

Gorham, E. and McFee, W.W., "Effects of Acid Deposition Upon Outputs from Terrestrial to Aquatic Ecosystems", Effects of Acid Rain Precipitation on Terrestrial Ecosystems, 1980, Plenum Press, New York, New York, pp. 465-480.

Graham, M.S. and Wood, C.M., "Toxicity of Environmental Acid to the Rainbow Trout: Interactions of Water Hardness, Acid Type, and Exercise", Canadian Journal of Zoology, Vol. 59, No. 8, Aug. 1981, pp. 1518-1526.

Grodzinska, K. et al., "First International Symposium on Acid Precipitation and the Forest Ecosystem", Water, Air, and Soil Pollution, Vol. 8, No. 1, May 1977, pp. 3-146.

Haines, B. and Waide, J., "Predicting Potential Impacts of Acid Rain on Elemental Cycling in a Southern Appalachian Deciduous Forest at Coweeta", Effects of Acid Rain Precipitation on Terrestrial Ecosystems, 1980, Plenum Press, New York, New York, pp. 335-339.

Haines, T.A., "Acidic Precipitation and Its Consequences for Aquatic Ecosystems: A Review", Transactions of the American Fisheries Society, Vol. 110, No. 6, Nov. 1981, pp. 669-707.

Harriman, R. and Morrison, B.R., "Ecology of Streams Draining Forested and Non-Forested Catchments in an Area of Central Scotland Subject to Acid Precipitation", Hydrobiologia, Vol. 88, No. 3, May 1982, pp. 251-263.

Harte, J., Holdren, J. and Tonnesson, K., "Potential for Acid-Precipitation Damage to Lakes of the Sierra Nevada, California", OWRT-A-081-CAL(1), Apr. 1983, Energy and Resources Group, University of California, Berkeley, California.

Havas, P. and Huttuneu, S., "Some Special Features of the Ecophysiological Effects of Air Pollution on Coniferous During the Winter", Effects of Acid Precipitation on Terrestrial Ecosystems, 1980, Plenum Press, New York, New York, pp. 123-131.

Heagle, A.S. et al., "Response of Soybeans to Simulated Acid Rain in the Field", Journal of Environmental Quality, Vol. 12, No. 4, Oct.-Dec. 1983, pp. 538-543.

Hendrey, G.R. et al., "Acid Precipitation: Some Hydrobiological Changes", Ambio, Vol. 5, No. 5-6, 1976, pp. 224-228.

Hendrey, G.R. et al., "Geological and Hydrochemical Sensitivity of the Eastern United States to Acid Precipitation", EPA-600/3-80-024, Jan. 1980, U.S. Environmental Protection Agency, Washington, D.C.

Hendrey, G.R. and Kaplan, E., "Identification of Fresh Waters Susceptible to Acidification", BNL-31000, 1982, Brookhaven National Laboratory, Upton, New York.

Hendrey, G.R. and Lipfert, F.W., "Acid Precipitation and the Aquatic Environment", May 1980, Brookhaven National Laboratory, Upton, New York.

Hendrey, G.R. and Vertucci, J.A., "Benthic Plant Communities in Acidic Lake Colden, New York: Sphagnum and the Algal Mat", Mar. 1980, Brookhaven National Laboratory, Upton, New York.

Henriksen, A. and Kirkhusmo, L.A., "Acidification of Groundwater in Norway", Nordic Hydrology, Vol. 13, No. 3, 1982, pp. 183-192.

Hileman, B., "Acid Precipitation", Environmental Science and Technology, Vol. 15, No. 10, Oct. 1981, pp. 1119-1124.

Hoffman, W.A., Lindberg, S.E. and Turner, R.R., "Precipitation Acidity: The Role of the Forest Canopy in Acid Exchange", Journal of Environmental Quality, Vol. 9, No. 1, 1980, pp. 95-101.

Hultberg, H. et al., "First International Symposium on Acid Precipitation and the Forest Ecosystem", Water, Air, and Soil Pollution, Vol. 7, No. 3, Mar. 1977, pp. 279-406.

Hutchinson, T.C., "Effects of Acid Leaching on Cation Loss from Soils", Effects of Acid Rain Precipitation on Terrestrial Ecosystems, 1980, Plenum Press, New York, New York, pp. 481-497.

Hutchinson, T.C. et al., "First International Symposium on Acid Precipitation and the Forest Ecosystem", Water, Air, and Soil Pollution, Vol. 7, No. 4, Apr. 1977, pp. 421-555.

Irving, P.M., "Acidic Precipitation Effects on Crops: A Review and Analysis of Research", Journal of Environmental Quality, Vol. 12, No. 4, Oct.-Dec. 1983, pp. 442-453.

Jacks, G. and Knutsson, G., "Susceptibility to Acidification of Ground Water in Different Parts of Sweden", KHM-TR-11, Oct. 1981, Statens Vattenfallsverk, Stockholm, Sweden.

Jacobson, J.S. "Experimental Studies on the Phototoxicity of Acidic Precipitation: The United States Experience," Effects of Acid Rain Precipitation on Terrestrial Ecosystems, 1980, Plenum Press, New York, New York, pp. 151-180.

Johnsen, I., "Regional and Local Effects of Air Pollution, Mainly Sulphur Dioxide, on Lichens and Bryophytes in Denmark", Effects of Acid Rain Precipitation on Terrestrial Ecosystems, 1980, Plenum Press, New York, New York, pp. 133-140.

Johnson, A.H. et al., "Recent Changes in Patterns of Tree Growth Rate in the New Jersey Pinelands: A Possible Effect of Acid Rain", Journal of Environmental Quality, Vol. 10, No. 4, Oct.-Dec. 1981, pp. 427-430.

Johnson, A.H. and Siccama, T.G., "Acid Deposition and Forest Decline", Environmental Science and Technology, Vol. 17, No. 7, July 1983, pp. 294A-305A.

Johnson, A.H. and Siccama, T.G., "Decline of Red Spruce in the Northern Appalachians: Assessing the Possible Role of Acid Deposition", Tappi, Vol. 67, No. 1, Jan. 1984, pp. 68-72.

Johnson, D.W., "Site Susceptibility to Leaching by H_2SO_4 in Acid Rainfall", Effects of Acid Rain Precipitation on Terrestrial Ecosystems, 1980, Plenum Press, New York, New York, pp. 525-535.

Johnson, D.W., "Acid Rain and Forest Productivity", 1981a, Oak Ridge National Laboratory, Oak Ridge, Tennessee.

Johnson, D.W., "The Natural Acidity of Some Unpolluted Waters in Southeastern Alaska and Potential Impacts of Acid Rain", Water, Air, and Soil Pollution, Vol. 16, No. 2, Aug. 1981b, pp. 243-252.

Johnson, D.W., "Effects of Acid Precipitation on Elemental Transport from Terrestrial to Aquatic Ecosystems", 1981c, Oak Ridge National Laboratory, Oak Ridge, Tennessee.

Johnson, D.W. and Richter, D.D., "Effects of Atmospheric Deposition on Forest Nutrient Cycles", Tappi, Vol. 67, No. 1, Jan. 1984, pp. 82-85.

Johnson, D.W., Turner, J. and Kelly, J.M., "Effects of Acid Rain on Forest Nutrient Status", Water Resources Research, Vol. 18, No. 3, June 1982, pp. 449-461.

Johnson, N.M., "Acid Rain: Neutralization Within the Hubbard Brook Ecosystem and Regional Implications", Science, Vol. 204, No. 4392, May 4, 1979, pp. 497-499.

Johnson, N.M. et al., "Acid Rain, Dissolved Aluminum and Chemical Weathering at the Hubbard Brook Experimental Forest, New Hampshire", Geochimica et Cosmochimica Acta, Vol. 45, No. 9, Sept. 1981, pp. 1421-1437.

Kaplan, E., Thode, H.C., Jr. and Protas, A., "Rocks, Soils and Water Quality--Relationships and Implications for Effects of Acid Precipitation on Surface Water in the Northeastern United States", Environmental Science and Technology, Vol. 15, No. 5, May 1981, pp. 539-544.

Killham, K., Firestone, M.K. and McColl, J.G., "Acid Rain and Soil Microbial Activity: Effects and Their Mechanisms", Journal of Environmental Quality, Vol. 12, No. 1, Jan.-Mar. 1983, pp. 133-137.

Klein, T.M., Kreitinger, J.P. and Alexander, M., "Nitrate Formation in Acid Forest Soils from the Adirondacks", Journal of the Soil Science Society of America, Vol. 47, No. 3, May-June 1983, pp. 506-508.

Klopatek, J.M., Harris, W.F. and Olson, R.J., "Regional Ecological Assessment Approach to Atmospheric Deposition: Effects on Soil Systems", 1979, Oak Ridge National Laboratory, Oak Ridge, Tennessee.

Kucera, V., "Effects of Sulfur Dioxide and Acid Precipitation on Metals and Anti-Rust Painted Steel", Ambio, Vol. 5, No. 5-6, 1976, pp. 248-254.

Lee, J.J. et al., "Effect of Simulated Sulfuric Acid Rain on Yield, Growth, and Foliar Injury of Several Crops", Environmental and Experimental Botany, Vol. 21, No. 2, 1981, pp. 171-185.

Lee, J.J., Neely, G.E. and Perrigan, S.C., "Sulfuric Acid Rain Effects on Crop Yield and Foliar Injury", EPA-600/3-80-016, Jan. 1980, U.S. Environmental Protection Agency, Washington, D.C.

Lee, J.J. and Weber, D.E., "The Effect of Simulated Acid Rain on Seedling Emergence and Growth of Eleven Woody Species", Forest Science, Vol. 3, No. 25, 1979, pp. 393-398.

Lee, J.J. and Weber, D.E., "Effects of Sulfuric Acid Rain on Major Cation and Sulfate Concentrations of Water Percolating Through Two Model Hardwood Forests", Journal of Environmental Quality, Vol. 11, No. 1, Jan.-Mar. 1982, pp. 57-64.

Likens, G.E. et al., "Biogeochemistry of a Forested Ecosystem", Report No. W80-00328, 1977, Cornell University-Section of Ecology and Systematics, Ithaca, New York.

Likens, G.E., Bormann, F.H. and Eaton, J.S., "Variations in Precipitation in Streamwater Chemistry at the Hubbard Brook Experimental Forest During 1964 to 1977", Effects of Acid Rain Precipitation on Terrestrial Ecosystems, 1980, Plenum Press, New York, New York, pp. 443-464.

Lindberg, S.E. and Harriss, R.C., "The Role of Atmospheric Deposition in an Eastern U.S. Deciduous Forest", Water, Air, and Soil Pollution, Vol. 16, No. 1, July 1981, pp. 13-31.

Lindberg, S.E., Shriner, D.S. and Hoffman, W.A., Jr., "Interaction of Wet and Dry Deposition with the Forest Canopy", 1981, Oak Ridge National Laboratory, Oak Ridge, Tennessee.

Logan, R.M., Derby, J.C. and Duncan, L.C., "Acid Precipitation and Lake Susceptibility in the Central Washington Cascades", Environmental Science and Technology, Vol. 16, No. 11, Nov. 1982, pp. 771-775.

Malmer, N., "Acid Precipitation: Chemical Changes in the Soil", Ambio, Vol. 5, No. 5-6, 1976, pp. 231-235.

Matziris, D.I. and Nakos, G., "Effect of Simulated Acid Rain on Juvenile Characteristics of Aleppo Pine (Pinus-Halepensis Mill)", Forest Ecology and Management, Vol. 1, No. 3, 1977, pp. 267-272.

Mayer, R. and Ulrich, B., "Input to Soil, Especially the Influence of Vegetation in Intercepting and Modifying Inputs--A Review", Effects of Acid Rain Precipitation on Terrestrial Ecosystems, 1980, Plenum Press, New York, New York, pp. 173-182.

McColl, J.G., "Effects of Acid Rain on Plants and Soils in California", ARB-R-81/148, Sept. 1981, California State Air Resources Board, Sacramento, California.

McKinley, V.L. and Vestal, J.R., "Effects of Acid on Plant Litter Decomposition in an Arctic Environment", Applied and Environmental Biology, Vol. 43, No. 5, May 1982, pp. 1188-1195.

McLaughlin, S.B. et al., "Interactive Effects of Acid Rain and Gaseous Air Pollutants on Natural Terrestrial Vegetation", 1983, Oak Ridge National Laboratory, Oak Ridge, Tennessee.

McLaughlin, S.B., West, D.C. and Blasing, T.J., "Measuring Effects of Air Pollution Stress on Forest Productivity: Perspectives, Problems, and Approaches", Tappi, Vol. 67, No. 1, Jan. 1984, pp. 74-80.

Mitchell, M.J., Landers, D.H. and Brodowski, D.F., "Sulfur Constituents of Sediments and Their Relationship to Lake Acidification", Water, Air, and Soil Pollution, Vol. 16, No. 3, Oct. 1981, pp. 351-359.

Mortvedt, J.J., "Impacts of Acid Deposition on Micronutrient Cycling in Agro-Ecosystems", Environmental and Experimental Botany, Vol. 23, No. 3, 1983, pp. 243-249.

National Technical Information Service, "Acid Precipitation: Effects on Terrestrial Ecosystems--1976-November, 1982 (Citations from the Energy Data Base)", Nov. 1982a, U.S. Department of Commerce, Springfield, Virginia.

National Technical Information Service, "Acid Precipitation: Effects on the Aquatic Ecosystem--1977-November, 1982 (Citations from the Energy Data Base)", Nov. 1982b, U.S. Department of Commerce, Springfield, Virginia.

Nelson, P.O. and Delwiche, G.K., "Sensitivity of Oregon's Cascade Lakes to Acid Precipitation", WRRI-85, Sept. 1983, Water Resources Research Institute, Oregon State University, Corvallis, Oregon.

Nilssen, J.P., "Acidification of a Small Watershed in Southern Norway and Some Characteristics of Acidic Aquatic Environments", Internationale Revue der Gesamten Hydrobiologie, Vol. 65, No. 2, 1980, pp. 177-207.

Nilsson, S.I., "Ion Adsorption Isotherms in Predicting Leaching Losses from Soils Due to Increased Inputs of 'Hydrogen Ions'--A Case Study", Effects of Acid Rain Precipitation on Terrestrial Ecosystems, 1980, Plenum Press, New York, New York, pp. 537-551.

Norton, S.A. et al., "Responses of Northern New England Lakes to Atmospheric Inputs of Acids and Heavy Metals", OWRT-A-048-ME(1), July 1981, Land and Water Resources Center, University of Maine, Orono, Maine.

Nriagu, J.O., "Isotopic Variation as an Index of Sulphur Pollution in Lakes Around Sudbury, Ontario", Nature, Vol. 273, No. 5659, May 18, 1978, pp. 223-224.

Olson, R.A., "Impacts of Acid Deposition on N and S Cycling", Environmental and Experimental Botany, Vol. 23, No. 3, 1983, pp. 211-223.

Olson, R.J., Johnson, D.W. and Shriner, D.S., "Regional Assessment of Potential Sensitivity of Soils in the Eastern United States to Acid Precipitation", ORNL/TM-8374, Dec. 1982, Oak Ridge National Laboratory, Oak Ridge, Tennessee.

Overrein, L.N., Seip, H.M. and Tollan, A., "Acid Precipitation: Effects on Forest and Fish", NP-2902584, Dec. 1980, Norges Landbrugshoegskole, Norway.

Petersen, L., "Podzolization: Mechanisms and Possible Effects of Acid Precipitation", Effects of Acid Rain Precipitation on Terrestrial Ecosystems, 1980a, Plenum Press, New York, New York, pp. 223-238.

Petersen, L., "Sensitivity of Different Soils to Acid Precipitation", Effects of Acid Rain Precipitation on Terrestrial Ecosystems, 1980b, Plenum Press, New York, New York, pp. 573-577.

Pfeiffer, M.H. and Festa, P.J., "Acidity Status of Lakes in the Adirondack Region of New York in Relation to Fish Resources", Aug. 1980, Bureau of

Fisheries, Department of Environmental Conservation, State of New York, Albany, New York.

Pitblado, J.R., Keller, W. and Conroy, N.I., "A Classification and Description of Some Northeastern Ontario Lakes Influenced by Acid Precipitation", Journal of Great Lakes Research, Vol. 6, No. 3, 1980, pp. 247-257.

Pough, F.H., "Mechanisms by Which Acid Precipitation Produces Embryonic Death in Aquatic Vertebrates", OWRT-A-077-NY(2), Mar. 1981, Cornell University, Ithaca, New York.

Puckett, L.J., "Acid Rain, Air Pollution, and Tree Growth in Southeastern New York", Journal of Environmental Quality, Vol. 11, No. 3, July-Sept. 1982, pp. 376-381.

Raynal, D.J., "Actual and Potential Effects of Acid Precipitation on a Forest Ecosystem in the Adirondack Mountains", NYSERDA-80-28, Dec. 1980, New York State Energy Research and Development Authority, Albany, New York.

Raynal, D.J., Roman, J.R. and Eichenlaub, W.M., "Response of Tree Seedlings to Acid Precipitation--II--Effect of Simulated Acidified Canopy Throughfall on Sugar Maple Seedling Growth", Environmental and Experimental Botany, Vol. 22, No. 3, 1982, pp. 385-392.

Reuss, J.D., "Chemical/Biological Relationships Relevant to Ecological Effects of Acid Rainfall", June 1975, Corvallis Environmental Research Laboratory, U.S. Environmental Protection Agency, Corvallis, Oregon.

Richter, D.D., Johnson, D.W. and Todd, D.E., "Atmospheric Sulfur Deposition, Neutralization, and Ion Leaching in Two Deciduous Forest Ecosystems", Journal of Environmental Quality, Vol. 12, No. 2, Apr.-June 1983, pp. 263-270.

Roberts, T.M. et al., "Effects of Sulphur Deposition on Litter Decomposition and Nutrient Leaching in Coniferous Forest Soils", Effects of Acid Rain Precipitation on Terrestrial Ecosystems, 1980, Plenum Press, New York, New York, pp. 381-393.

Rorison, I.H., "The Effects of Soil Acidity on Nutrient Availability and Plant Response", Effects of Acid Rain Precipitation on Terrestrial Ecosystems, 1980, Plenum Press, New York, New York, pp. 283-304.

Scherbatskoy, T. and Klein, R.M., "Response of Spruce and Birch Foliage to Leaching by Acidic Mists", Journal of Environmental Quality, Vol. 12, No. 2, Apr.-June 1983, pp. 189-195.

Schindler, D.W. et al., "Effects of Acidification on Mobilization of Heavy Metals and Radionuclides from the Sediments of a Freshwater Lake", Canadian Journal of Fisheries and Aquatic Sciences, Vol. 37, No. 3, Mar. 1980, pp. 373-377.

Schindler, D.W. and Turner, M.A., "Biological Chemical and Physical Responses of Lakes to Experimental Acidification", Water, Air, and Soil Pollution, Vol. 18, No. 1-3, 1982, pp. 259-271.

Schnitzer, M., "Effect of Low pH on the Chemical Structure and Reactions of Humic Substances", Effects of Acid Rain Precipitation on Terrestrial Ecosystems, 1980, Plenum Press, New York, New York, pp. 203-222.

Schofield, C.L., "Acid Precipitation: Effects on Fish", Ambio, Vol. 5, No. 5-6, 1976, pp. 228-231.

Schofield, C.L., "Effects of Acid Rain on Lakes", Acid Rain: Proceedings of ASCE National Convention, 1979, American Society of Civil Engineers, New York, New York, pp. 55-69.

Sears, S.O. and Langmuir, D., "Sorption and Mineral Equilibria Controls on Moisture Chemistry in a C-Horizon Soil", Journal of Hydrology, Vol. 56, No. 3/4, Apr. 1982, pp. 287-308.

Shinn, J.H., "Critical Survey of Measurements of Foliar Deposition of Airborne Sulfates and Nitrates", Air Pollution Control Association Meeting, Houston, Texas, Mar. 1978, Lawrence Livermore Laboratory, Livermore, California.

Shriner, D.S., "Effects of Simulated Acidic Rain on Host-Parasite Interactions in Plant Diseases", Phytopathology, Vol. 68, No. 2, Feb. 1978, pp. 213-219.

Shriner, D.S. and Cowling, E.B., "Effects of Rainfall Acidification on Plant Pathogens", Contract No. 2-7405-ENG-26, 1978, NATO Institute on Acid Rain Effects, Toronto, Canada.

Sigma Research, Inc., Proceedings: Ecological Effects of Acid Precipitation, Feb. 1982, Richland, Washington.

Singh, B.R., Abrahamsen, G. and Stuanes, A., "Effect of Simulated Acid Rain on Sulfate Movement in Acid Forest Soils", Journal of the Soil Science Society of America, Vol. 44, No. 1, Jan.-Feb. 1980, pp. 75-80.

Sparks, D.L. and Curtis, C.R., "An Assessment of Acid Rain on Leaching of Elements from Delaware Soils into Groundwater", OWRT-A-053-DEL(1), 1983, College of Agricultural Sciences, University of Delaware, Newark, Delaware.

Sposito, G., Page, A.L. and Frink, M.E., "Effects of Acid Precipitation on Soil Leachate Quality--Computer Calculations", EPA-600/3-80-015, Jan. 1980, U.S. Environmental Protection Agency, Washington, D.C.

Sprules, W.G., "Midsummer Crustacean Zooplankton Communities in Acid-Stressed Lakes", Fisheries Research Board of Canada Journal, Vol. 32, No. 3, Mar. 1975, pp. 389-399.

Stednick, J.D. and Johnson, D.W., "Natural Acidity of Waters in Podzolized Soils and Potential Impacts from Acid Precipitation", 1982, Oak Ridge National Laboratory, Oak Ridge, Tennessee.

Strayer, R. and Alexander, M., "Effects of Simulated Acid Rain on Glucose Mineralization and Some Physicochemical Properties of Forest Soils", Journal of Environmental Quality, Vol. 18, No. 4, Dec. 1981, pp. 460-465.

Strayer, R.F., Chyi-Jiin, L. and Alexander, M., "Effect of Simulated Acid Rain on Nitrification and Nitrogen Mineralization in Forest Soils", Journal of Environmental Quality, Vol. 10, No. 4, Oct.-Dec. 1981, pp. 547-551.

Tamm, C.O., "Acid Precipitation: Biological Effects in Soil and on Forest Vegetation", Ambio, Vol. 5, No. 5-6, 1976, pp. 235-239.

Taylor, F.B. and Symons, G.E., "Effects of Acid Rain on Water Supplies in the Northeast", Journal of the American Water Works Association, Vol. 76, No. 3, Mar. 1984, pp. 34-41.

Tetra Tech, Inc., "Integrated Lake-Watershed Acidification Study (ILWAS): Contributions to the International Conference on the Ecological Impact of Acid Precipitation", May 1981, Lafayette, California.

Torrenueva, A.L., "Effects of Acid-Rain on Terrestrial Ecosystems", Research Review of Ontario Hydro, No. 2, May 1981, pp. 49-56.

Tschupp, E.J., "Effect on Groundwater Resources of Precipitation Modified by Air Pollution", 55th Annual Meeting of the American Geophysical Union, Apr. 8-12, 1974, Washington, D.C.

Tukey, H.B., Jr., "Some Effects of Rain and Mist on Plants, with Implications for Acid Precipitation", Effects of Acid Rain Precipitation on Terrestrial Ecosystems, 1980, Plenum Press, New York, New York, pp. 141-150.

Turk, J.T. and Adams, D.B., "Sensitivity to Acidification of Lakes in the Flat Tops Wilderness Area, Colorado", Water Resources Research, Vol. 19, No. 2, Apr. 1983, pp. 346-350.

Tveite, B. and Abrahamsen, G., "Effects of Artificial Acid Rain on the Growth and Nutrient Status of Trees", Effects of Acid Rain Precipitation on Terrestrial Ecosystems, 1980, Plenum Press, New York, New York, pp. 305-318.

Tyler, G., "Leaching Rates of Heavy Metal Ions in Forest Soil", Water, Air, and Soil Pollution, Vol. 9, No. 2, Feb. 1978, pp. 137-149.

Ulrich, B., Mayer, R. and Khanna, P.K., "Chemical Changes Due to Acid Precipitation in a Loess-Derived Soil in Central Europe", Soil Science, Vol. 130, No. 4, Oct. 1980, pp. 193-199.

U.S. Fish and Wildlife Service, "Effects of Acid Precipitation on Aquatic Resources: Results of Modeling Workshops", FWS/OBS-80/40.12, July 1982, Eastern Energy and Land Use Team, Kearneysville, West Virginia.

Vangenechten, J.H., "Interrelations Between pH and Other Physiochemical Factors in Surface Waters of the Campine of Antwerp, (Belgium): With Special Reference to Acid Moorland Pools", Archiv fur Hydrobiologie, Vol. 90, No. 3, Nov. 1980, pp. 265-283.

Vaughan, H.H., Underwood, J.K. and Ogden, J.G., III, "Acidification of Nova Scotia Lakes I: Response of Diatom Assemblages in the Halifax Area", Water, Air, and Soil Pollution, Vol. 18, No. 1-3, 1982, pp. 353-361.

Wainwright, M., Supharungsun, S. and Killham, K., "Effect of Acid Rain on the Solubility of Heavy Metal Oxides and Fluorspar (CaF$_2$) Added to Soil", Science of the Total Environment, Vol. 24, No. 1, May 1982, pp. 85-90.

Webb, A.H., "Weak Acid Concentrations and River Chemistry in the Tovdal River, Southern Norway", Water Research, Vol. 16, No. 5, May 1982, pp. 641-648.

Wetstone, G.S. and Foster, S.A., "Acid Precipitation: What is It Doing to Our Forests", Environment, Vol. 25, No. 4, May 1983, pp. 10-12, 38-40.

Wiklander, L., "The Sensitivity of Soils to Acid Precipitation", Effects of Acid Rain Precipitation on Terrestrial Ecosystems, 1980, Plenum Press, New York, New York, pp. 553-567.

Wood, T. and Bormann, F.H., "Increases in Foliar Leaching Caused by Acidification of an Artificial Mist", Ambio, Vol. 4, No. 4, 1975, pp. 169-172.

Wright, R.F. and Gjessing, E.T., "Acid Precipitation: Changes in the Chemical Composition of Lakes", Ambio, Vol. 5, No. 5-6, 1976, pp. 219-224.

CHAPTER 5

CONTROL STRATEGIES FOR ACID RAIN

As a result of the growing body of knowledge about the causes and consequences of acid rain, an increased emphasis is being given to control strategies to minimize both causes and consequences. Control strategies include both source emission reduction measures and environmental clean-up and restoration programs. Many policy considerations are associated with decisions on control strategies. This chapter summarizes each of these issues following a background section highlighting a summary of the pertinent abstracts contained in Appendix F.

BACKGROUND INFORMATION

This chapter is based on information from 28 references whose abstracts are contained in Appendix F. Table 25 provides a brief comment on each of these references in the order in which they appear in Appendix F. Not every reference cited in Table 25 will be specifically addressed in this chapter.

Table 25: References in Appendix F Dealing with Control Strategies

Author(s) (Year)	Comments
Ball and Menzies (1982a)	Capital and operating costs for flue gas desulfurization (FGD) systems retrofitted to industrial boilers.
Ball and Menzies (1982b)	Appendices of capital and operating costs for FGD systems.
Ball, Muela and Meling (1982)	Identification of control techniques for reducing SO_2 emissions from industrial sources east of the Mississippi River.
Bengtsson, Dickson and Nyberg (1980)	Costs and effectiveness of Swedish program for liming lakes and rivers.
Blake (1981)	Costs, techniques, and effectiveness of liming ponds in New York.
Bomberger and Phillips (1979)	Description of control technologies for reducing SO_x and NO_x emissions from power plants.
Catalono and Makansi (1983)	Technical and policy issues associated with SO_2 controls.

Table 25: (continued)

Author(s) (Year)	Comments
Chadwick (1983)	Control strategies for reducing SO_x and NO_x emissions from fuel use.
Chawla and Varma (1981-1982)	Systems approach for developing a control strategy for acid rain.
DePinto and Edzwald (1982)	Study of various chemical control measures for recovery of Adirondack acid lakes.
Electric Power Research Institute (1981)	Proceedings of symposium on control of NO_x from stationary combustion.
Ellis and Golomb (1981)	Efficacy of neutralization as a lake restoration technique.
Fortin and McBean (1983)	Linear programming-based model for examining management alternatives for acid rain abatement.
Fraser et al. (1982)	Feasibility of liming for mitigating surface water acidification.
Gilbert (1982)	Economic basis for source control.
Golomb (1983)	Prediction of changes in rain acidity for various source emission reduction scenarios.
Gorham (1982)	General proposal for reducing acid rain.
Hastings and Schaefer (1980)	Status of combustion technologies for controlling NO_x emissions.
McLean (1981)	SO_x and NO_x control strategies based on their relative contributions to the acid rain problem.
PEDCo-Environmental, Inc. (1981)	Cost estimates for SO_2 and NO_x control strategies for top 50 coal-fired emission sources.
Rohde (1981)	Potential effects on aquatic biology from liming lakes.
Streets et al. (1983)	Analysis of proposed legislation to control acid rain in the United States.

Table 25: (continued)

Author(s) (Year)	Comments
Sverdrup (1983)	Mathematical model for determining the effectiveness of lake liming.
Szabo, Shah and Abraham (1982)	Detailed evaluation of the costs and effectiveness of conventional SO_x and NO_x controls for coal-fired utility boilers.
U.S. Environmental Protection Agency (1980)	Description of new and improving technologies for control of SO_x emissions.
Weisenfeld and Kreiss (1982)	Policy issues associated with source controls and source-receptor relationships.
Wright and Henriksen (1983)	General planning model for estimating improvements in Norwegian lakes from reductions in SO_x emissions in Europe.
Zawadzki Ltd. (1981)	Policy issues associated with use of low sulfur coal or cleaned coal to reduce SO_x emissions.

SOURCE EMISSION REDUCTION MEASURES

Since sulfur dioxide and oxides of nitrogen are the two major precursors in acid rain formation, source emission reduction measures are typically focused on control technologies for these two pollutants. Table 26 lists 11 references from Appendix F which address the effectiveness and costs of technologies for accomplishing sulfur dioxide and oxides of nitrogen control.

Table 26: References in Appendix F which Describe Source Emission Reduction Measures

Author(s) (Year)	Comments
Ball and Menzies (1982a)	Capital and operating costs for flue gas desulfurization (FGD) systems retrofitted to industrial boilers.
Ball and Menzies (1982b)	Appendices of capital and operating costs for FGD systems.

Table 26: (continued)

Author(s) (Year)	Comments
Ball, Muela and Meling (1982)	Identification of control techniques for reducing SO_2 emissions from industrial sources east of the Mississippi River.
Bomberger and Phillips (1979)	Description of control technologies for reducing SO_x and NO_x emissions from power plants.
Chadwick (1983)	Control strategies for reducing SO_x and NO_x emissions from fuel use.
Electric Power Research Institute (1981)	Proceedings of symposium on control of NO_x from stationary combustion.
Gilbert (1982)	Economic basis for source control.
Hastings and Schaefer (1980)	Status of combustion technologies for controlling NO_x emissions.
PEDCo-Environmental, Inc. (1981)	Cost estimates for SO_2 and NO_x control strategies for top 50 coal-fired emission sources.
Szabo, Shah and Abraham (1982)	Detailed evaluation of the costs and effectiveness of conventional SO_x and NO_x controls for coal-fired utility boilers.
U.S. Environmental Protection Agency (1980)	Description of new and improving technologies for control of SO_x emissions.

Chadwick (1983) denoted three stages in fuel usage where choices could be made which would minimize the atmospheric emissions of sulfur dioxide and oxides of nitrogen. These stages and choices include: (1) before burning, choose fuel with minimal nitrogen and sulfur contents; (2) during burning, use staged, catalytic or coal-limestone combustion; and (3) during venting of waste gas, use flue gas desulfurization and ammonia injection. This section of Chapter 5 will focus on the options during waste gas venting, particularly flue gas desulfurization for sulfur dioxide control.

Flue Gas Desulfurization

Flue gas desulfurization (FGD) technology is the SO_2 emission control approach that is most common today. FGD systems remove SO_2 from power plant combustion gases via absorption in an alkaline solution. Since these systems are used after the combustion process, they can be retrofitted to existing boilers. Therefore, they can provide an effective and efficient means of meeting SO_2 emission standards without modifying or replacing existing boilers. FGD systems can be classified into two general types: (1) non-regenerable or throwaway systems, in which the sulfur material generated through scrubbing or absorption is disposed of as a waste product; and (2) regenerable or recovery systems, in which sulfur materials, such as gypsum, elemental sulfur, liquid SO_2, and sulfuric acid, are marketed as saleable products. Frequently mentioned processes for FGD include lime or limestone scrubbing, dual (double) alkali scrubbing, magnesia scrubbing, and the Wellman-Lord process. About two out of three FGD systems that are now operational, under construction, or planned for the next five years in the utility industry in the United States use limestone as the absorption reagent.

The lime/limestone scrubbing process extracts SO_2 from the flue gas by reacting it with lime (CaO) or limestone ($CaCO_3$) in a slurry to form a precipitate. The relevant chemical reactions are:

Lime:
$$SO_2 + Ca(OH)_2 \rightarrow H_2O + CaSO_3 \tag{1}$$

Limestone:
$$SO_2 + CaCO_3 \rightarrow CO_2 + CaSO_3 \tag{2}$$

$$2\ CaSO_3 + O_2 \rightarrow 2\ CaSO_4 \tag{3}$$

The power plant flue gas, from which fly ash has been previously removed in an electrostatic precipitator or fabric filter, and the slurry come into contact in the scrubber, and it is there that the first two reactions take place. From the scrubber, the reacted slurry flows to a reaction tank where fresh lime or limestone is added to make the calcium sulfite and sulfate precipitate (as hydrates). Part of the slurry leaving the reaction tank is pumped back to the scrubber. The remainder is sent to a solid/liquid separator, which may be a centrifuge, filter, or holding pond. Solids (sludge) from the separator--consisting of hydrated calcium sulfite and sulfate, unreacted lime or limestone, and some fly ash--are removed for disposal. The liquid-output stream of the separator is split, one portion being sent back to the reaction tank and the remaining portion to the scrubber, thus completing the cycle. Makeup water is added to the scrubber-bound portion to compensate for evaporation losses and water lost with the sludge.

Oxides of Nitrogen Control

Several technologies are available for reducing the oxides of nitrogen emissions from fuel combustion. The reduction of NO_x emissions may involve combustion modification, NO_x removal from stack gases, dry and wet NO_x removal processes, and fluidized bed combustion (Bomberger and Phillips, 1979). Considerable research work is being done on combustion technologies (Hastings and Schaefer, 1980). The Electric Power Research Institute has published the Proceedings of a symposium on stationary combustion oxides of nitrogen control (Electric Power Research Institute, 1981). The symposium consisted of over 50 presentations describing recent advances in NO_x control technology, including applications for pulverized coal-fired utility boilers.

Cost Considerations

Several studies have been made of the cost of control programs for sulfur dioxide and oxides of nitrogen. A detailed breakdown of cost estimates for SO_2 and NO_x control strategies applied to the top 50 coal-fired SO_2-NO_x emitters associated with utility boilers has been prepared (PEDCo-Environmental, Inc., 1981). Information is available on the costs for eastern and western low-sulfur coal strategies, costs for the 90 percent and 60 percent (partial scrubbing) SO_2 control attained through the use of wet flue gas desulfurization, costs for dry flue gas desulfurization (applicable to only three plants), and NO_x control costs for low-NO_x burners and overfire air. All costs are in 1980 dollars, and treatment of annual operating and maintenance and fixed charges is as uniform as possible to maximize comparisons on the same or similar basis. In addition, Ball and Menzies (1982a; and 1982b) have provided a consistent set of capital investment and operating costs for FGD systems retrofitted to existing industrial boilers.

Szabo, Shah and Abraham (1982) have provided a detailed evaluation of the cost and effectiveness of conventional controls for emissions of sulfur oxides (SO_x) and nitrogen oxides (NO_x) from coal-fired utility boilers. The cost and control efficiency data are based on analyses of the 50 U.S. utility plants emitting these pollutants in the greatest quantities in 1979 (the 50 highest emitters for each pollutant, with some overlap of plants emitting large amounts of both). The evaluation was based on the premise that coal-fired utility power plants in the midwestern United States are the major contributors to the acid rain problem in the northeastern United States. One conclusion was that reducing SO_x and NO_x emissions from midwestern coal-fired power plants may not significantly reduce the acidity of rain, even at the cost of billions of dollars for controls. In fact, local sources of SO_x and NO_x, chiefly oil-fired boilers and automobiles in the northeast, may contribute more significantly to the acid rain occurring there than previously realized.

ENVIRONMENTAL CLEAN-UP AND RESTORATION

Liming of acidified surface bodies of water has been practiced as a means of environmental clean-up and restoration. While this approach can be effective, it only addresses one aspect of the environmental damages from acid rain, namely, aquatic ecosystem effects. Table 27 lists 7 references from Appendix F which relate to liming of lakes.

Table 27: References in Appendix F which Address Liming of Lakes

Author(s) (Year)	Comments
Bengtsson, Dickson and Nyberg (1980)	Costs and effectiveness of Swedish program for liming lakes and rivers.
Blake (1981)	Costs, techniques, and effectiveness of liming ponds in New York.
DePinto and Edzwald (1982)	Study of various chemical control measures for recovery of Adirondack acid lakes.

Table 27: (continued)

Author(s) (Year)	Comments
Ellis and Golomb (1981)	Efficacy of neutralization as a lake restoration technique.
Fraser et al. (1982)	Feasibility of liming for mitigating surface water acidification.
Rohde (1981)	Potential effects on aquatic biology from liming lakes.
Sverdrup (1983)	Mathematical model for determining the effectiveness of lake liming.

In 1976, the Swedish government initiated liming lakes and rivers to improve the quality of water polluted by acid rain (Bengtsson, Dickson and Nyberg, 1980). During 1977-1979, 700 lakes and rivers were treated directly with a total of 120,000 tons of lime. A single application may be effective for five to ten years. Ecological studies showed that liming acid waters greatly increased the populations of zooplankton, phytoplankton, and fish. Sweden has now treated over 1,000 lakes (Rohde, 1981). In Canada, Ellis and Golomb (1981) indicated that lake neutralization experiments have been only partly successful in restoring low-pH lakes to a nonacid-stressed oligotrophic state.

Blake (1981) reported on the experience of liming acid surface waters in New York. Application of hydrated lime is limited to open-water periods for economic reasons. Hydrated lime is superior to the less potent agricultural lime, calcium carbonate, although the latter permits a longer time between treatments. No matter which type of lime is used, the lime is dumped from bags from a slowly moving boat. More lime is spread in shallow water areas than over deep water areas so that the lime is placed where the fish live and it is exposed more effectively to wave action and currents. The use of lime in remote areas is more difficult due to the high cost of transport. Costs for liming remote ponds range as high as $297 per acre. Experience with the dry dispersal method has shown that lime can be moved at a rate of 5 tons per hour. The average cost of this method should be about $100 per acre. It was concluded that for both accessible and remote ponds, liming is an effective and economically feasible tool which can be used to counteract the adverse impacts of acid precipitation and maintain selected fisheries.

Fraser et al. (1982) summarized a study of liming as a technique for mitigating surface water acidification. The study addressed the types and combinations of alkaline materials, the input location, delivery modes, and approximate cost of some of the techniques. An assessment of potential chemical, physical and biological changes resulting from short- or long-term lime applications was made. Finally, recommendations were suggested for additional research needed to evaluate liming as a technique for mitigating the effects from surface water acidification.

POLICY CONSIDERATIONS

There are many policy considerations associated with source emission reduction and environmental clean-up strategies. These strategies may be used singly or in combination. Table 28 identifies 10 references from Appendix F which address various policy issues. One of the difficult issues is associated with the need for decision-making in the absence of complete information on the causes, consequences, and necessary corrective actions for the acid precipitation problem. This need is heightened by the perception that quick actions are needed to begin to address the problem (Catalano and Makansi, 1983). Another issue is associated with recognizing the environment as a system and accounting for pollutant transfer between air, water, and soil (Chawla and Varma, 1981-1982). Pollution control efforts directed toward specific components such as air must also recognize the side-effects of these efforts on other components such as water and soil.

Table 28:   References in Appendix F which Address Policy Issues Related to Acid Rain Control Strategies

Author(s) (Year)	Comments
Catalono and Makansi (1983)	Technical and policy issues associated with SO_2 controls.
Chawla and Varma (1981-1982)	Systems approach for developing a control strategy for acid rain.
Fortin and McBean (1983)	Linear programming-based model for examining management alternatives for acid rain abatement.
Golomb (1983)	Prediction of changes in rain acidity for various source emission reduction scenarios.
Gorham (1982)	General proposal for reducing acid rain.
McLean (1981)	SO_x and NO_x control strategies based on their relative contributions to the acid rain problem.
Streets et al. (1983)	Analysis of proposed legislation to control acid rain in the United States.
Weisenfeld and Kreiss (1982)	Policy issues associated with source controls and source-receptor relationships.
Wright and Henriksen (1983)	General planning model for estimating improvements in Norwegian lakes from reductions in SO_x emissions in Europe.

Table 28: (continued)

Author(s) (Year)	Comments
Zawadzki Ltd. (1981)	Policy issues associated with use of low sulfur coal or cleaned coal to reduce SO_x emissions.

Acid rain control strategies have typically been oriented to source control of sulfur dioxide emissions. McLean (1981) justifies this approach by noting that the contributions of nitric acid and sulfuric acid in atmospheric deposition to the acidification of the terrestrial and aquatic ecosystems are not equivalent and that, except possibly during the spring thaw, nitric acid contributes considerably less to the acidification of the ecosystem. Therefore, though there is little doubt that tighter control strategies are needed to diminish the effects of acid rain on ecosystems, the existing control strategies, which have put more emphasis on the control of emissions of sulfur oxides than nitrogen oxides, have a reasonable scientific basis given present limited knowledge of their effects on ecosystems.

Control of sulfur dioxide emissions represents a major component of programs to reduce the damage from acid precipitation in Norway (Wright and Henriksen, 1983). Norway, along with Sweden and Finland, have proposed that emissions of sulfur in Europe be reduced by 30 percent by 1993. Using data from a 1974-75 survey of major-ion chemistry and fish populations in 683 lakes in southern Norway, a 30 percent reduction in sulfur deposition will restore chemical conditions such that 22 percent of the lakes now experiencing fisheries problems should be able to support fish. The method used to make these predictions was derived from empirical relationships between sulfur deposition, water chemistry and fisheries status for several thousand lakes and rivers in Europe and North America. The method takes into account lake-to-lake differences in sensitivity, changes in base cation concentrations, and the buffering of dissolved aluminum at pH readings less than 5.4 and of bicarbonate at pH levels greater than 5.4.

A major determinant in any control strategy for acid rain is its costs and benefits, or cost-effectiveness. Fourteen bills introduced into the 97th Congress of the United States addressed acid rain or the long-range transport of air pollutants. Streets et al. (1983) analyzed the emission reductions and costs required by the five major bills. The required emissions reductions ranged between 6.5 and 12.6 million tons of sulfur dioxide per year, at a cost of $2.6 to $5.4 billion per year.

Weisenfeld and Kreiss (1982) reported the conclusions of a technical panel convened to examine the source-receptor relationship in acid precipitation. They noted that while a decrease in acid deposition is certainly a worthy goal for pursuit, the ecological consequences with or without such decreases are uncertain. Even more difficult to predict, however, is how best to limit deposition of acidifying species by the control of emissions. Although control costs using various strategies may be estimated to about a factor of two, assessment of the benefits to be derived (in terms of deposition abatement) from the imposition of such controls remains a matter of significant uncertainty. The authors noted that it is imperative that a far better understanding of the source-receptor relationship be gained in order to optimize the benefits of a major emissions reduced program so that the likely benefits of alternative control strategies may be estimated with reasonable accuracy.

SELECTED REFERENCES

Ball, J.G. and Menzies, W.R., "Acid Rain Mitigation Study--Volume I: FGD Cost Estimates", EPA-600/2-82-070A, Sept. 1982a, U.S. Environmental Protection Agency, Research Triangle Park, North Carolina.

Ball, J.G. and Menzies, W.R., "Acid Rain Mitigation Study--Volume II: FGD Cost Estimates (Appendices)", EPA-600/2-82-070B, Sept. 1982b, U.S. Environmental Protection Agency, Research Triangle Park, North Carolina.

Ball, J.G., Muela, C.A. and Meling, J.L., "Acid Rain Mitigation Study--Volume III: Industrial Boilers and Processes", EPA-600/2-82-070C, Sept. 1982, U.S. Environmental Protection Agency, Research Triangle Park, North Carolina.

Bengtsson, B., Dickson, W. and Nyberg, P., "Liming Acid Lakes in Sweden", Ambio, Vol. 9, No. 1, 1980, pp. 34-36.

Blake, L.M., "Liming Acid Ponds in New York", New York Fish and Game Journal, Vol. 28, No. 2, 1981, pp. 208-214.

Bomberger, D.C. and Phillips, R.C., "Technological Options for Mitigation of Acid Rain", Acid Rain: Proceedings of ASCE National Convention, 1979, American Society of Civil Engineers, pp. 132-166.

Catalono, L. and Makansi, J., "Acid Rain: New SO_2 Controls Inevitable", Power, Vol. 127, No. 9, Sept. 1983, pp. 25-33.

Chadwick, M.J., "Acid Depositions and the Environment", Ambio, Vol. 12, No. 2, 1983, pp. 80-82.

Chawla, R.C. and Varma, M.M., "Pollutant Transfer Between Air, Water and Soil: Criteria for Comprehensive Pollution Control Strategy", Journal of Environmental Systems, Vol. 11, No. 4, 1981-82, pp. 363-374.

DePinto, J.V. and Edzwald, J.K., "An Evaluation of the Recovery of Adirondack Acid Lakes by Chemical Manipulation", OWRT-B-095-NY(1), June 1982, Office of Water Research and Technology, U.S. Department of Interior, Washington, D.C.

Electric Power Research Institute, "Joint Symposium on Stationary Combustion NO_x Control: Proceedings", EPRI-WS-79-220(V.1), May 1981, Palo Alto, California.

Ellis, J.D. and Golomb, A., "Lake Acidity and Its Neutralization", Research Review of Ontario Hydro, No. 2, May 1981, pp. 67-74.

Fortin, M. and McBean, E.A., "Management Model for Acid Rain Abatement", Atmospheric Environment, Vol. 17, No. 11, 1983, pp. 2331-2336.

Fraser, J. et al., "Feasibility Study to Utilize Liming as a Technique to Mitigate Surface Water Acidification", EPRI-EA-2362, Apr. 1982, General Research Corporation, McLean, Virginia.

Gilbert, A.H., "Economics of Acid Rain: An Invisible Hand of Control", International Journal of Environmental Studies, Vol. 18, No. 2, Feb. 1982, pp. 85-90.

Golomb, D., "Acid Deposition-Precursor Emission Relationship in the Northeastern U.S.A.: The Effectiveness of Regional Emission Reduction", Atmospheric Environment, Vol. 17, No. 7, 1983, pp. 1387-1390.

Gorham, E., "What To Do About Acid Rain", Technology Review, Vol. 85, No. 7, Oct. 1982, pp. 59-66.

Hastings, A. and Schaefer, M., "Controlling Nitrogen Oxides", EPA-600/8-80-004, Feb. 1980, U.S. Environmental Protection Agency, Washington, D.C.

McLean, R.A., "The Relative Contributions of Sulfuric and Nitric Acids in Acid Rain to the Acidification of the Ecosystem--Implications for Control Strategies", Journal of the Air Pollution Control Association, Vol. 31, No. 11, Nov. 1981, pp. 1184-1187.

PEDCo-Environmental, Inc., "Acid Rain: Control Strategies for Utility Boilers--Volume II--Appendices", DOE-METC-82-42-V.2, May 1981, Cincinnati, Ohio.

Rohde, W., "Reviving Acidified Lakes", Ambio, Vol. 10, No. 4, 1981, pp. 195-196.

Streets, D.G. et al., "Analysis of Proposed Legislation to Control Acid Rain", ANL/EES-TM-209, Jan. 1983, Argonne National Laboratory, Argonne, Illinois.

Sverdrup, H., "Lake Liming", Chemistry Series, Vol. 22, No. 1, 1983, pp. 12-18.

Szabo, M., Shah, Y. and Abraham, J., "Acid Rain: Control Strategies for Coal-Fired Utility Boilers--Volume I", DOE/METC-82-42-V.1, May 1982, U.S. Department of Energy, Washington, D.C.

U.S. Environmental Protection Agency, "Controlling Sulfur Oxides", EPA-600/8-80-029, Aug. 1980, Washington, D.C.

Weisenfeld, J.R. and Kreiss, W.T., "Source-Receptor Relationship in Acid Precipitation: Implications for Generation of Electric Power from Coal", PD-LJ-82-268R, 1982, Physical Dynamics, Inc., La Jolla, California.

Wright, R.F. and Henriksen, A., "Restoration of Norwegian Lakes by Reduction in Sulphur Deposition", Nature, Vol. 305, No. 5933, Sept. 1983, pp. 422-424.

Zawadzki Ltd., "Preliminary Study: Use of Low-Sulfur Coal and Coal Cleaning in Control of Acid Rain", DOE/MC/14784-T1, May 1981, U.S. Department of Energy, Washington, D.C.

CHAPTER 6

DRY DEPOSITION, PRECIPITATION METALS AND ORGANICS

Atmospheric pollutants can return to the earth's surface by wet or dry deposition. This chapter gives attention to dry deposition of various pollutants, including sulfates, metals and nutrients. In addition, wet deposition or precipitation can also return metals and organics to the earth's surface in addition to acid rain components such as sulfates and nitrates. This chapter will also address precipitation metals and organics.

DRY DEPOSITION

This portion of Chapter 6 is based on information from the 13 references whose abstracts are found in Appendix G. Table 29 provides a brief comment on each of these references in the order they appear in Appendix G. Not every reference listed in Table 29 will be specifically addressed in this portion of Chapter 6. As a background, Garland (1979) discusses a range of issues relating to dry deposition, including the deposition process, methods of measurement, deposition velocities, and uptake by the sea.

Table 29: References in Appendix G Dealing with Dry Deposition

Author(s) (Year)	Comments
Dolske and Sievering (1979)	Dry deposition of trace elements over southern Lake Michigan.
Dolske and Sievering (1980)	Dry deposition of nutrients over southern Lake Michigan.
Droppo and Doran (1979)	Dry deposition removal rates of various atmospheric constituents.
Garland (1979)	General information on deposition processes and velocities, and methods of measurement.
Hicks (1979)	Deposition velocities of sulfate particles.
Hicks, Wesely and Durham (1982)	Discussion and critique of methods for measuring dry deposition.
Ibrahim, Barrie and Fanaki (1983)	Mathematical model of particle deposition to a smooth surface.

Table 29: (continued)

Author(s) (Year)	Comments
Lindberg, Turner and Lovett (1982)	Dry deposition of metals and acids in a Tennessee forest area.
Robinson, Lamb and Chockalingam (1982)	Dry deposition rates for ozone.
Sehmel (1980)	Discussion of published numerical values of particle and gas dry deposition rates.
Sheih, Wesely and Hicks (1979)	Deposition velocities of SO_2 and sulfate particles in eastern North America.
Sickles, Bach and Spiller (1983)	Comparison of techniques for determining dry deposition fluxes for nitrates and sulfates.
Sievering et al. (1979)	Dry deposition of metals and nutrients in southern Lake Michigan.

Several studies have been conducted on the rates of dry deposition. For example, Droppo and Doran (1979) presented data applicable to the computation of dry removal rates for both radioactive and nonradioactive materials, with the latter including iodine, sulfur dioxide, ozone, oxides of nitrogen, and nitric oxide particles. Sehmel (1980) has published numerical values of particle and gas dry deposition velocities. The deposition velocities for particles range over three orders of magnitude, and the deposition velocities for gases range over four orders of magnitude.

Sulfate particles present in air are often acidic (Hicks, 1979). These particles, which are typically small and hygroscopic, might be expected to attach themselves to foliage, thus imparting a strong but very localized dose of acid. Application of deposition velocities in the range presently advocated for sulfate particles suggests acid fluxes by dry deposition that average about two orders of magnitude less than those probably resulting from rainfall. However, Hicks (1979) noted that this should not be interpreted as an indication that dry deposition effects can be neglected, since it is clear that acid particles might reside on surfaces for considerable times, perhaps until washed off by rain or sufficiently diluted by dewfall.

Dry deposition has been measured and/or calculated for several locations in the United States. For example, Sheih, Wesely and Hicks (1979) computed surface deposition velocities of sulfur dioxide and sulfate particles over the eastern half of the United States, southern Ontario, and nearby oceanic regions from equations developed in recent field experiments, for use in studies of regional-scale atmospheric sulfur pollution. Surface roughness scale lengths and resistances to pollutant uptake by the surface were estimated from consideration of land-use characteristics and the likely biological status of the vegetation.   Midsummer conditions were assumed, but other seasons can be easily considered.   Average deposition velocities for grid cells corresponding to half-degree increments of longitude and latitude are presented for a range of atmospheric stabilities.

Dry deposition in the southern Lake Michigan area has been reported by several authors (Dolske and Sievering, 1979; Dolske and Sievering, 1980; and Sievering et al., 1979).   A midlake and nearshore trace element and nutrients data base with associated meteorology capable of establishing a climatology for mass transfer to Lake Michigan was collected during 1977 and 1978 (Sievering et al., 1979).   Significant data for Al, Ca, Cu, Fe, Mg, Mn, Pb, Ti, Zn, total P, NO_3 and SO_4 were obtained.   A strong linear dependence upon atmospheric thermal stability in the variability of all 12 aerosol constituents was found, but no linear dependence upon wind speed was found.   Bulk deposition velocities as a function of overlake climatology were used to calculate dry deposition atmospheric loadings to Lake Michigan.

Dolske and Sievering (1979) estimated annual dry deposition loadings to southern Lake Michigan for several elements.   For four of the elements, Fe, Mn, Pb, and Zn, dry deposition loadings to the southern basin alone of at least 500,000, 30,000, 250,000, and 100,000 kg/yr were found.   For Fe and Mn, these loadings represent about 15 percent of the total of all inputs to Lake Michigan. For Zn and Pb, about one-third to one-half of the annual loading from all sources is from dry deposition of atmospheric aerosols.   In addition, Dolske and Sievering (1980) indicated that dry deposition accounted for 15 percent or more of all atmospheric nutrient inputs of total phosphorus and nitrate-nitrite nitrogen and for 3 percent of total aerosol mass over southern Lake Michigan. Dry deposition loadings for this area of the lake was 150,000-180,000 kg per year for P and 3.5 million to 5.1 million kg per year for N.   The aerosol P and N was strongly associated with fine particles, less than 1.0 micrometers in diameter.

Lindberg, Turner and Lovett (1982) quantified the atmospheric deposition of water-leachable fractions of trace metals in the Walker Branch watershed in Tennessee.   They found that atmospheric wet and dry deposition contributed significantly to the annual flux of these elements to the forest floor; deposition supplied from 14 percent (Mn) to approximately 40 percent (Cd, Zn) to approximately 99 percent (Pb) to this flux.   Dry deposition constituted a major fraction of the total annual atmospheric input of Cd and Zn (approximately 20 percent), Pb (approximately 55 percent), and Mn (approximately 90 percent).

One of the problem issues associated with dry deposition is related to its measurement.   Sickles, Bach and Spiller (1983) have compared several techniques for determining the dry deposition flux of nitrates and sulfates. Direct flux estimates were made by using actual leaf surfaces and foliar wash, and by exposing and washing three surrogate surfaces: bucket, petri dish, and cellulose filter.   Indirect flux estimates were made using high-volume sampling along with meteorological measurements.   The time scale for the direct methods was nominally one month.   The indirect method used 24-hour particulate sampling and three-hour meteorological observations to calculate average flux values for the exposure periods of the direct methods.   This permitted comparison of direct and indirect methods on a similar time scale.

PRECIPITATION METALS AND ORGANICS

Metals and organics can be washed from the atmosphere as a result of wet deposition processes. This portion of Chapter 6 is based on information from the 68 references whose abstracts are found in Appendix H. Table 30 provides a brief comment on each of these references in the order they appear in Appendix H. Not every reference listed in Table 30 will be specifically addressed in this portion of Chapter 6. The information in Appendix H can be considered in terms of geographical studies of metals in wet deposition, metals in the terrestrial environment, metals in the aquatic environment, organics in wet deposition, and organics in the aquatic environment.

Table 30:   References in Appendix H Dealing with Precipitation Metals and Organics

Author(s) (Year)	Comments
Andersen (1978)	Dry and wet deposition of metals in Copenhagen, Denmark.
Andren, Lindberg and Bates (1975)	Wet deposition of metals in Oak Ridge, Tennessee.
Aten, Bourke and Walton (1983)	Metals in rainwater in Geneva, New York.
Beavington (1975)	Metals in herbage near an industrial site in Australia.
Berry and Wallace (1974)	General information on trace elements in the environment resulting from fossil fuel usage.
Budd et al. (1981)	Aluminum in precipitation, surface and ground waters in the New Jersey Pine Barrens.
Bufalini et al. (1979)	Rainwater concentrations of hydrogen peroxide in North Carolina.
Carpenter (1927)	Lethal action of soluble metallic salts on fishes.
Costeque and Hutchinson (1972)	Metals in soil and vegetation near a Canadian metal smelter.
Crecelius, Johnson and Hofer (1974)	Metals contamination of soil near a copper smelter in Washington.
Dorn et al. (1972)	Environmental cycling of metals near a lead smelter.

Table 30: (continued)

Author(s) (Year)	Comments
Dorn et al. (1975)	Metals in vegetation near a cooling tower operation.
Eisenreich, Looney and Thornton (1980)	Dry and wet deposition of organics in the Great Lakes area.
Eisenreich, Looney and Thornton (1981)	Annual deposition of airborne trace organics to the five Great Lakes.
Eriksson (1981)	Aluminum in ground water from acid precipitation in western Sweden.
Evans and Rigler (1980)	Atmospheric deposition of lead in a rural Canadian lake.
Franzin, McFarlane and Lutz (1979)	Atmospheric fallout of metals near a Canadian metal smelter.
Freedman and Hutchinson (1980)	Metals in forest soils and vegetation near a Canadian metal smelter.
Glotfelty (1978)	Atmospheric transport and removal of pesticides.
Hanson, Norton and Williams (1982)	Lead and zinc in New England lake sediments as a function of acid precipitation.
Harder et al. (1980)	Rainfall input of toxaphene and other pesticides to a South Carolina estuary.
Heit et al. (1981)	Polycyclic aromatic hydrocarbon levels in two Adirondack lakes.
Hemenway (1982)	Heavy metals in acid precipitation in Vermont.
Hemphill and Pierce (1974)	Vegetative accumulation of heavy metals near lead smelters in southeastern Missouri.
Henrikson and Wright (1978)	Concentrations of four heavy metals in the waters of 213 Norwegian lakes.
Hirao and Patterson (1974)	Lead deposition and environmental distribution in the High Sierra mountains.

Table 30: (continued)

Author(s) (Year)	Comments
Hoffman, Lindberg and Turner (1980)	Plasticizers and chlorohydrocarbons in rainfall.
Hopkins, Wilson and Smith (1972)	Lead levels in plants adjacent to a highway near Austin, Texas.
Hutchinson (1971)	Traffic-based metals in soils and vegetation in the Toronto area.
Hutchinson and Whitby (1973)	Copper and nickel in soils and vegetation near Canadian smelters.
Jackson et al. (1980)	Trace metal behavior in two Canadian soft-water lakes.
Kahl (1982)	Fate of atmospherically deposited metals in two acid-stressed lakes in Maine.
Klein and Russell (1973)	Soil and vegetation enrichment with heavy metals around a coal-fired power plant.
Komai et al. (1972)	Atmospherically deposited zinc in soils near Osaka, Japan.
LaRiviere et al. (1978)	Bibliography on the cycling of trace metals in fresh water ecosystems.
Lindberg et al. (1981)	Atmospheric deposition of heavy metals in a deciduous forest in the eastern United States.
Little and Martin (1972)	Zinc, lead, and cadmium in soil and vegetation around a British smelting complex.
Little and Martin (1974)	Monitoring of airborne metal pollution using Sphagnum moss.
Moeller and Stanley (1970)	Airborne contributions to lead in pasture grasses.
Montagnini, Naufeld and Uhl (1984)	Heavy metals in an Amazonian rainforest in southwestern Venezuela.
Muhlbaier and Tisue (1981)	Atmospheric inputs to cadmium mass balance for southern Lake Michigan.
Munshower (1972)	Cadmium cycling in a grassland ecosystem following its airborne deposition.

Table 30: (continued)

Author(s) (Year)	Comments
Murphy and Rzeszutko (1978)	Polychlorinated biphenyls (PCBs) in rainfall samples in the Lake Michigan basin.
Navarre, Ronneau and Priest (1980)	Contribution of heavy metals to Belgian agricultural soils from wet and dry deposition processes.
Nieboer et al. (1972)	Heavy metal content of lichens near a Canadian smelter area.
Nriagu, Wong and Coker (1982)	Metals accumulation in lake sediments near a Canadian smelter area.
Ohta et al. (1977)	Washout of chlorinated hydrocarbons from the ambient air.
Oliver et al. (1974)	Lead in snow and runoff samples from Ottawa, Canada.
Page, Ganje and Joshi (1971)	Lead in soil and vegetation near some major southern California highways.
Palmer and Kucera (1980)	Lead in soil and vegetation near smelters and mines in eastern Missouri.
Patterson et al. (1976)	Lead sources and cycling in sea water.
Ratcliffe (1975)	Biological indicators for monitoring atmospheric lead deposition.
Ratsch (1974)	Heavy metals in soil and vegetation in the vicinity of a copper smelter.
Reid and Crabbe (1980)	Drift and dry deposition of insecticides.
Ruhling and Tyler (1968)	Lead in soil and vegetation near three large roads in Sweden.
Ruhling and Tyler (1970)	Sorption and retention of heavy metals in a woodland moss and its usage as a biological indicator.
Shirahata, Elias and Patterson (1980)	Pond sediment lead from atmospheric deposition.
Siccama and Smith (1980)	Metals in forest floor material in central Massachusetts.

Table 30: (continued)

Author(s) (Year)	Comments
Soederlund (1975)	Atmospheric fallout of metals in the Baltic Sea.
Swank and Henderson (1976)	Atmospheric input of selected cations and anions in forests in North Carolina and Tennessee.
Taylor, Hanna and Parr (1979)	Chromium deposition and environmental cycling from cooling tower drift.
Ter Haar (1971)	Contribution of lead in plants from atmospheric deposition of lead.
Tucker and Preston (1984)	Procedures for estimating atmospheric deposition properties of organic chemicals.
Tyler and Westman (1979)	Influence of metal deposition on soil biological processes in northern Sweden.
Ward, Brooks and Reeves (1974)	Lead in soils and vegetation near a New Zealand highway.
Welch and Dick (1975)	Lead in soils and mice near a highway.
Zajic and Grodzinska (1982)	Heavy metals in snow near a steel mill in southern Poland.
Zimdahl (1975)	Environmental cycling of atmospherically deposited lead.

Metals in Wet Deposition

Several studies have been made of different geographical areas subject to the wet deposition of metals. Table 31 lists 14 references from Appendix H which address these studies from the United States, Canada, and Europe. Hemenway (1982) reported on the aluminum, cadmium, lead, and vanadium in seven separate rainfall events in Vermont. The samples were collected sequentially on a subevent basis. The subsamples for each event were filtered through a 0.1 um membrane filter in an attempt to separate particulate matter from soluble species. Filtrate concentrations were as high as 243 ug/l, 53 ug/l, 91 ug/l and 8.3 ug/l for Al, V, Pb, and Cd, respectively, for the initial subsample. The only metal investigated associated with the suspended particulate was Al with concentrations of up to 350 ug/l in the initial subsamples.

Table 31: References from Appendix H Addressing Metals in Wet Deposition

Author(s) (Year)	Comments
Andersen (1978)	Dry and wet deposition of metals in Copenhagen, Denmark.
Andren, Lindberg and Bates (1975)	Wet deposition of metals in Oak Ridge, Tennessee.
Aten, Bourke and Walton (1983)	Metals in rainwater in Geneva, New York.
Evans and Rigler (1980)	Atmospheric deposition of lead in a rural Canadian lake.
Franzin, McFarlane and Lutz (1979)	Atmospheric fallout of metals near a Canadian metal smelter.
Hemenway (1982)	Heavy metals in acid precipitation in Vermont.
Lindberg et al. (1981)	Atmospheric deposition of heavy metals in a deciduous forest in the eastern United States.
Muhlbaier and Tisue (1981)	Atmospheric inputs to cadmium mass balance for southern Lake Michigan.
Navarre, Ronneau and Priest (1980)	Contribution of heavy metals to Belgian agricultural soils from wet and dry deposition processes.
Oliver et al. (1974)	Lead in snow and runoff samples from Ottawa, Canada.
Siccama and Smith (1980)	Metals in forest floor material in central Massachusetts.
Soederlund (1975)	Atmospheric fallout of metals into the Baltic Sea.
Swank and Henderson (1976)	Atmospheric input of selected cations and anions in forests in North Carolina and Tennessee.
Zajic and Grodzinska (1982)	Heavy metals in snow near a steel mill in southern Poland.

Lead, zinc, copper, dry weight, and organic matter content of the forest floor were measured in 10 white pine stands in central Massachusetts in 1962, and again in 1978, thus providing quantitative estimates of these parameters at two points in time 16 years apart (Siccama and Smith, 1980). A net increase in lead of 30 mg/m^2/year was observed, with this being approximately 80 percent of the estimated total annual input of this element via precipitation in this region during the 16-year period. There were no statistically significant changes in total zinc and copper content of the forest floor. Since the forest floor total dry weight increased signficantly (42 percent), zinc and copper concentration decreased significantly. The results of this study emphasize the importance of determining both concentration and amount of trace elements in soil studies.

A preliminary mass balance for the southern basin of Lake Michigan, an area of 18,000 square kilometers, has been developed (Muhlbaier and Tisue, 1981). The overall input rate of Cd exceeds the estimated loss rate by a factor of 2.5. Sources of Cd input in tons per year are as follows: rain, 4.3 (40.1 percent); dry deposition, 2.2 (20.8 percent); erosion, 1.0 (9.4 percent); and tributaries, 3.1 (29.2 percent), for a total of 10.6 tons. Sedimentation is the major (93 percent) sink in the system, 3.8 tons per year, with outflow of water at a Cd concentration of 20 ng per liter eliminated 0.28 tons per year or 6.9 percent.

Franzin, McFarlane and Lutz (1979) described a study of atmospheric fallout in the vicinity of a base metal smelter at Flin Flon, Manitoba, Canada. The fallout was monitored with bulk precipitation collections over one year and by winter snow samples collected from the surfaces of frozen lakes. Simple correlation analysis of the data obtained from both types of collection indicated that the smelter was a major source of Zn, Cd, Pb, As, and Cu in bulk precipitation in the Flin Flon area.

Metal deposition with precipitation has also been reported in Belgium, the Baltic Sea, Denmark, and Poland. For example, Navarre, Ronneau and Priest (1980) have noted that wet plus dry deposition processes contribute more heavy metals to Belgian agricultural soils than do chemical fertilizers. This was based on sampling conducted for one year in 19 purely rural sites as far removed as possible from obvious contamination sources. Metals clearly attributable to human origin were zinc, arsenic, selenium, mercury, lead, cadmium, and antimony. Soederlund (1975) described the importance of atmospheric fallout of nitrogen, phosphorus, and lead to the material balance of these elements in the Baltic Sea.

Atmospheric dry and wet deposition (bulk precipitation) of the heavy metals Cu, Pb, Zn, Ni, V and Fe over the Copenhagen area in Denmark has been measured by sampling in plastic funnels from 17 stations during a 12-month period (Andersen, 1978). These metals were also measured in lichens and bryophytes in the study areas. There was a linear correlation between bulk precipitation and heavy metal concentration in lichens and bryophytes. Finally, samples of snow were collected on three dates in 1977 and 1978 at 27 sites in open and forest stands east of the large steel mill at Cracow, Poland (Zajac and Grodzinska, 1982). The concentrations of heavy metals, light metals, and sulfate decreased with increasing distance from the source, but snow acidity increased.

Metals in the Terrestrial Environment

A number of monitoring studies have confirmed the presence of metals in the terrestrial environment resulting from wet deposition. Table 32 lists 29 references from Appendix H that focus on the occurrence of wet deposition-related metals in the terrestrial environment. Berry and Wallace (1974) summarized chemical and biological information about the nature, occurrence and effects of trace elements in the environment, particularly as related to their ecological behavior. They found that, in general, the ecology of trace elements is very delicately balanced, and small additions of some trace elements from polluting sources can significantly alter an existing ecosystem. Some animals in particular are very sensitive to changes in concentration of trace elements, and the limits between adequacy and toxicity are very often narrow. Trace elements, at both low and high levels, are also of great importance to the health of man. The pathways and rates of movement of trace elements between each component of any soil-water-plant-animal system need to be better understood.

Table 32: References from Appendix H Addressing Metals in the Terrestrial Environment

Author(s) (Year)	Comments
Beavington (1975)	Metals in herbage near an industrial site in Australia.
Berry and Wallace (1974)	General information on trace elements in the environment resulting from fossil fuel usage.
Costeque and Hutchinson (1972)	Metals in soil and vegetation near a Canadian metal smelter.
Crecelius, Johnson and Hofer (1974)	Metals contamination of soil near a copper smelter in Washington.
Dorn et al. (1972)	Environmental cycling of metals near a lead smelter.
Dorn et al. (1975)	Metals in vegetation near a cooling tower operation.
Freedman and Hutchinson (1980)	Metals in forest soils and vegetation near a Canadian metal smelter.
Hemphill and Pierce (1974)	Vegetative accumulation of heavy metals near lead smelters in southeastern Missouri.
Hirao and Patterson (1974)	Lead deposition and environmental distribution in the High Sierra mountains.

Table 32: (continued)

Author(s) (Year)	Comments
Hopkins, Wilson and Smith (1972)	Lead levels in plants adjacent to a highway near Austin, Texas.
Hutchinson (1971)	Traffic-based metals in soils and vegetation in the Toronto area.
Hutchinson and Whitby (1973)	Copper and nickel in soils and vegetation near Canadian smelters.
Klein and Russell (1973)	Soil and vegetation enrichment with heavy metals around a coal-fired power plant.
Komai et al. (1972)	Atmospherically deposited zinc in soils near Osaka, Japan.
Little and Martin (1972)	Zinc, lead, and cadmium in soil and vegetation around a British smelting complex.
Montagnini, Neufeld and Uhl (1984)	Heavy metals in an Amazonian rainforest in southwestern Venezuela.
Moeller and Stanley (1970)	Airborne contributions to lead in pasture grasses.
Munshower (1972)	Cadmium cycling in a grassland ecosystem following its airborne deposition.
Nieboer et al. (1972)	Heavy metal content of lichens near a Canadian smelter area.
Page, Ganje and Joshi (1971)	Lead in soil and vegetation near some major southern California highways.
Palmer and Kucera (1980)	Lead in soil and vegetation near smelters and mines in eastern Missouri.
Ratsch (1974)	Heavy metals in soil and vegetation in the vicinity of a copper smelter.
Ruhling and Tyler (1968)	Lead in soil and vegetation near three large roads in Sweden.
Taylor, Hanna and Parr (1979)	Chromium deposition and environmental cycling from cooling tower drift.
Ter Haar (1971)	Contribution of lead in plants from atmospheric deposition of lead.

Table 32: (continued)

Author(s) (Year)	Comments
Tyler and Westman (1979)	Influence of metal deposition on soil biological processes in northern Sweden.
Ward, Brooks and Reeves (1974)	Lead in soils and vegetation near a New Zealand highway.
Welch and Dick (1975)	Lead in soils and mice near a highway.
Zimdahl (1975)	Environmental cycling of atmospherically deposited lead.

Considerable information has been developed on lead in the terrestrial environment. For example, lead uptake and translocation by plants and its subsequent effects are reviewed by Zimdahl (1975). Lead is taken up by many plants primarily via their roots. Large amounts of lead are deposited on plant foliage, but most remains as a topical deposit. Nevertheless, foliar uptake is also a demonstrated fact. Soil lead levels above 1000 ppm are assumed to be required to cause observable plant effects. Environmental variables, plant age, and species are important determinants of lead uptake. Increasing soil lead availability increases plant uptake; whereas increasing soil phosphorus, organic content, and pH decrease plant uptake. Mosses are excellent indicators of lead from aerial sources since they only absorb minerals from precipitation and settled dust. Moeller and Stanley (1970) have presented a model for determining the relationship of airborne lead and soil lead concentrations to lead concentrations in pasture grasses. On the basis of several published studies, it was concluded that not more than 15 micrograms Pb/gm dry weight in forage crops is due to the uptake of lead from soil even at soil lead contents up to about 700 micrograms/gm and possibly 3000 micrograms/gm. Amounts of lead in herbage substantially greater than 15 micrograms/gm are therefore due to aerial fallout.

Studies of metals in the terrestrial environment have been reported in several states in the United States, including Kentucky, Tennessee, Missouri, Texas, Montana, Washington, and California. For example, the transfer and fate of chromium from cooling tower drift to terrestrial ecosystems has been quantified at a uranium enrichment facility at Paducah, Kentucky (Taylor, Hanna and Parr, 1979). Chromium concentrations in fescue grass decreased with increasing distance from the cooling tower, ranging from 251 ± 19 ppm at 15 meters to 0.52 ± 0.07 ppm at 1500 meters.

An inventory of plant and soil materials in the environs of the Oak Ridge Gaseous Diffusion Plant provided quantitative evidence for the transfer of chromium and zinc to vegetation from a cooling tower operation (Dorn et al., 1975). Chromium concentrations in foliage ranged from several orders of magnitude above background levels to near background at 1609 m from the tower. Concentrations of extractable zinc ranged from 5 to 72 ppm at distances of 180 and 15 m, respectively, indicating no increase above background at 180 m. Potted tobacco plants placed downwind from the plume accumulated chromium, reaching equilibrium and attaining maximum concentrations (237 $\pm$ 18 ppm) after 4 to 6 weeks of exposure. While the tobacco plants (highly sensitive to chromium) developed symptoms of toxicity to drift, similar effects were not observed for native plant species at similar concentrations and downwind distances.

Metals in the environs of lead, zinc, and copper smelters have been monitored in Missouri, Montana and Washington. The accumulation of lead and other heavy metals in or on vegetation near lead smelters, mines, and mills in southeastern Missouri was investigated from 1971 to 1974 (Hemphill and Pierce, 1974). An area 12 by 25 mi in the New Lead Belt, containing four mines and mills and one smelter, as well as an area of 14 by 14 mi outside the New Lead Belt, with a smelter at the center, were sampled. Post oak (Quercus stellata) and shortleaf pine (Pinus echinata) foliage collected within 0.5 mi of a smelter accumulated maximum lead levels of 8,125 ppm and 11,750 ppm, respectively. Elevated lead levels in white oak and blueberry leaves were detected at distances greater than 7 mi from the smelters and at distances of 4 mi from the mines and mills. Similar patterns of elevated levels of cadmium, copper, and zinc were found. Palmer and Kucera (1980) determined lead concentrations of both plant tissue (Platanus occidentalis L.) and soils (total and available) for various sampling locations within a 6 mi radius of two lead smelters and two lead mines located in eastern Missouri. Lead contents of washed foliage ranged from 1.3 to 1,120 ppm, while values in twigs, which were generally lower, ranged from 1.8 to 320 ppm. Nitric acid-extractable lead from soils ranged from 7 to 62,000 ppm, while 3 percent acetic acid-extractable lead varied from 1 to 20,400 ppm. The largest lead levels for both soils and plants occurred at the smelter sites.

The distribution, compartmentation, and transfer of cadmium has been studied in a grassland ecosystem where airborne cadmium from a zinc smelter had accumulated for approximately 50 years until permanent closure in 1969 (Munshower, 1972). The soil reservoir of introduced Cd is restricted to the top 2 to 5 cm, is geographically distributed in relation to the smelter and prevailing winds, and normally is not transferred to lower soil horizons. Analyses of plants grown in controlled Cd concentrations in nutrient solutions and soil cultures, and of plants collected in the field, established that Cd was absorbed and translocated by plants in a predictable manner. Grasshopper Cd concentrations, and the liver and kidney Cd concentrations of cattle, swine, and columbian ground squirrels, demonstrated Cd accumulation above plant Cd levels as a function of the animal's age. The presence of Cd in the kidney of red fox, badger, and weasel documented Cd transfer from herbivores to carnivores. A mathematical model of the ecosystem was developed depicting the system as a closed, self-contained unit. The half-life for the introduced Cd in the grassland ecosystem appears to be in excess of 1,000 years, and shorter in cultivated lands.

Stack dust from a large copper smelter near Tacoma, Washington has contaminated soil in the area with arsenic, antimony, and lead (Crecelius, Johnson and Hofer, 1974). Within 3 mi of the smelter, 380 ppm (dry weight) arsenic, 200 ppm antimony, and 540 ppm lead were measured in the surface soil (0-3 cm). Standard orchard leaves in the area were also collected, and the following mean concentrations and standard deviations were determined for orchard leaves: 10.5 $\pm$ 0.5 ppm arsenic, and 3.5 $\pm$ 0.5 ppm antimony.

The usage of tetraethyl lead in gasoline prompted several studies of lead in soils and plants adjacent to highways. For example, plant samples were taken along a highway near Austin, Texas in August, 1970 (Hopkins, Wilson and Smith, 1972). It was found that the levels of lead decreased with increasing distance from the highway. Plants taken 32 m northwest of the road contained less than one-half as much lead as plants adjacent to the road. More lead was found in plants on the northwest side of the road than on the southeast due to prevailing winds.

The lead contents of 27 varieties of consumer crops and plants growing near some major southern California highways were reported by Page, Ganje and Joshi (1971). The amounts of lead in surface and subsurface soils and in suspended air particulates at or near the locations where the plant samples were obtained were also recorded. Exposed tissues of plants grown very close to highways contained more lead than similar tissues of plants grown some distance from the highways. This effect was most apparent at distances less than about 150 m from the highway. The direction of the prevailing wind also significantly affected lead concentrations in plants near highways; without exception, lead in plants on the leeward side of the road exceeded that in plants on the windward side. In soils and suspended particulates, lead concentrations were influenced by distance from highway and direction of prevailing winds. These results all demonstrate that the lead accumulations were caused primarily by aerial deposition and not--at least to any great extent--by absorption by the plant from lead-contaminated soil.

The Sudbury region of Ontario in Canada has large deposits of nickel, iron, and copper, and thus a number of smelting plants which produce sulfur dioxide and heavy metal pollution. Several studies have been made of metals in the terrestrial environment around these smelters. Costeque and Hutchinson (1972) have reported that soil contamination from metals has a pattern indicative of an airborne smelter source. A significant reduction of pH occurred in the soil within 1½ mi of the smelter. A pH of 2.2 was recorded in one instance, suggesting the presence of free sulfuric acid. Nickel levels were 2835 ppm at 0.5 mi from the smelter, 1522 ppm at 4 to 5 mi, 306 ppm at 12 mi, and 83 ppm at 31 mi. Copper followed the same pattern from 1528 ppm to 31 ppm; cobalt decreased from 127 ppm to 19 ppm. Nieboer et al. (1972) have indicated that the pattern of heavy metal content in lichen species in the area of a copper smelter in Sudbury could be correlated with distance from the smelter. All seven species of lichens studied contained copper, iron, zinc, nickel, manganese, and lead. A pollution model was developed and compared to field results. The simple dilution lichen metal content was related to the reciprocal of the distance from the pollution source.

There have been ecological consequences of metal deposition in the Sudbury basin region of Canada. For example, Hutchinson and Whitby (1973) reported that the growth of the radicles of tomato and lettuce seedlings in water extracts of soils collected at various distances from a smelter was severely inhibited at distances up to 8 km from the smelter. Freedman and Hutchinson (1980) reported lower rates of leaf decomposition, lower populations in soil micofungi and micoarthropods, and lower rates of soil metabolic activity at forest sites near a smelter.

The distribution of airborne zinc, lead, and cadmium from a lead and zinc smelting plant in the Avonmouth area of Severnside, Great Britain, has been determined (Little and Martin, 1972). Levels of zinc, lead, and cadmium in elm leaves collected in October, 1971, ranged from 8000, 5000, and 50 ppm dry matter close to a smelting complex, to values of about 200, 100, and less than 0.25 ppm, respectively, at distances of 6 to 9 mi from the factory.

Tyler and Westman (1979) determined to what extent soil biological processes are influenced in the coastal region of Vaesterbotten, northern Sweden, by deposition of metals and acid from a large smelter. Elevated levels of several heavy metals in the topsoil of forested areas are measurable at least 40 to 50 mi from the main source. The pollution has a measurable influence on important biological reactions in conifer forest soil over an area of at least 900 to 1400 mi^2, particularly evident northwards and southwards from the smelter. Of the main polluting metals, cadmium and copper have the highest simple correlation coefficients with biological variables such as soil respiration rate and phosphatase activity. However, the authors noted that soil pH contributes roughly as much as all the metals together to account for the biological variability of the soil reactions studied, according to stepwise regression analysis. The variation in pH will partly be explained by natural differences between sites and partly by soil acidification due to deposition of acid pollutants.

Samples of white clover and paspalum have been taken from an industrial area (copper smelter and steel works) and a rural area of Australia and analyzed for copper, zinc, and iron contents (Beavington, 1975). Mean levels of copper, zinc, and iron for all herbage collected in the industrial area were 38.0, 70.4, and 723 ppm as compared with levels of 9.8, 28.6, and 129 ppm, respectively, for the rural samples. Zinc and copper levels correlated significantly with distance from the smelter, and iron levels with distance from the steelworks.

Metals in the Aquatic Environment

Metals associated with wet deposition have also been monitored in the aquatic environment. Table 33 lists 11 references from Appendix H that focus on the occurrence of wet deposition-related metals in surface and ground waters. A bibliography addressing the cycling of trace metals in fresh water ecosystems is available (LaRiviere et al., 1978).

Table 33: References from Appendix H Addressing Metals in the Water Environment

Author(s) (Year)	Comments
Budd et al. (1981)	Aluminum in precipitation, surface and ground waters in the New Jersey Pine Barrens.
Carpenter (1927)	Lethal action of soluble metallic salts on fishes.
Eriksson (1981)	Aluminum in ground water from acid precipitation in western Sweden.

Table 33: (continued)

Author(s) (Year)	Comments

Hanson, Norton and
Williams (1982)

Lead and zinc in New England lake
sediments as a function of acid precipitation.

Henrikson and Wright
(1978)

Concentrations of four heavy metals in
the waters of 213 Norwegian lakes.

Jackson et al. (1980)

Trace metal behavior in two Canadian
soft-water lakes.

Kahl (1982)

Fate of atmospherically deposited metals
in two acid-stressed lakes in Maine.

LaRiviere et al. (1978)

Bibliography on the cycling of trace
metals in fresh water ecosystems.

Nriagu, Wong and
Coker (1982)

Metals accumulation in lake sediments
near a Canadian smelter area.

Patterson et al. (1976)

Lead sources and cycling in sea water.

Shirahata, Elias and
Patterson (1980)

Pond sediment lead from atmospheric deposition.

Aluminum in surface and ground water has been recognized as resulting from atmospheric depositions as well as terrestrial and aquatic sediment sources. For example, aluminum appears in waters of the New Jersey Pine Barrens in the United States in the following volume weighted averages: 105 ug/l in precipitation; 345 ug/l in streams; and 468 ug/l in ground water (Budd et al., 1981). These levels are 10 times greater than in most terrestrial waters, except for the acidified waters of New England and New York. The pH of precipitation in this study averaged 4.0; of ground water, 4.6; and of streams, 4.1. The total (acid reactive) aluminum in wet or dry deposition was 140 mg per sq meter per year over the study period, May 1978 to May 1980. Stream and ground water outputs were 149 and 110 mg per sq meter per year, respectively.

As part of a lake regional survey in Norway, the concentrations of Zn, Pb, Cu, and Cd were measured in surface- and bottom-water samples collected from representative, small, pristine lakes (136 in southern Norway samples in October 1974, 58 resampled in March 1975, and 77 in northern Norway sampled in March 1975) (Henriksen and Wright, 1978). The lakes, a statistically representative sample of small lakes in Norway, were chosen such that their watersheds are undisturbed. Heavy-metal concentrations in these lakes thus reflect only natural inputs and anthropogenic inputs via the atmosphere. The generally low concentrations (Zn 0.5-12.0 ug/l; Pb 0-2.0 ug/l; Cu 0-210 ug/l; and Cd 0.1-0.5 ug/l) measured in lakes in central and northern Norway provide estimates of natural "background" levels. These estimates may be too high because they include the global-scale deposition of heavy metals from the atmosphere which has increased as a result of industrial activities. Concentrations of Zn and Pb in lakes in southernmost and southeastern Norway lie above these "background" levels, apparently because of atmospheric deposition associated with the acidic precipitation that falls over southern Scandinavia. Increased heavy-metal concentrations in acid lakes may also be due to increased mobilization of metals due to acidification of soil- and surface-waters.

Organics in Wet Deposition

The atmospheric content of various synthetic organic chemicals can be transported to the earth's surface via wet or dry deposition. For example, Flotfelty (1978) noted that pesticides may be removed from the atmosphere by surface deposition, washout, or chemical reaction. Hoffman, Lindberg and Turner (1980) determined some organic components of rain in samples collected on an event basis above and below the canopy of a deciduous forest. Plasticizers and chlorohydrocarbons that were identified were attributed to sources external to the vegetation. Organic acids and high molecular weight "waxy" components identified were attributed to the foliage itself. Approximately equal amounts of material were ascribed to each of these sources.

Factors reducing the concentrations of several chlorinated hydrocarbons on both rainy and clear days have been determined by Ohta et al. (1977). Analyses of 1,1,1-trichloroethane ($C_2H_3Cl_3$), carbon tetrachloride (CCl_4), trichloroethylene (C_2HCl_3), and tetrachloroethylene (C_2Cl_4) in the ambient air were made every half hour using an automatic gas sample inlet system for about seven weeks. Factors influencing the concentrations of these chemicals were: amount emitted from the original source; variations in wind directions and velocities; rain scavenging or washout; and decomposition by direct or indirect photochemical reactions.

A rationale for procedures to estimate the climatological mean dry deposition velocity and precipitation scavenging ratio of organic chemicals has been developed (Tucker and Preston, 1984). Separate formulae are presented for vapor and sorbed fractions using their mass-weighted average to describe the deposition parameters for the total airborne concentration. The deposition of sorbed fractions is controlled solely by particle size. The dry deposition velocity for the vapor fraction is derived by semiempirical analysis, and is found to depend on Henry's Law constant, molecular diffusion coefficients in air and water, and the chemical's reaction rate in water.

Organics in the Aquatic Environment

One source of synthetic organics in an aquatic environment is atmospheric deposition. Murphy and Rzeszutko (1978) analyzed rainfall samples from the southern Lake Michigan area for polychlorinated biphenyls (PCBs). The precipitation weighted mean concentration of 35 samples of rain was 111 parts per trillion. This would result in the deposition of 4800 kg/yr of PCBs to the lake from precipitation. Presently available evidence on other sources of PCBs to the lake indicates that precipitation is now the major source. Eisenreich, Looney and Thornton (1981) have noted that the total sources of PCBs in Lake

Superior can be estimated at 8,000-9,700 kg per year, with 6,600-8,300 kg of this coming from the atmosphere.

The degree to which anthropogenic trace elements and polycyclic aromatic hydrocarbons have been deposited in remote lakes known to be affected by acid rainfall in the Adirondack State Park in New York has been studied by Heit et al. (1981). Sagamore Lake is a brown water lake due to high levels of humic materials, while Woods Lake is generally clear. Sediment cores were taken from each of the lakes during March, 1978. Several conclusions were reached. Except for perylene, the prime source of all parental PAHs measured and the majority of trace elements appears to be combustion. All combustion products had primarily an anthropogenic origin. Levels of all parental PAHs except perylene and levels of several of the metals significantly increased in the surface sediments of both lakes compared to their background concentrations. PAH concentrations in Woods Lake were quite high, approaching levels reported for more heavily populated and industrialized areas. Metals were present at 3-4 times higher concentrations in Woods Lake than in Sagamore Lake. Metals and PAH levels decreased in concentration with depth to background levels in both lakes. This baseline depth corresponded to about 30 years ago. Lead has increased to the greatest extent of any of the metals considered relative to their baseline quality. Long distance atmospheric transport and region-wide deposition of anthropogenically derived elements and PAHs into these remote lakes appears to be more significant than input from local sources.

SELECTED REFERENCES

Andersen, A., "Atmospheric Heavy Metal Deposition in the Copenhagen Area", Environmental Pollution, Vol. 17, No. 2, 1978, pp. 133-151.

Andren, A.W., Lindberg, S.E. and Bates, L.C., "Atmospheric Input and Geochemical Cycling of Selected Trace Elements in Walker Branch Watershed", Report No. ESD-728, June 1975, Oak Ridge National Laboratory, Oak Ridge, Tennessee.

Aten, C.F., Bourke, J.B. and Walton, J.C., "Heavy Metal Content of Rainwater in Geneva, New York During Late 1982", Bulletin of Environmental Contamination and Toxicology, Vol. 31, No. 5, Nov. 1983, pp. 574-581.

Beavington, F., "Some Aspects of Contamination of Herbage with Copper, Zinc, and Iron", Environmental Pollution, Vol. 8, No. 1, Jan. 1975, pp. 65-71.

Berry, W.L. and Wallace, A., "Trace Elements in the Environment: Their Role and Potential Toxicity as Related to Fossil Fuels. A Preliminary Study", Report No. UCLA-12-946, Jan. 1974, National Technical Information Service, U.S. Department of Commerce, Springfield, Virginia.

Budd, W.W. et al., "Aluminum in Precipitation, Streams, and Shallow Groundwater in the New Jersey Pine Barrens", Water Resources Research, Vol. 17, No. 4, Aug. 1981, pp. 1179-1183.

Bufalini, J.J. et al., "Hydrogen Peroxide Formation from the Photooxidation of Formaldehyde and Its Presence in Rainwater", Journal of Environmental Science and Health, Vol. 14, No. 2, 1979, pp. 135-141.

Carpenter, K.E., "The Lethal Action of Soluble Metallic Salts on Fishes", British Journal of Experimental Biology, Vol. 4, No. 4, 1927, pp. 378-390.

Costeque, L.M. and Hutchinson, T.C., "The Ecological Consequences of Soil Pollution by Metallic Dust from the Sudbury Smelters", Proceedings of Institute of Environmental Sciences Annual Technical Meeting, 1972, Institute of Environmental Sciences, Mt. Prospect, Illinois, pp. 540-545.

Crecelius, E.A., Johnson, C.J. and Hofer, G.C., "Contamination of Soils Near a Copper Smelter by Arsenic, Antimony, and Lead", Water, Air, and Soil Pollution, Vol. 3, No. 3, Sept. 1974, pp. 337-342.

Dolske, D.A. and Sievering, H., "Trace Element Loading of Southern Lake Michigan by Dry Deposition of Atmospheric Aerosol", Water, Air, and Soil Pollution, Vol. 12, No. 4, Nov. 1979, pp. 485-502.

Dolske, D.A. and Sievering, H., "Nutrient Loading of Southern Lake Michigan by Dry Deposition of Atmospheric Aerosol", Journal of Great Lakes Research, Vol. 6, No. 3, 1980, pp. 184-194.

Dorn, R.C. et al., "Study of Lead, Copper, Zinc, and Cadmium Contamination of Food Chains of Man", EPA-R3-73-034, Dec. 1972, University of Missouri, Columbia, Missouri.

Dorn, R.C. et al., "Environmental Contamination by Lead, Cadmium, Zinc, and Copper in a New Lead-Producing Area", Environmental Research, Vol. 9, No. 2, Apr. 1975, pp. 159-172.

Droppo, J.G. and Doran, J.C., "Dry Deposition of Air Pollutants", 1979, Battelle Pacific Northwest Laboratories, Richland, Washington.

Eisenreich, J.J., Looney, B.B. and Thornton, J.D., "Assessment of Airborne Organic Contaminants in the Great Lakes Ecosystem", Nov. 1980, Department of Civil and Mineral Engineering, University of Minnesota, Minneapolis, Minnesota.

Eisenreich, J.J., Looney, B.B. and Thornton, J.D., "Airborne Organic Contaminants in the Great Lakes Ecosystem", Environmental Science and Technology, Vol. 15, No. 1, Jan. 1981, pp. 30-38.

Eriksson, E., "Aluminum in Groundwater: Possible Solution Equilibria", Nordic Hydrology, Vol. 12, No. 1, 1981, pp. 43-50.

Evans, R.D. and Rigler, F.H., "Calculation of the Total Anthropogenic Lead in the Sediments of a Rural Ontario Lake", Environmental Science and Technology, Vol. 14, No. 2, Feb. 1980, pp. 216-218.

Franzin, W.G., McFarlane, G.A. and Lutz, A., "Atmospheric Fallout in the Vicinity of a Base Metal Smelter at Flin Flon, Manitoba, Canada", Environmental Science and Technology, Vol. 13, No. 12, Dec. 1979, pp. 1513-1525.

Freedman, B. and Hutchinson, T.C., "Smelter Pollution Near Sudbury, Ontario, Canada, and Effects on Forest Litter Decomposition", Effects of Acid Rain

Precipitation on Terrestrial Ecosystems, 1980, Plenum Press, New York, New York, pp. 395-424.

Garland, J.A., "Dry Deposition of Gaseous Pollutants", 1979, AERE, Harwell, England.

Glotfelty, D.E., "The Atmosphere as a Sink for Applied Pesticides", Journal of the Air Pollution Control Association, Vol. 28, No. 9, Sept. 1978, pp. 917-922.

Hanson, D.W., Norton, S.A. and Williams, J.S., "Modern and Paleolimnological Evidence for Accelerated Leaching and Metal Accumulation in Soils in New England, Caused by Atmospheric Deposition", Water, Air, and Soil Pollution, Vol. 18, No. 1-3, 1982, pp. 227-230.

Harder, H.W. et al., "Rainfall Input of Toxaphene to a South Carolina Estuary", Estuaries, Vol. 3, No. 2, June 1980, pp. 142-147.

Heit, M. et al., "Anthropogenic Trace Elements and Polycyclic Aromatic Hydrocarbon Levels in Sediment Cores from Two Lakes in the Adirondack Acid Lake Region", Water, Air, and Soil Pollution, Vol. 15, No. 4, May 1981, pp. 441-464.

Hemenway, D.R., "Heavy Metals in Acid Precipitation", Dec. 1982, Vermont Water Resources Research Center, University of Vermont, Burlington, Vermont.

Hemphill, D.D. and Pierce, J.O., "Accumulation of Lead and Other Heavy Metals by Vegetation in the Vicinity of Lead Smelters and Mines and Mills in Southeastern Missouri", 2nd Annual Trace Contaminants Conference Proceedings, 1974, Pacific Grove, California, pp. 325-332.

Henriksen, A. and Wright, R.F., "Concentrations of Heavy Metals in Small Norwegian Lakes", Water Research, Vol. 12, 1978, pp. 101-112.

Hicks, B.B., "Dry Deposition of Acid Particles to Natural Surfaces", 1979, Argonne National Laboratory, Argonne, Illinois.

Hicks, B.B., Wesely, M.L. and Durham, J.L., "Critique of Methods to Measure Dry Deposition--Concise Summary of Workshop", EPA-600/D-82-155, Jan. 1982, U.S. Environmental Protection Agency, Research Triangle Park, North Carolina.

Hirao, Y. and Patterson, C.C., "Lead Aerosol Pollution in the High Sierra Overrides Natural Mechanisms which Exclude Lead from a Food Chain", Science, Vol. 184, No. 4140, May 31, 1974, pp. 989-992.

Hoffman, W.A., Lindberg, S.E. and Turner, R.R., "Some Observations of Organic Constituents in Rain Above and Below a Forest Canopy", Environmental Science and Technology, Vol. 14, No. 8, Aug. 1980, pp. 999-1000.

Hopkins, J.M., Wilson, R.H. and Smith, B.N., "Lead Levels in Plants", Naturwissenschaften, Vol. 59, No. 9, 1972, pp. 421-422.

Hutchinson, T.C., "The Occurrence of Lead, Cadmium, Nickel, Vanadium, and Chloride in Soils and Vegetation of Toronto in Relation to Traffic Density",

Canadian Botanical Association and the American Institute of Biological Sciences, Joint Meeting, June 20-24, 1971, Alberta, Canada.

Hutchinson, T.C. and Whitby, L.M., "A Study of Airborne Contamination of Vegetation and Soils by Heavy Metals from the Sudbury Copper-Nickel Smelters, Canada", Report EL-3, Nov. 1973, Department of Botany and the Institute of Environmental Sciences and Engineering, Toronto University, Toronto, Canada.

Ibrahim, M., Barrie, L.A. and Fanaki, F., "Experimental and Theoretical Investigation of the Dry Deposition of Particles to Snow, Pine Trees and Artificial Collectors", Atmospheric Environment, Vol. 17, No. 4, 1983, pp. 781-788.

Jackson, T.A. et al., "Experimental Study of Trace Metal Chemistry in Soft-Water Lakes at Different pH Levels", Canadian Journal of Fisheries and Aquatic Sciences, Vol. 37, No. 3, Mar. 1980, pp. 387-402.

Kahl, J.S., "Metal Input and Mobilization in Two Acid-Stressed Lake Watersheds in Maine", Dec. 1982, Land and Water Resources Center, University of Maine, Orono, Maine.

Klein, D.H. and Russell, P., "Heavy Metals: Fallout Around a Power Plant", Environmental Science and Technology, Vol. 7, No. 4, Apr. 1973, pp. 357-358.

Komai, Y. et al., "Accumulation of Heavy Metals and Their Behavior in Soil and Plants Due to Air Pollution. Part I. Accumulation of Zinc in Soil in the South Part of Osaka Prefecture", Japan Social Science Meeting, 1972, Osaka Prefecture, Osaka, Japan, p. 132.

LaRiviere, M.G. et al., "Bibliography on Cycling of Trace Metals in Freshwater Ecosystems", Report No. PNL-2706, July 1978, National Technical Information Service, U.S. Department of Commerce, Springfield, Virginia.

Lindberg, S.E., Turner, R.R. and Lovett, G.M., "Processes of Atmospheric Deposition of Metals and Acids to Forests", 1982, Oak Ridge National Laboratory, Oak Ridge, Tennessee.

Lindberg, S.E. et al., "Atmospheric Deposition of Heavy Metals and Their Interaction with Acid Precipitation in a North American Deciduous Forest", 1981, Oak Ridge National Laboratory, Oak Ridge, Tennessee.

Little, P. and Martin, M.H., "Biological Monitoring of Heavy Metal Pollution", Environmental Pollution, Vol. 6, No. 1, Jan. 1974, pp. 1-19.

Little, P. and Martin, M.H., "A Survey of Zinc, Lead, and Cadmium in Soil and Natural Vegetation Around a Smelting Complex", Environmental Pollution, No. 3, July 1972, pp. 241-254.

Moeller, P.K. and Stanley, R.L., "Origin of Lead in Surface Vegetation", Report 87, July 1970, California State Department of Public Health, Berkeley, California.

Montagnini, F., Neufeld, H.S. and Uhl, C., "Heavy Metal Concentrations in Some Non-Vascular Plants in an Amazonian Rainforest", Water, Air, and Soil Pollution, Vol. 21, No. 1-4, Jan. 1984, pp. 317-321.

Muhlbaier, J. and Tisue, G.T., "Cadmium in the Southern Basin of Lake Michigan", Water, Air, and Soil Pollution, Vol. 15, No. 1, 1981, pp. 45-59.

Munshower, F.F., "Cadmium Compartmentation and Cycling in a Grassland Ecosystem in the Deer Lodge Valley, Montana", Thesis (Ph.D.), June 1972, Montana University, Missoula, Montana.

Murphy, T.J. and Rzeszutko, C.P., "Polychlorinated Biphenyls in Precipitation in the Lake Michigan Basin", July 1978, De Paul University, Chicago, Illinois

Navarre, J.L., Ronneau, C. and Priest, P., "Deposition of Heavy Elements on Belgian Agricultural Soils", Water, Air, and Soil Pollution, Vol. 14, 1980, pp. 207-213.

Nieboer, E. et al., "Heavy Metal Content of Lichens in Relation to Distance from a Nickel Smelter in Sudbury Ontario", Lichenologist, Vol. 5, 1972, pp. 292-304.

Nriagu, J.O., Wong, H.K. and Coker, R.D., "Deposition and Chemistry of Pollutant Metals in Lakes Around the Smelters at Sudbury, Ontario", Environmental Science and Technology, Vol. 16, No. 9, 1982, pp. 551-560.

Ohta, T. et al., "Washout Effect and Diurnal Variation for Chlorinated Hydrocarbons in Ambient Air", Atmospheric Environment, Vol. 11, No. 10, 1977, pp. 985-988.

Oliver, B.G. et al., "Chloride and Lead in Urban Snow", Journal of Water Pollution Control Federation, Vol. 46, No. 4, Apr. 1974, pp. 766-771.

Page, A.L., Ganje, T.J. and Joshi, M.S., "Lead Quantities in Plants, Soil, and Air Near Some Major Highways in Southern California", Hilgardia, Vol. 41, No. 1, July 1971, pp. 1-31.

Palmer, K.T. and Kucera, C.L., "Lead Contamination of Sycamore and Soil from Lead Mining and Smelting Operations in Eastern Missouri", Journal of Environmental Quality, Vol. 9, No. 1, 1980, pp. 106-111.

Patterson, C.C. et al., "Transport of Pollutant Lead to the Oceans and Within Ocean Ecosystems", NSF/INTL Decade of Ocean Exploration Pollution Survey Report, Jan. 1976, Savannah, Georgia, pp. 23-29.

Ratcliffe, J.M., "An Evaluation of the Use of Biological Indicators in an Atmospheric Lead Survey", Atmospheric Environment, Vol. 9, No. 6-7, June/ July 1975, pp. 623-629.

Ratsch, H.C., "Heavy-Metal Accumulation in Soil and Vegetation from Smelter Emissions", EPA-660/3-74-012, Aug. 1974, U.S. Environmental Protection Agency, Washington, D.C.

Reid, J.D. and Crabbe, R.S., "Two Models of Long-Range Drift of Forest Pesticide Aerial Spray", Atmospheric Environment, Vol. 14, No. 9, 1980, pp. 1017-1025.

Robinson, E., Lamb, B. and Chockalingam, M.P., "Determination of Dry Deposition Rates for Ozone", EPA-600/3-82-042, Apr. 1982, U.S. Environmental Protection Agency, Research Triangle Park, North Carolina.

Ruhling, A. and Tyler, G., "An Ecological Approach to the Lead Problem", Botan Notiser, Vol. 121, 1968, pp. 321-342.

Ruhling, A. and Tyler, G., "Sorption and Retention of Heavy Metals in the Woodland Moss Hylocomium Splendens", Oikos, Vol. 21, No. 1, 1970, pp. 92-97.

Sehmel, G.A., "Particle and Gas Dry Deposition: A Review", Atmospheric Environment, Vol. 14, No. 9, 1980, pp. 983-1011.

Sheih, C.M., Wesely, M.L. and Hicks, B.B., "Estimated Dry Deposition Velocities of Sulfur Over the Eastern United States and Surrounding Regions", Atmospheric Environment, Vol. 13, No. 10, Oct. 1979, pp. 1361-1368.

Shirahata, H., Elias, R.W. and Patterson, C.C., "Chronological Variations in Concentrations and Isotopic Compositions of Anthropogenic Atmospheric Lead in Sediments of a Remote Subalpine Pond", Geochimica et Cosmochimica Acta, Vol. 44, No. 2, 1980, pp. 149-162.

Siccama, T.G. and Smith, W.H., "Changes in Lead, Zinc, Copper, Dry Weight and Organic Matter Content of the Forest Floor of White Pine Stands in Central Massachusetts Over 16 Years", Environmental Science and Technology, Vol. 14, No. 1, Jan. 1980, pp. 54-56.

Sickles, J.E., Bach, W.D. and Spiller, L.L., "Comparison of Several Techniques for Determining Dry Deposition Flux", EPA-600/D-83-057, June 1983, U.S. Environmental Protection Agency, Research Triangle Park, North Carolina.

Sievering, H. et al., "An Experimental Study of Lake Loading by Aerosol Transport and Dry Deposition in the Southern Lake Michigan Basin", EPA-905/4-79-016, July 1979, U.S. Environmental Protection Agency, Chicago, Illinois.

Soederlund, R., "Some Preliminary Views on the Atmospheric Transport of Matter to the Baltic Sea", Report No. AC-31, Mar. 1975, Stockholm University, Stockholm, Sweden.

Swank, W.T. and Henderson, G.S., "Atmospheric Input of Some Cations and Anions to Forest Ecosystems in North Carolina and Tennessee", Water Resources Research, Vol. 12, No. 3, June 1976, pp. 541-546.

Taylor, F.G., Hanna, S.R. and Parr, P.D., "Cooling Tower Drift Studies at the Paducah, Kentucky Gaseous Diffusion Plant", Cooling Tower Institute Annual Meeting, Houston, Texas, 1979, Oak Ridge National Laboratory, Oak Ridge, Tennessee.

Ter Haar, G.L., "The Effect of Lead Antiknocks on the Lead Content of Crops", Journal of Washington Academy of Science, Vol. 61, No. 2, 1971, pp. 114-120.

Tucker, W.A. and Preston, A.L., "Procedures for Estimating Atmospheric Deposition Properties of Organic Chemicals", Water, Air, and Soil Pollution, Vol. 21, No. 1-4, Jan. 1984, pp. 247-260.

Tyler, F. and Westman, L., "Effects of Heavy Metal Pollution on Decomposition in Forest Soils--VI--Metals and Sulfuric Acid", May 1979, National Swedish Environment Protection Board, Stockholm, Sweden.

Ward, N.I., Brooks, R.R. and Reeves, R.D., "Effect of Lead from Motor-Vehicle Exhausts on Trees Along a Major Thoroughfare in Palmerston North, New Zealand", Environmental Pollution, Vol. 6, No. 2, Feb. 1974, pp. 149-158.

Welch, W.R. and Dick, D.L., "Lead Concentrations in Tissues of Roadside Mice", Environmental Pollution, Vol. 8, No. 1, Jan. 1975, pp. 15-21.

Zajac, P.K. and Grodzinska, K., "Snow Contamination by Heavy Metals and Sulphur in Cracow Agglomeration (Southern Poland)", Water, Air, and Soil Pollution, Vol. 17, No. 3, 1982, pp. 269-280.

Zimdahl, R.L., "Entry and Movement in Vegetation of Lead Derived from Air and Soil Sources", Paper 75-18.3, Air Pollution Control Association Annual Meeting, June 15-20, 1975, Boston, Massachusetts.

APPENDIX A

GENERAL INFORMATION ON ACID RAIN

Ahern, J. and Baird, C., "Acid Precipitation in Southeastern Wyoming", Sept. 1983, Water Resources Research Institute, University of Wyoming, Laramie, Wyoming.

Snowfall, snowpack, and rainfall samples were collected in Laramie, Wyoming and in the Snowy Range west of Laramie from March to June 1981 to determine the occurrence and sources of acid precipitation in southeast Wyoming. Electrodes measured different pH values in the samples; however, fast-response electrodes yielded higher and apparently more accurate pH measurements. The pH values in the Laramie precipitation and snowpack were typically greater than 5.0, but all the Snowy Range snowpack pH values were less than 5.0. The lower pH values in the Snowy Range snowpack were caused by higher concentrations of the acid-forming nitrate and lower concentrations of the neutralizing calcium. Two organic species, formate and acetate, were detected in the Laramie samples, but had no significant influence on the acidity of the samples.

Alexander, T., "New Fears Surround the Shift to Coal", Fortune, Vol. 98, No. 10, Nov. 1978, pp. 50-56.

The environmental and health costs of converting to a coal-based economy are dangerously high. Coal production and consumption cause many more deaths from occupational exposure, air pollution, and accidents than oil, gas, or nuclear energy cause. Mining causes lasting ecological and aesthetic damage. Acid mine drainage, which has sterilized 10,000 miles of Appalachian streams, has no known remedy. Sulfates from coal burning are long-range agents, not merely local pollutants. Sulfates contribute to acidic rain, which has critically polluted lakes and rivers in Sweden, New York, and elsewhere. Sludges from coal ash scrubbers are harmful wastes, and must be disposed of safely without their leaching into ground waters. The hazards of nuclear energy appear negligible compared with coal's proved hazards.

Ambler, M., "Power Emissions May Reduce AG Productivity", High Country News, Vol. 10, No. 15, July 28, 1978, pp. 1-4.

Studies of the vegetation and animals in the vicinity of the Colstrip coal-fired power plant in Montana indicate that subtle changes may be occurring as the vegetation absorbs sulfur dioxide emitted from the power plant. Acid rains have increased within a 40-60 mile radius of power plants in the Eastern U.S. and are likely to become a problem in the west as most coal-fired power plants are being built. Some of the subtle clues researchers have to long-term changes include: a decrease in the number of beetles present; a 2 percent reduction in the protein content of native grasses; a less diverse population of plants; and a shorter growing season for grasses. Livestock that eat the vegetation may also be affected: frequent calf deaths could be another indication of change. Researchers feel that the present standards for SO_2 levels are inadequate. When Colstrip studies are complete in 1980, they should help to evaluate the standards and to predict effects of coal development on agriculture more readily than is presently possible.

Anderson, D.E. and Landsberg, H.E., "Detailed Structure of pH in Hydrometeors", Environmental Science and Technology, Vol. 13, No. 8, Sept. 1978, pp. 992-994.

The pH values of various forms of atmospheric precipitation, dew, frost, and fog were studied. Automatic precipitation collector-pH analyzers were developed for sampling of liquid and solid forms of precipitation. Data were collected under a wide variety of weather conditions at College Park, Maryland, near a major power plant, and from an area without nearby sources of pollution. Results from these 3 sites indicate that the precipitation pH, with a mean of 4.7, is not normally distributed when measured per millimeter of precipitation. In many events a strong variation of pH was evident. In the course of the study, pH values ranged from 2.7 to 6.9 units. Sequential sampling upwind and downwind of a major power plant indicated that concentrations of nitrate, sulfur, trace metals (Na, Pb, V) and H^+ were higher downwind. Observations of dew and frost at 3 different sites indicate that the pH is normally distributed about 5.7 units. Values of fog water pH collected at Catoctin Mountain (near Camp David in Northern Maryland) averaged 5.3 units.

Assam, W.R.H. and Corrada, L.A., "Bibliography on Precipitation Chemistry", 1976, Utrecht University, Netherlands.

This bibliography of 1470 references on atmospheric precipitation chemistry covers the period 1950 to 1975.

Babich, H. and Davis, D.L., "Acid Precipitation--Part 1--Causes and Consequences", Environment, Vol. 22, No. 4, May 1980, pp. 6-13, 40-41.

The causes of acid precipitation are detailed and its consequences for the abiotic and biotic components of the terrestrial and aquatic ecosystems and for natural and man-made materials are described. Acid precipitation is defined as rain or snow with pH values of less than 5.6. The incidence and severity of acid precipitation falling over the United States, Canada, and most of western Europe had increased significantly within the last 20 to 25 years. Acid precipitation is associated with industrial and automotive emissions of sulfur oxides, nitrogen oxides, and gaseous hydrogen chloride resulting from the combustion of fossil fuels. These pollutants may remain in the atmosphere for several days, during which they may be transported large distances, sometimes exceeding 500 miles, before being deposited on water or land surfaces. Acid precipitation may have a toxic effect on indigenous plants and animals by altering the chemical composition of the land or aquatic environments on which it is deposited. Soil type is a factor in determining the susceptibility of terrestrial environments to acid precipitation. The pH of some soils may be lowered, or the leaching of calcium, magnesium, or heavy metals from the soil may be accelerated. In addition, acid rain may adversely affect the reproductive success of some terrestrial animals and cause a variety of plant growth and foliage damage problems. The pH of lakes or streams may be lowered, resulting in the release of toxic heavy metals, the disappearance of fish populations, a reduction in the microbial decomposition processes, and a general lowering of the fertility of the aquatic ecosystem. Acid precipitation has also been shown to promote the corrosion of metals, particularly iron and iron-containing alloys, and the deterioration of stone and marble structures.

Balson, W.E., Boyd, D.W. and North, D.W., "Acid Deposition: Decision Framework--Volume 1--Description of Conceptual Framework and Decision-Tree Models", Aug. 1982, Decision Focus, Inc., Palo Alto, California.

Acid precipitation and dry deposition of acid materials have emerged as an important environmental issue affecting the electric utility industry. This report presents a framework for the analysis of decisions on acid deposition. The decision framework is intended as a means of summarizing scientific information and uncertainties on the relation between emissions from electric utilities and other sources, acid deposition, and impacts on ecological systems. The methodology for implementing the framework is that of decision analysis, which provides a quantitative means of analyzing decisions under uncertainty. The decisions of interest include reductions in sulfur oxide and other emissions thought to be precursors of acid deposition, mitigation of acid deposition impacts through means such as liming of waterways and soils, and choice of strategies for research. The report first gives an overview of the decision framework and explains the decision analysis methods with a simplified caricature example. The state of scientific information and the modeling assumptions for the framework are then discussed for the three main modules of the framework: emissions and control technologies; long-range transport and chemical conversion in the atmosphere; and ecological impacts. The report then presents two versions of a decision tree model that implements the decision framework. The basic decision tree addresses decisions on emissions control and mitigation in the immediate future and a decade hence, and it includes uncertainties in the long-range transport and ecological impacts. The research emphasis decision tree addresses the effect of research funding on obtaining new information as the basis for future decisions. Illustrative data and calculations using the decision tree models are presented.

Beamish, R.J. and Van Loon, J.C., "Precipitation Loading of Acid and Heavy Metals to a Small Acid Lake Near Sudbury, Ontario", Fisheries Research Board of Canada Journal, Vol. 34, No. 5, May 1977, pp. 649-659.

Acidity levels in Lumsden Lake, Sudbury, Ontario, were measured to determine if precipitation was the source of high pH levels and high heavy metal content. The pH of Lumsden Lake was closely related to the measured amount of acid entering the Lake from bulk precipitation. High concentrations of sulfate, manganese, zinc, and nickel were found in Lumsden Lake. Atmospheric fallout also contributed substantial quantities of nickel and copper to the watershed.

Bilonick, R.A. and Nichols, D.G., "Temporal Variations in Acid Precipitation over New York State--What the 1965-1979 USGS Data Revealed", Atmospheric Environment, Vol. 17, No. 6, 1983, pp. 1063-1072.

During the period from 1965 to 1979, the USGS collected rainfall samples throughout New York State; pH, sulfate, nitrate and calcium levels were measured among other species. The statistical nature of the variation of these quantities with time was examined in this study. The time series analysis method of Box and Jenkins was applied; the results indicate that the long-term mean levels of hydrogen, sulfate, nitrate and calcium ions in the bulk (wet plus dry) precipitation samples were essentially constant. Use of the Box-Jenkins method was appropriate because samples that were collected close together in time were clearly more alike than those far apart, indicating autocorrelation. An analysis that fails to take this dependency into account could lead to incorrect conclusions.

Bowersox, V.C. and DePena, R.G., "Analysis of Precipitation Chemistry at a Central Pennsylvania Site", Journal of Geophysical Research, Vol. 85, No. C10, Oct. 1980, pp. 5614-5620.

Data collected at the multistate atmospheric power production pollution study (MAPPPS) precipitation chemistry network were analyzed in detail. Precipitation was sampled on a storm-by-storm basis at a rural central Pennsylvania site during the period January 1, 1977, to December 31, 1978. Samples were analyzed for sulfate, nitrate, ammonium, and hydronium ion concentrations. Sulfate concentration varied seasonally, with high values in summer and low values in winter. Sulfate is of interest as the major determinant of precipitation acidity, particularly in rain. The principal contributor to H_3O was sulfuric acid, but the acidity of snow was determined principally by NO_3. The NH_4 in precipitation can be explained by assuming equilibrium between cloud water and gaseous ammonia in air. There was a good stoichiometric balance of the four major inorganic ions. Knowledge of the nature of acid-determining components in rain can be used to help establish strategies to control the quality of precipitation.

Brocksen, R.W., "Acid Precipitation: The Complexity of a Perceived Problem", Feb. 1980, Electric Power Research Institute, Palo Alto, California.

A brief overview of the acid precipitation problem as viewed by the Electric Power Research Institute is presented. Research programs conducted or planned by EPRI are described.

Brosset, C., "Acid Particulate Air Pollutants in Sweden", EPA International Environmental Document, Report 30781A-Sweden, Feb. 1975, U.S. Environmental Protection Agency, Washington, D.C.

For the most part, the acid ammonium sulfate aerosols episodically entering the air of Sweden are generated by emissions from remote sources and transported to Scandinavia under certain weather conditions. The chemical nature of the particle-borne sulfur; formation mechanisms for black and white sulfur episodes; interaction of the sulfur episodes with the local pollution; change in the acidity content in particles during transport over land; and systematic measurements of the sulfur concentration in particles in Sweden are examined. Particulate pollution, which arises when the long-distance transported aerosols are mixed with local emissions, has been insufficiently investigated.

Brosset, C., Andreasson, K. and Ferm, M., "The Nature and Possible Origin of Acid Particles Observed at the Swedish West Coast", Atmospheric Environment, Vol. 6, No. 7, Sept. 1975, pp. 631-643.

During March 1-August 31, 1973, the particle content of ammonium and sulfate ions was measured on the western coast of Sweden. The investigations were completed by phase analysis, determination of particle volume distribution, and trajectory estimates of air masses. The occurrence of two types of acid particles is discussed. The relationship between their acid properties and their formation is explained.

Brosset, C. and Ferm, M., "Man-Made Airborne Activity and Its Determination", 1976, Swedish Water and Air Pollution Research Laboratory, Goteborg, Sweden.

Man-made airborne acidity has been defined as an unneutralized part of H_2SO_4 and HNO_3. It has been shown that in most practical cases man-made air-borne acidity can be determined using Gran's Titration Method. The theory, possibilities and limitations of this method are discussed.

Brown, R.J., "Precipitation Washout--1964-June 1980 (A Bibliography with Abstracts)", June 1980, National Technical Information Service, U.S. Department of Commerce, Springfield, Virginia.

The reports cited in this bibliography cover the removal of pollutants, radioactive isotopes, and dust by rain and snow. The theory, modeling, sampling and effects of precipitation washout are presented. This updated bibliography contains 154 abstracts.

Budiansky, S., "Understanding Acid Rain", Environmental Science and Technology, Vol. 15, No. 6, June 1981, pp. 623-624.

Acid rain is a complex phenomenon which involves almost every component of the biosphere. The intensity and effects of acid rain may be affected by varying weather, varying soils, the presence of other pollutants, and species differences. Investigations of the effects of meteorological factors on amounts of chemical wet deposition have identified a number of systematic relationships. Cold fronts and squall lines, thunderstorms, and summer rainfall were all found to be associated with increased concentrations of hydrogen ions and sulfates. A considerable proportion of the available data on acid rain, both current and historical, is disjointed and based on inadequate sampling procedures. Much of the data does not include direct measurements of pH. Methods used to measure dry deposition are particularly inadequate, relying on such surrogate collectors as plastic buckets, which are incapable of measuring dry gaseous deposition. Studies of the effects of acid rain on individual species often ignore the overall effect of pollutants on the ecosystem as a whole and the effects of multiple pollutants on an individual species. Also, the amount of acidic material accumulated and absorbed into the soil will be affected by such subsequent events as runoff and snow melts.

Burns, D.A., Galloway, J.N. and Hendrey, G.R., "Acidification of Surface Waters in Two Areas of the Eastern United States", Water, Air, and Soil Pollution, Vol. 16, No. 3, Oct. 1981, pp. 277-285.

Some acidification of surface waters has occurred in two undisturbed, upland watersheds in New Hampshire's White Mountains and North Carolina's Blue Ridge Mountains. The New Hampshire waters had a pH of 6.06 in 1979 vs. 6.66 in 1936-39. The North Carolina waters declined in pH from 6.77 in 1961-64 to 6.51 in 1979. Recent determinations of Ca, Mg, Na, and K were compared with bicarbonate alkalinity. The current ratios of bicarbonate alkalinity to the sum of the base cations were 0.312 for New Hampshire and 0.562 for North Carolina. These may be compared with ratios from western lakes receiving no acid precipitation (0.776-0.942) and some acid lakes in Connecticut (0.333), Maine (0.427), and Scandinavia (0.064). Acidification was more pronounced in New Hampshire waters, corresponding with historical data that New Hampshire has been receiving precipitation of average annual pH below 4.5 since at least 1955, whereas North Carolina precipitation in 1955 had a pH of 4.7-5.2, and in 1963, 5.0.

Carter, L.J., "Uncontrolled SO_2 Emissions Bring Acid Rain", Science, Vol. 204, No. 4398, June 15, 1979, pp. 1179-1182.

Sulfur dioxide and the fine sulfate mist into which SO_2 is readily transformed are the principal percursors of acid rain. Acid rain may reduce forest yields, poison soils, and cause the death of numerous species of fish. Although new EPA standards will reduce SO_2 emissions from new coal-fired plants to half of what would have been allowed under the preexisting standards, no similar restrictions have been placed on existing plants. The problem represented by the laxity of the legislation addressing existing power plants, and the environmental implications of the massive sulfur emissions that these plants will produce, are discussed.

Chamberlain, J. et al., "Physics and Chemistry of Acid Precipitation", Nov. 1981, SRI International, Arlington, Virginia.

Studies were conducted on the sources of emissions having the potential to raise the acidity of precipitation, their transformation in the air and in hydro-meteors and droplets to strong acids, the transport mechanisms and the statistical nature of acid precipitation. Findings indicate that the acidity of precipitation has increased in the eastern United States over the past three decades; the largest percentage increase has occurred in the Southeast. Data on SO_2 and NO_x emissions in various regions of the U.S. are presented. Information is presented under the following section headings: atmospheric chemistry; droplet chemistry; stack heights and acid rain; transport models of pollutants; comments on the use of statistics in acid precipitation analysis; possible observations and experiments in acid precipitation research; and partial bibliography on acid precipitation.

Church, T.M., "Chemistry of Acid Rain at Lewes, Delaware as Part of the Precipitation Chemistry Network of MAP3S", Contract W-7405-ENG-48, 1978, Delaware University-College of Marine Studies, Lewes, Delaware.

Rain water was collected and analyzed from July 1977 to the present at an Atlantic coastal site at the southern terminus of Delaware Bay near Lewes, Delaware, as part of a network established by the Multi-State Atmospheric Power Production Pollution Study (MAP3S). The pH of rain water goes through a minimum as low as 3.6 in the summer and a maximum as high as 5.8 in the winter, paralleling a maximum in mean monthly weighted sea salt sulfate in summer and a minimum in winter. Sea salt sulfate contributions are usually small (less than 10 percent) even at the coastal site. The contribution of sodium, magnesium, and chloride in Lewes rain, however, is dominated by sea salt perhaps of local origin. The nature of rain acidity at Lewes is due also to sulfuric and nitric acids apparently from the remote combustion of fossil fuels. Local contributions of basic sea salt components such as bicarbonate are insufficient to cause any apparent neutralization of the acid rain. Quantitatively, the composition of Lewes acid rain is highly typical of other MAP3S sites except for higher sea salt contributions. The annual net deposition of nonsea salt sulfate is 12.8 KgS/hectare-year, which is at the upper end of ranges in the MAP3S network suggesting rain falling at Lewes has accumulated sulfate from other sources. One potential source of nonmarine sulfate is local biogenic contributions from exposed salt marshes.

Coghill, C.V. and Likens, G.E., "Acid Precipitation in the N.E. United States", Water Resources Research, Vol. 10, No. 6, Dec. 1974, pp. 1133-1137.

Acid precipitation in New York and New England is evaluated in this paper. The distribution of acid precipitation encompasses most of the northeastern U.S. This pattern apparently has existed since about 1950-55, but the intensity of acid deposition has increased since then. Analysis of prevailing winds indicates that much of the acidity originates as a general source over industrial areas in the Midwest.

Cowling, E.B., "Acid Precipitation in Historical Perspective", Environmental Science and Technology, Vol. 16, No. 2, Feb. 1982, pp. 110A-123A.

This article reviews several steps in the transformation of the concepts of acid precipitation from the domain of scientific curiosity to the domain of public concern and debate. An historical resume is presented in tabular form of the progress that has been made toward understanding the phenomena connected with man's influence on the chemical climate of the earth and their biological consequences. Contributions are presented in the chronological order of their occurrence, whether or not the work was recognized or accepted at the time. The names of the major scientists, the country in which the research was conducted, and the principal contribution to science or public affairs are presented, with appropriate references when possible. Many features of the acid rain phenomenon were first discovered in the mid-1800s in and around the city of Manchester, England. Contemporary concepts about acid precipitation and its environmental effects originated in three seemingly unrelated fields of science: limnology, agriculture and atmospheric chemistry. Concern about acid precipitation and its ecological effects in North America developed first in Canada and later in the United States, initially centering on the effects of sulfur dioxide exposure and associated acid precipitation and heavy metal deposition in the vicinity of metal smelting and sintering operations, especially near Sudbury, Ontario.

Cowling, E.B., "Discovering the Causes, Consequences, and Implications of Acid Rain and Atmospheric Deposition", Tappi Journal, Vol. 66, No. 9, Sept. 1983, pp. 43-46.

Areas receiving high acidic precipitation (average annual pH between 4.1 and 4.6) include many regions in which surface waters and aquatic life are sensitive to or are presently being affected by acid precipitation. The release of pollutants in meltwater during warm periods in winter or in early spring can cause dramatic consequences of acid input into aquatic ecosystems. In certain industrial regions of the world substantial damage to forests and crops is caused by dry deposition of toxic gases. Ozone, sulfur dioxide, oxides of nitrogen, fluoride, and hydrogen chloride are some of the worst offenders. The consequences of acid inputs to soils vary greatly, depending on the rates and recent history of atmospheric acid inputs, the character of the vegetation, natural rates of acid formation in the soil, and the physical/chemical properties of the soil. Acid precipitation increases the solubility and mobility of many cations in the soil. This increases the concentration of toxic metal cations, including potassium, aluminum, manganese, and zinc in the soil solution. It also increases the leaching of nutrient cations, including calcium and magnesium. These toxic and nutrient ions are transferred from soils into surface waters and ground waters. Certain aspects of the acid precipitation problem are still not well understood and additional work is needed to develop possible solutions to the question.

Curtis, K.E., "Emissions of Acid Rain Precursors in North America", Research Review Ontario Hydro, May 1981, pp. 5-10.

Precipitation falling over large areas of eastern North America is highly acidic, with sulphur and nitrogen compounds dominating this acidity. Current information was reviewed to identify the major sources of the precursors of acid rain in North America. A brief description is also given of the work done to characterize the emissions from Ontario Hydro's fossil-fuel-fired generating stations. Anthropogenic emissions of SO_2 and NO_x were found to exceed those from natural sources, having increased significantly since 1940. In Canada and Ontario, industrial processes and transportation are by far the major sources of SO_2 and NO_x, respectively. However, electric utilities are major emitters of both these acid gases in the United States.

Davies, T.D., "Dissolved Sulphur Dioxide and the Sulphate in Urban and Rural Precipitation", Atmospheric Environment, Vol. 13, 1979, pp. 1275-1285.

Sulphur dioxide washout and total sulphate scavenging by precipitation at a rural and urban site have been measured in Norfolk (easternmost part of the U.K.) over a period of one year. The basic sampling period was one day although, on occasions, more intensive sampling was undertaken. Estimates have been made of the role of sulphur dioxide washout in the total sulphur removal process. Because the "same" rainfall was collected in the urban and rural locations, attempts have been made to compare city and rural sulphur dioxide washout and sulphate scavenging rates in relation to precipitation amount, intensity, atmospheric sulphur dioxide, rainfall pH and weather-type. There is a greater difference between urban and rural sulphur dioxide washout levels than between the respective precipitation sulphate levels. The field measurements of sulphur dioxide solubilities agree reasonably well with some previous theoretical work.

Dawson, G.A., "Ionic Composition of Rain During Sixteen Consecutive Showers", Atmospheric Environment, Vol. 12, No. 10, 1978, pp. 1991-2001.

Concentrations of inorganic constituents were measured during 16 summer showers occurring over Tucson, Arizona, in 1976 and 1977. Data were subjected to statistical analysis to determine whether sources or associations could be inferred. The method cannot be used to infer sources--at least not in the clean conditions of the southwestern U.S. Observed concentration changes are determined primarily by precipitation processes.

Deland, M.R., "Acid Rain", Environmental Science & Technology, Vol. 14, No. 6, June 1980, p. 657.

A 1975 survey of 214 lakes in the Adirondack region of New York found over 50 percent with a pH which imperiled fish survival (less than 5.0) and that 82 were in fact fishless. The acid rain problem is spreading from the northeast to other regions in the U.S. In Europe, Scottish scientists in 1974 recorded a rainstorm with the acid equivalent of vinegar. In 1979 President Carter directed that $10 million per year be allocated for research to solve the acid rain problem. EPA Administrator Costle urges that the transition be made from research to action. Regulations for emissions from coal and petroleum are being urged for immediate implementation.

DeWalle, D.R. et al., "Nature and Extent of Acid Snow Packs in Pennsylvania", Feb. 1983, Institute for Research on Land and Water Resources, Pennsylvania State University, University Park, Pennsylvania.

Snowpack chemistry was evaluated with repeated surveys at 30 sampling sites across Pennsylvania during 1979-81. The mean snowpack pH of 4.25 for 280 samples was largely due to H(+) from nitric acid. Snowpack sulfates were high, especially in the southwest sector of the state where SO_2 emissions were also high, but did not originate only from sulfuric acid. Rain-on-snow and coastal storm snowpacks produced higher snowpack pH. Localized increases in snowpack pH were found due to additions of dust from limestone and dolomite quarry operations, especially in the central Ridge and Valley Province. Acid loading on the landscape from snow in Pennsylvania is controlled by the total amount of snowfall as well as variations in concentrations of chemical constituents. Lead concentrations in snowpacks occasionally exceeded maximum contaminant levels prescribed for drinking water.

Dillon, P.J. et al., "Acid Precipitation in South-Central Ontario: Recent Observations", Fisheries Research Board of Canada Journal, Vol. 35, No. 6, June 1978, pp. 809-816.

Evidence is presented to suggest that acidic rain either already has caused or soon will cause a severe environmental problem in south-central Ontario. The chemical analysis of precipitation and lake samples for pH values in areas of Ontario east of Georgian Bay, south of Lake Nipissing, and north of the Precambrian Shield-Paleozoic boundary indicated the existence of a potentially serious environmental problem. This conclusion is supported by the low and possibly declining buffering capacities of some lakes in the region. The pH and sulfate concentrations of precipitation and bulk samples collected at 11 sites in south-central Ontario from May 1976-April 1977 are tabulated.

Drabloes, D. and Tollan, A., "Proceedings of International Conference on the Ecological Impact of Acid Precipitation", Oct. 1980, Sandefjord, Norway.

Separate abstracts were prepared for 134 of the papers presented in these proceedings. Investigations into: transport and deposition, vegetation and soils, water quality, aquatic biota, impacts on soils and indirect effects on vegetation, effects on water quality, effects on aquatic biota, and various integrated studies are all discussed.

Duncan, L.C., "Acid Precipitation in the Washington Cascades", Sept. 1981, Department of Chemistry, Central Washington University, Ellensburg, Washington.

Five sites maintained across the Washington Cascades from January-June 1981 sampled precipitation for determinations of its potential acidity. The precipitation collected at each site was acidic in that the average pH values were substantially below pH 5.6, a level expected for waters containing only equilbrium concentrations of dissolved carbon dioxide. The pH values varied somewhat with location. Sulfate was the dominant negative ion at all sites, even after correction for sea-salt input. The mean strong acid composition over the sites ranged from 50 to 57 weight % H_2SO_4. Precipitation falling on the Cascades east and windward of the urban Puget Sound area was notably acidic, with a volume average pH of 4.71 at the Cascade Crest (Snoqualmie Pass). Strong acid content in precipitation increased in the summer months. In addition, 29 watershed source lakes lying along the Cascade Crest 3 to 20 miles north of the site collectors were sampled in the summer of 1981; all lakes were found to be highly sensitive (but not acidified), with alkalinities ranging from 4 to 190 micro-equivalents/liter (mean value of 50.4). The lakes were nonetheless, considered to be highly susceptible to acid input, being in a region of thin top soil and resistant strata.

Eaton, J.S., Likens, G.E. and Bormann, F.H., "Wet and Dry Deposition of Sulfur at Hubbard Brook", Effects of Acid Precipitation on Terrestrial Ecosystems, Plenum Press, New York, New York, 1980, pp. 69-75.

This paper is a summary of the biogeochemistry of sulfur dry deposition of gases and aerosols in the Hubbard Brook Experimental Forest. An excellent diagram of the deposition and reactions of sulfur in a northern hardwood forest ecosystem are given. Seasonal quantities and qualities of deposition are given as well as a description of field collection methods.

Ellis, B.A., Verfaillie, J.R. and Kummerow, J., "Nutrient Gain from Wet and Dry Atmospheric Deposition and Rainfall Acidity in the Southern California Chaparral", Oecologia, Vol. 60, No. 1, 1983, pp. 118-121.

Ionic concentration and annual deposition of NO3(-)-N, NH4(+)-N, Ca(2+), and Mg(2+) from bulk precipitation and dry atmospheric deposition were studied for one year in southern California. Data were collected from an inland chapparal site at 1,300 m elevation, 75 km from the coast. The annual depositions of NO3(-)-N, and NH4(+)-N amounted to 96.3 and 56.0 mg/sq m ground area/yr, respectively. The corresponding values for calcium and magnesium ions were 207.4 and 57.4 mg/sq m/yr. The average pH of rainwater was 3.74 (range 3.37 to 4.75), thus documenting acid rain for an inland site in California, distant from urban sources of air contamination. An estimate of nitrogen gains and losses indicated that the time between recurrent chapparal fires should be about 60 years in order to maintain a balanced nitrogen budget.

Ember, L., "States Anguish Over Acid Rain Problem", Chemical and Engineering News, Apr. 21, 1980, pp. 22-23.

Thirty states (representatives) recently met to exchange information on controlling the acid rain problem. Some possible options to control acid rain are: (1) set tighter tall-stack regulations; (2) improve enforcement and enforcement monitoring; (3) set tighter parameters on atmospheric models to yield more stringent emissions limitations; (4) set an upper limit on pollutants; (5) require universal coal washing at coal-using facilities; (6) establish SO_2 and NO_2 taxes; (6) encourage least emissions rather than least cost dispatching of electric utilities, thereby retiring older, less efficient plants; and (7) encourage favorable tax breaks for facilities that reduce emissions. Opposition by coal and electric industries to regulatory actions to curb acid rain stems from expenditures (by these companies) needed to reduce their emissions which are principally responsible for acid rain production. An agreement with Canada to control acid rain sources is not likely before 1982.

Ember, L.R., "Acid Rain Focus on International Cooperation", Chemical and Engineering News, Dec. 3, 1979, pp. 15-17.

To fight transboundary air pollution, 33 nations sign a pact in Geneva; the U.S. and Canada agree to speed efforts to reduce acid precipitation. Canada and the U.S. will sign a bilateral air pollution treaty sometime in 1981. Acid rain damage to freshwater lakes and architectural structures in the U.S. may be as high as 2 billion dollars. On August 2, 1979, the President of the United States created the Interagency Acid Rain Coordination Committee, co-chaired by EPA and the Department of Agriculture, and called for a 10 year acid rain research program to be funded at $10 million the first year.

Evans, L.S., "Considerations of an Air-Quality Standard to Protect Terrestrial Vegetation from Acidic Precipitation", 1981, Laboratory of Plant Morphogenesis, Manhattan College, Bronx, New York.

Studies on the effects of acidic precipitation, which is here defined as wet or frozen deposition with a hydrogen ion concentration greater than 2.5 mu eq per liter, are reviewed. At the present time there is an inadequate amount of information that shows decreases in crop growth except for one field study. Most studies with plants (crops and forests) are inadequate for standard setting because they are not conducted in the field with adequate randomization of plots coupled with rigorous statistical analyses. Although visible injury to foliage has been documented in a variety of greenhouse studies, no experimental evidence demonstrates loss of field crop value or reduction in plant productivity due to visible foliar injury. Acidic precipitation can contribute nutrients to vegetation and could also influence leaching rates of nutrients from vegetation. Although these processes occur, there are no data that show changes in nutrient levels in foliage that relate to crop or natural ecosystem productivity. Experimental results show that fertilization of ferns is inhibited by current levels of acidic precipitation in the northeastern United States. However, the overall impacts of inhibited fertilization on perpetuation of the species or ecosystem productivity have not been evaluated. Simulated acidic precipitation has been shown to effect plant pathogens in greenhouse and field experiments. Simulated acidic precipitation inhibited pathogen activities under some circumstances and promoted pathogen activities under other circumstances. No conclusion can be drawn about the effects of current levels of precipitation acidity on plant pathogen-host interactions. From these data it must be concluded that research on the effects of acidic precipitation on terrestrial vegetation is too meager to draw any conclusions with regard to an air quality standard.

Evans, L.S. et al., "Acidic Precipitation: Considerations for an Air Quality Standard", Water, Air, and Soil Pollution, Vol. 16, No. 4, 1981, pp. 469-509.

This paper discusses the relationship between precipitation acidity and receptor effects. First, the existence of acidic precipitation is documented, including temporal and spatial variability. A review of receptor characteristics is presented, along with responses to acidic precipitation, describing links between acidic precipitation and environmental effects where such links are known. Information gaps are highlighted. Finally the possibility of establishing a critical threshold for precipitation acidity is evaluated, which can serve as a meaningful standard to protect terrestrial and/or aquatic biota. The chief anions accounting for the acidity in rainfall are nitrate and sulfate. Greater nutritional benefits are derived from acid rain by agricultural rather than forest systems, when considering soils alone. Acid deposition has caused acidification of fresh waters of southern Scandinavia, southwestern Scotland, southeastern Canada, and northeastern United States. Biological effects of acidification of fresh waters are detectable below a pH of 6.0. As lake and stream pH levels decreased below pH 6.0, many species of plants, invertebrates, and vertebrates were progressively eliminated. Current research indicates that establishing a maximum permissible value for the volume weighted annual hydrogen ion concentration of precipitation at 25 microeq/liter may protect the most sensitive areas from permanent lake acidification.

Feeley, J.A. and Liljestrand, H.M., "Source Contributions to Acid Precipitation in Texas", Atmospheric Environment, Vol. 17, No. 4, 1983, pp. 807-814.

Wet-precipitation-only sampling at Austin, Prairie View and Highlands, Texas has been conducted to characterize the main sources of acid rain chemical composition. These results are compared with those of previous investigators to identify spatial distributions of source contributions. A chemical mass balance model is used to determine the spatial distributions of background crustal source alkalinity and sea salt contributions.

Forland, E.J. and Gjessing, Y.T., "Snow Contamination from Washout-Rainout and Dry Deposition", Atmospheric Environment, Vol. 9, No. 3, Mar. 1975, pp. 339-353.

The deposition of air pollutants in the Bergen Region, western Norway, was investigated by a simple and inexpensive method. Dry snow was sampled from the ground at 46 locations just after a snowfall, and again three days later. Consistent and logical patterns for the samples were found for sulfate, magnesium, calcium, zinc, pH, and conductivity. The rate of dry deposition in rural areas was significantly smaller than the rate of washout-rainout.

Frey, J.H., "Acid Precipitation, June, 1976-May, 1980 (Citations from the Energy Data Base)", June 1980, National Technical Information Service, U.S. Department of Commerce, Springfield, Virginia.

Reports concerning the causes, effects, sources, and controls of acid precipitation and acidification are contained in this retrospective bibliography. Contains 138 citations.

Fuzzi, S., Orsi, G. and Mariotti, M., "pH of Fog", Journal of Aerosol Science, Vol. 14. No. 3, 1983, pp. 298-301.

This paper reports on the problem of fog water acidity, based on a study carried out in a country site of the Po Valley in Italy. During the winter months, for about 30 percent of the time, visibility is lower than 1 km and, in some cases, conditions of a few meters visibility are reached. Fog liquid water samples were collected during three fog events, followed step by step from formation to dissipation. The pH of the samples was measured by means of miniature glass electrodes. The samples were also bulk, but the collection time (20-30 min) can be considered short with respect to the entire fog event, so that these measurements describe the pH trend during the fog evolution.

Galloway, J.N., "Sulfur Deposition in the Eastern United States", Air Pollution Control Association Mid-Atlantic States Section Semi-Annual Technical Conference, Philadelphia, Pennsylvania, Apr. 13-14, 1978, pp. 101-115.

The temporal and spatial trends of sulfur in wet deposition in the eastern U.S. are investigated in regard to increasing rates of deposition. Sulfur deposition due to fossil fuel combustion is five times greater than that due to natural causes in the eastern U.S. Increasing sulfur deposits in precipitation as sulfuric acid have caused acidification of precipitation over most of the eastern U.S. Increased nitric acid in precipitation causes increased precipitation acidity. Presently, it is not possible to quantitate the contributions of HNO_3 and H_2SO_4 to acidity in precipitation.

Galloway, J.N. and Cowling, E.B., "The Effects of Precipitation on Aquatic and Terrestrial Ecosystems: A Proposed Precipitation Chemistry Network", Air Pollution Control Association Journal, Vol. 28, No. 3, Mar. 1978, pp. 229-236.

The changing chemistry of precipitation and its effects on terrestrial and aquatic ecosystems are discussed. A network of monitoring stations is needed in the U.S. and Canada to permit prudent management of emissions. The magnitude of the effect of chemicals in precipitation depends on two related parameters: magnitude of the input of chemical components relative to the sensitivity of the area to the inputs.

Galloway, J.N. and Likens, G.E., "Acid Precipitation: The Importance of Nitric Acid", Atmospheric Environment, Vol. 15, No. 6, 1981, pp. 1081-1085.

Using the information on temporal trends of sulfate and nitrate ions accumulated from the long-term precipitation data collected at Hubbard Brook Experimental Forest, New Hampshire, temporal trends were determined in the maximum contributions of sulfuric acid and nitric acid to the acidity of the precipitation. Nitric acid in acid rain was introduced primarily by increased emissions of NO_x from fossil fuel combustion. Its contribution to the acidity of precipitation is becoming increasingly significant. The maximum contribution of H_2SO_4 to the acidity of precipitation is 73% in the summer and 59% in the winter. For HNO_3, it is 31% in the summer and 61% in the winter. This analysis indicated that the importance of H_2SO_4 has decreased by about 30% relative to HNO_3 and HNO_3 has increased by about 50% relative to H_2SO_4 for the years studied. These conclusions should be applicable to most of the eastern United States.

Galloway, J.N., Likens, G.E. and Edgerton, E.S., "Acid Precipitation in the Northeastern United States: pH and Acidity", Science, Vol. 194, No. 4266, Nov. 12, 1976, pp. 722-725.

Detailed chemical analyses reveal that acid precipitation in the northeastern U.S. is caused by two strong mineral acids: sulfuric and nitric. There is a large array of other proton sources in precipitation, including weak and bronsted acids. Although these other acids contribute to the total acidity of precipitation, their influence on the free acidity of acid precipitation is minimal.

Galloway, J.N. and Whelpdale, D.M., "Atmospheric Sulfur Budget for Eastern North America", Atmospheric Environment, Vol. 14, No. 4, 1980, pp. 409-417.

Atmospheric sulfur budgets for the eastern United States, eastern Canada and eastern North America are formulated, based on the fluxes: natural and man-made emissions, wet and dry deposition, and atmospheric inflows and outflows. The uncertainty in each of these terms is examined, and a best estimate is chosen for the budget calculations.

Galvin, P.J. et al., "Transport of Sulfate to New York State", Environmental Science and Technology, Vol. 12, No. 5, Aug. 1979, pp. 580-584.

Trajectory analysis provides evidence that high sulfate concentrations observed during summer high-pressure systems at three rural sites in New York State are products of sulfur dioxide emissions to the south and southwest of New York State. The highest concentrations occur in air that first stagnates over the area surrounding the Ohio River Valley and then is advected into New York State as the high-pressure system begins to move eastward. The amount of oxides of nitrogen that emissions data indicate should accompany these high sulfate concentrations is not present in the particulate samples as inorganic nitrate.

Gannon, J., "Acid Rain Fallout: Pollution and Politics", National Parks and Conservation Magazine, Vol. 52, No. 10, Oct. 1978, pp. 17-23.

Acid rain kills fish, stunts plants, and corrodes buildings. Acid rain has wiped out commercial salmon fishing in Norway and Sweden and has destroyed sport fishing in parts of Canada, the U.S. and Scandinavia. The burning of fossil fuels and the smelting of ores produce sulfur and nitrogen oxides, which are converted to acids in the atmosphere. The problem of acid rain is transnational and transcontinental. There is a high correlation between acid levels in lakes and mercury levels in fish. Acid rain, mercury pollution, and the construction of a new coal-fired power plant threaten the existence of an Ojibway Indian Fishing Village on Lac la Croix in Quebec, Ontario. Governmental efforts to control acid rain are insufficient.

GCA Corporation, Acid Rain Information Book, May 1981, Bedford, Massachusetts.

Acid rain is one of the most widely publicized environmental issues of the day. The potential consequences of increasingly widespread acid rain demand that this phenomenon be carefully evaluated. Review of the literature shows a rapidly growing body of knowledge, but also reveals major gaps in understanding that need to be narrowed. This document discusses major aspects of the acid rain phenomenon, points out areas of uncertainty, and summarizes current and projected research by responsible government agencies and other concerned organizations.

General Accounting Office, "The Debate Over Acid Precipitation--Opposing Views--Status of Research", Sept. 1981, Washington, D.C.

The use of coal as a substitute for imported oil is raising concern over the possible impact of acid precipitation on the environment and human health. Precipitation can become acidified when sulfur and nitrogen oxides emitted by fossil-fueled powerplants, vehicles, and other man-made or natural sources are chemically changed in the atmosphere and return to earth as acid compounds. Environmental organizations and some agencies and interest groups contend that more stringent emission controls are needed immediately. On the other hand, some agencies and many industries, particularly in the coal and utility sectors, argue that much more research is needed to determine if achievable emissions reductions could significantly diminish the extent of acid precipitation, and if the benefits of such regulations would be worth their potentially high cost.

Gibson, J.H., "A National Program for Assessing the Problem of Atmospheric Deposition (Acid Rain)--A Report to the Council on Environmental Quality", Report NC-141, Dec. 1978, National Atmospheric Deposition Program, Natural Resources Ecology Laboratory, Colorado State University, Fort Collins, Colorado.

This booklet discusses present and future needs and trends in acid rain research.

Gibson, J.H., Proceedings of Workshop on Data Management Needs for Atmospheric Deposition, Aug. 1980, Natural Resources Ecology Laboratory, Colorado State University, Fort Collins, Colorado.

With the recent organization of the National Atmospheric Deposition Program atmospheric chemistry monitoring network it became evident that there was a need for a data management system to store and make readily available the extensive data base that would be generated. Such a system needed to be designed to handle data from other U.S. atmospheric monitoring networks, as well as non-U.S. networks as for example the Canadian CANSAP program. A workshop was organized to address this question of data management requirements. The attendees representing major U.S. and Canadian studies agreed that a data management function was critical if assessment of the data with respect to the effects of atmospheric deposition was to be successful. In addition, the data are critical to those involved in the studies of atmospheric chemistry and transport. It was agreed that a single system could serve many programs and a statement was prepared stating the required characteristics and capabilities of the data management function. It was recommended that development and support of such a system should be the responsibility of a federal agency and the offer by the EPA to assume this responsibility was strongly endorsed.

Glover, G.M. et al., "Ion Relationships in Acid Precipitation and Stream Chemistry", Effects of Acid Precipitation on Terrestrial Ecosystems, Plenum Press, New York, New York, 1980, pp. 95-109.

The authors are employed at the General Electricity Research Laboratories in Leatherhead, Surrey, England. In this study of the composition of precipitation and the fate and reactions it undergoes after deposition the authors conclude that low concentrations of sulfur dioxide are unlikely to produce significant changes in rain pH in remote areas. They attribute changes in the pH of surface water to the presence of weak acids.

Glustrom, L. and Stolzenberg, J., "Acid Rain: A Background Report", July 1982, Wisconsin Legislative Council, Madison, Wisconsin.

This Staff Brief was prepared for the Wisconsin Legislative Council's Special Committee on Acid Rain to provide an introduction to the issue of acid rain. It is divided into four parts. Part I provides an overview on the controversies surrounding the measurement, formation and effects of acid rain. As described in Part I, the term acid rain is used to describe the deposition of acidic components through both wet deposition (e.g., rain or snow) and dry deposition (e.g., direct contact between atmospheric constituents and the land, water or vegetation of the earth). Part II presents background information on state agency activities relating to acid rain in Wisconsin, describes what is known about the occurrence of, susceptibility to and effects of acid rain in Wisconsin, and provides information related to man-made sources of sulfur and nitrogen oxides in Wisconsin. Part III describes major policies and regulations relating to acid rain which have been or are being developed jointly by the United States and Canadian governments, by the United States government and by the State of Wisconsin. Part IV briefly discusses possible areas for Committee action.

Grennfelt, P., Bengston, C. and Skaerby, L., "Estimation of the Atmospheric Input of Acidifying Substances to a Forest Ecosystem", July 1978, Swedish Water and Air Pollution Research Laboratory, Goeteborg, Sweden.

Acidifying substances are compounds capable of increasing the hydrogen ion activity in the ecosystem. They are deposited to vegetation and soil in different forms (gases, particles, rain) and the deposition mechanisms are extremely complex. However, to evaluate possible effects of the deposition of acidifying substancs to an ecosystem, the quality and quantity of atmospheric input must be known, as well as the pathways by which the deposition takes place. The paper gives a brief description of the occurrence of acidifying substances in the atmosphere, the essential pathways, and the mechanisms for deposition of anthropogenic nitrogen and sulphur compounds. The paper also presents a rough estimate of the total input, that is, a deposition budget, of these compounds to a coniferous forest ecosystem in southern Sweden.

Gunnerson, C.G. and Willard, B.E., (editors), Acid Rain, 1979, American Society of Civil Engineers, New York, New York.

This book is composed of papers presented at the proceedings of a session sponsored by the Research Council on Environmental Impact Analysis of the ASCE National Convention in Boston, Massachusetts in April, 1978. It is in paperback version, may be read quickly, and provides an excellent bibliography at the end of each paper.

Haines, B., "Forest Ecosystem SO_4-S Input-Output Discrepancies and Acid Rain: Are They Related?", Oikos, Vol. 41, No. 1, 1983, pp. 139-143.

The inputs and outputs of SO_4-S have been estimated for seven terrestrial ecosystems in the Americas. Five of the systems, an agricultural watershed, two deciduous forests in the Southeastern United States, and a Central and a South American rainforest had an excess of inputs over outputs. The reported discrepancies were between +4 and +11 kg/ha/yr. Hypotheses to explain the excess of inputs over outputs include: accumulation of S in biomass; accumulation of S in soil; conversion of SO_4-S to organic S compounds which leave the system in drainage water; conversion of SO_4-S to volatile S compounds which leave as gases; and estimation errors. Acid rain occurs at both rain forest sites. If the S were volatilized out of the forest, oxidized in the atmosphere to SO_4, then washed out of the atmosphere by rain, the resulting quantity of H_2SO_4 would be sufficient to account for the rainfall acidity observed in the field in Central America.

Hales, J.M. and Slinn, W.G.N., "Natural Recovery Processes in Polluted Atmospheres", Proceedings of the Second National Conference on Complete Water Reuse: Water's Interface with Energy, Air, and Soils, 1975, American Institute of Chemical Engineers, New York, New York, pp. 338-354.

The purpose of this paper is to present a brief survey of current capabilities for assessing atmospheric recovery processes as they affect the quality of surface and ground waters. The fact that delivery from the atmosphere can constitute a significant portion of pollutant burdens in surface waters is well documented, accordingly, it is important for scientists specializing in this area to be fully cognizant of the processes responsible for this coupling relationship between the earth's surface and the atmosphere.

Handa, B.K., Kumar, A. and Goel, D.K., "Chemical Composition of Rain Water Over Lucknow in 1980", Mausam, Vol. 33, No. 4, 1982, pp. 485-488.

Rainwater collected over Lucknow, India, from June to December 1980 was analyzed. This area, several hundred km from the sea, is affected by industrial and domestic air pollution, gases from anaerobic decomposition in agricultural lands and swampy areas, emanations from surface water bodies, and other airborne pollutants from adjacent areas. There was no evidence of acid rain; pH values found ranged from 6.55 to 8.15. Chemical composition of the rain varied with each shower, but it was predominantly of the bicarbonate type. Mean values for constituents were: electrical conductivity, 24.9 miscrosiemen/cm; bicarbonates, 83 mg/l; chlorides, 1.14 mg/l; nitrates, 2.12 mg/l; Ca, 1.70 mg/l; Mg, 0.43 mg/l; Na, 0.59 mg/l; K, 0.32 mg/l; and silica, 0.24 mg/l. Mean trace element levels (micrograms per liter) were: Fe, 26.5; Zn, 9.74; Sr, 3.63; Cu, 3.08; Mn, 2.45; Mo, 1/10; and Ag, Co, Cd, Cs, Li, and Rb, all less than 1.0. The highest Cd value in a sample was 2.5 micrograms per liter.

Hendry, C.D. and Brezonik, P.L., "Chemistry of Precipitation at Gainesville, Florida", Environmental Science and Technology, Vol. 14, No. 7, July 1980, pp. 843-849.

Chemical analyses of wet-only and bulk precipitation collected from June 1976 to July 1977 indicate that rain is the predominant deposition mechanism for SO_4^{2-}, NH_4^+, NO_3^-, Mg^{2+}, and K^+, but dry fallout is of comparable importance to rainfall for deposition of Na^+, Cl^-, Ca^{2+}. Heavy metal levels (especially Zn) were highly variable in both types of samples. The (volume-weighted) average concentrations of total nitrogen, NH_4^+, and NO_3^- in bulk precipitation were 0.82, 0.12, and 0.23 mg of N/L, respectively. Averages for ortho- and total phosphate were 24 and 85 ug of P/L respectively, in bulk precipitation and 19 and 34 ug/l in wet-only samples. Highest nutrient concentrations occurred in spring and early summer. Total loadings of N and P from bulk precipitation (1.15 g of N and 0.12 g of P/m²-year) are above permissible rates (relative to eutrophication) for shallow lakes. The volume-weighted mean pH of wet-only precipitation was 4.53; bulk precipitation had a slightly higher pH. Lowest pH values occurred in the spring and summer. Acidity titrations and ionic balances indicate that rainfall acidity resulted mainly from sulfuric acid (69 percent) and nitric acid (23 percent).

Hibbard, Jr., W.R., "Acid Precipitation: A Critique of Present Knowledge and Proposed Action", June 1982, Virginia Center for Coal and Energy Research, Virginia Polytechnic Institute, Blacksburg, Virginia.

This study surveys the evidence, describes the situation, and proposes recommendations to reduce acid precipitation in a cost effective way. Acid precipitation is damaging to the ecology and is present in many areas. Possible sources of acid precipitation are identified, but their specific role and effects have not yet been resolved in an unambiguous manner. Research is needed to determine the mechanism of acid transformation, transport and disposition so that the role of various sources can be evaluated and the most cost-effective counter measures chosen. The effects of actual acid precipitation on forests, vegetation, fish, lakes and ecology are real, but not as bad as originally thought nor does the acidification seem to be changing very quickly. Current legislation, focused on sulfur dioxide emissions from coal burning utilities, will cost $2 to $6 billion per year and will be paid for primarily by utility customers in 10 states. The benefits of this legislation are not at all certain. Sulfur dioxide reduction in Japan did not control acid precipitation, for example. It is highly probable that local sources are a significant cause of acid precipitation, and these sources are not addressed by the legislation. With hindsight, we can now say that local sources were the major problem in both Japan and the Netherlands. At present, further legislation aimed only at coal burning and distant sources will be expensive and probably ineffective. Research results that will target the major causes and identify the effective remedies are needed. This will lower the uncertainty and the cost.

Hileman, B., "Acid Deposition", Environmental Science and Technology, Vol. 16, No. 6, June 1982, pp. 323A-327A.

Papers presented at the American Chemical Society's symposium on acid precipitation (March 29 to April 2, 1982) are summarized. It was shown that a pH isopleth of 4.5 encloses an area of the U.S. and Canada that is in general symmetrical about and displaced somewhat to the northeast of states with high sulfur dioxide emissions. Across the north central lake states the acidity concentration gradient is almost the same as that which existed in Scandinavia when lakes there became acidified. The extent of injury caused by acid rain in lakes is much greater than the killing of fish, and other negative effects precede fish disappearance. A study of about 2,000 observations on the tolerance of aquatic organisms to acidity revealed that fish and amphibians are the most tolerant and leeches and mollusks the least. Experimental acidification of a small precambrian shield lake showed that fathead minnows died out first at a pH of about 5.5, while lake trout and white suckers showed no stress until a pH of about 5.1. In 12 seepage lakes, 6 of which were acidic, species richness of fish was much less in the acidic lakes, and minnows were significantly fewer here. In surface waters a reduction in pH is accompanied by high manganese concentrations and indicates the release of other heavy metals into the water. Increased manganese and aluminum concentrations were found in fish in acidified lakes. Some models for water chemistry were presented, but as yet these tend to describe an existing state rather than to predict what acid inputs would cause a given change in chemistry. Several papers were given on the chemical processes of acidification of precipitation and on long-range transport.

Hileman, B., "1982 Stockholm Conference on Acidification of the Environment", Environmental Science and Technology, Vol. 17, No. 1, Jan. 1983a, pp. 15A-18A.

Recently West Germany and Switzerland took up the position of the Scandinavian countries in viewing acid rain as a problem which should warrant control actions. The problem of forest destruction by acid rain was discussed intensively at the 1982 Stockholm Conference on Acidification of the Environment. The conference consisted of three international meetings. The first was on the ecological effects of acid deposition, and the second was on methods to control sulfur and nitrogen oxide emissions. The third conference was a ministerial meeting of government representatives. The first two expert conferences concluded that forest damage in an estimated one million hectares of Central Europe seems to be related to gaseous pollutants and soil impoverishment, and that trees will not survive beyond 30 to 50 years of age. The United States and Great Britain were the only two countries that contended immediate reductions in sulfur dioxide and nitrogen oxide emissions were not in order until more research is performed. The ministers' conference adopted a consensus statement asserting that further concrete action is needed to control trans-boundary air pollution and called for concerted international programs for the reduction of sulfur emissions. The consensus statement, however, is nonbinding, and did not receive the support of all polluting nations.

Hileman, B., "Acid Rain: A Rapidly Shifting Scene", <u>Environmental Science and Technology</u>, Vol. 17, No. 9, 1983b, pp. 401A-405A.

Some changes are noted in the concern shown by top levels of the United States government regarding the problem of acid rain. A recent government report indicates that the problem is serious enough to warrant a search for immediate solutions, that emissions of sulfur dioxide and nitrogen oxides are at least 10 times greater from human activities than arises from natural happenings, that the areas receiving the highest deposition are within and downwind of the major source regions, and that some lakes in the major receptor areas have become more acidic in the past two decades. A National Research Council report agrees that the current increase of acidic substances in the environment cannot be arising from natural causes. The report concluded that there is no evidence for a strong nonlinearity in the relationship between long-term average emissions and deposition. The conclusion is based on analysis of historical trends, a comparison between the historical molar ratio of sulfur dioxide and nitrogen oxides in emissions to the molar ratio of sulfates and nitrates in deposition, and theoretical calculations based on lab studies of the chemical reactions involved in conversion of the oxides to the sulfates and nitrates. Confidence in current mathematical models describing the movement of acid-forming pollutants over long distances is not high. Two types of models, the STEM series and the air shed model, incorporate a significant number of gas phase reactions and cloud processes into an Eulerian grid calculation. But these models have not been compared with each other or with simpler schemes.

Hill, F.B. et al., "Development of a Technique for Measurement of Biogenic Sulfur Emission Fluxes", Report No. CONF-771102-2, July 1977, Brookhaven National Laboratory, Upton, New York.

Atmospheric sulfur compounds of biogenic origin are thought to constitute a significant fraction of the atmospheric sulfur burden. Determination of fluxes of these compounds into the atmosphere is desirable in order to permit accurate assessment of the relative roles of anthropogenic and biogenic sources in contributing to such phenomena as the atmospheric sulfate burden and acidity in precipitation among others. In the present paper a number of steps in the development of an emission flux measurement technique for use of the technique in a Long Island salt marsh are presented. Experimental fluxes are compared to estimates of biogenic fluxes derived from global sulfur budgets and from a simple mass transfer model. Comparison is also made with anthropogenic emissions expressed as fluxes. Further steps in the development of the technique are suggested.

Hinds, W.T., "Incursion of Acid Deposition into Western North America", Environmental Conservation, Vol. 10, No. 1, 1983, pp. 53-58.

Acidification processes in western North America will probably increase in scope and visibility over the foreseeable future. Localized acidification seems more probable for the next few decades than does a generalized acidified deposition over the western States. At least part of current acid deposition downwind of urban areas is associated with automobile emissions. However, increases in coal combustion for power generation might be expected to increase sulfur dioxide concentrations, along with several related acidifying agents. Interactions of air pollutants with orographic precipitation and snowmelt may bring some areas to public attention much earlier than others, depending upon site-specific matters of climate, topography, soils, water-bodies, and biota. Lakes and streams on windward slopes facing large urban or industrial sources of sulfur and nitrogen oxides may be most seriously at risk. Avoiding acidification may be much easier than amelioration, mitigation, or regulation. This suggests that susceptible landscapes and ecosystems should be identified as soon as possible. Then concerned public and private institutions can agree on specific areas that need protection from acid deposition.

Hitchcock, D.R., "Biogenic Contributions to Atmospheric Sulfate Levels", Proceedings of the Second National Conference on Complete Water Reuse: Water's Interface with Energy, Air, and Soils, 1975, American Institute of Chemical Engineers, New York, New York, pp. 291-310.

An investigation of sulfate levels by bacteria, sea water, fossil fuels, and man is described. Previous works are discussed. Data are given for the eastern part of the USA. Seasonal variations in levels are presented.

Hoban, T.M., "Acid Rain and the Interstate Provisions of the Clean Air Act", International Journal of Environmental Studies, Vol. 18, No. 2, Feb. 1982, pp. 109-116.

The Clean Air Act as originally conceived and as amended through 1977 fails to deal directly with the causes of acid rain because of the mechanism used by the Act. Amendments in 1977 intended to address the problem of "interstate pollution" likewise failed to come to grips with this new phenomenon. The Clean Air Act is to be reconsidered this summer at which time Congress should remember that acid rain is merely one type of long range pollution, that there are likely to be others, and that legislation should be as flexible as possible to better deal with similar future threats to the environment.

Hornbeck, J.W., "Acid Rain: Facts and Fallacies", Journal of Forestry, Vol. 79, No. 7, July 1981, pp. 438-443.

The sources and effects of acid rain are complex environmental phenomena. Although the major sources are thought to be atmospheric emissions, primarily from urban sources, the exact locations of sources and the mechanisms of transformation of emissions to acid rain are not entirely understood. In the absence of long-term records, it is difficult to analyze changes in rainfall acidity and whether this problem is intensifying or spreading. Long range transport of airborne emissions by weather systems has made acid rain a potential threat to aquatic and forest ecosystems in rural areas. Acid rain contains not only substances that donate hydrogen ions, such as nitric and sulfuric acid, but also dissolved ions of common and heavy metals and trace elements. Aquatic ecosystems are the most vulnerable to acid rain in the short run. Acidification of streams and lakes may lead to decreased fish populations and increased concentrations of toxic metals. Acid rain may accelerate the replacement and leaching of basic cations and speed the rate of acidification of soils, particularly shallow, coarse textured forest soils which are already lacking in basic cations. This gradual acidification of the soil could increase leaching of nutrient ions, slow microbiological processes, reduce the variety and populations of soil fauna, and increase the mobility of some toxic elements. Acid rain also appears to be depositing heavy metals on forest soils. Long range effects of acid rain may include reduction of forest productivity and alteration of species composition or diversity.

Interagency Task Force on Acid Precipitation, "Interagency Task Force on Acid Precipitation: First Annual Report to the President and the Congress of the United States", Jan. 1982a, U.S. Department of Energy, Washington, D.C.

The Acid Precipitation Act of 1980 (Title VII of P.L. 96-294) created the Interagency Task Force on Acid Precipitation (the Task Force) to plan and implement the National Acid Precipitation Assessment Program (NAPAP). The Act mandates that the Task Force issue an Annual Report to the President and the Congress on the status of the national program. This first Annual Report describes progress in establishing the National Program since passage of the Act. The report outlines the scope and policy focus of the research underway, Task Force accomplishments, and describes future activities.

Interagency Task Force on Acid Precipitation, "National Acid Precipitation Assessment Plan", June 1982b, U.S. Department of Energy, Washington, D.C.

The Plan has five parts. Part I describes the authorizing legislation and the organization of the Plan itself. Previous federal planning efforts in the current acid rain research programs are discussed. Part II presents a general overview of current understanding of the phenomenon and consequences of acid rain. Subjects discussed include: evidence of trends in rainfall acidity; possible sources of acid precipitation; atmospheric chemistry and transport; monitoring of acid deposition; the effects of acid precipitation; the relationship of acid precipitation to human health; and control technologies and mitigation of impacts. Part III identifies nine categories of information needed to enhance the ability to make sound energy, environmental, economic, and resource policy decisions as: (a) natural sources; (b) man-made sources; (c) atmospheric processes; (d) deposition monitoring; (e) aquatic impacts; (f) terrestrial impacts; (g) effects on materials and cultural resources; (h) control technologies; and (i) assessment and policy analysis. Possible indirect effects on human health are addressed in both categories (e) and (f). Part IV presents the research proposed to address these information needs. The management and coordination of the Program are described in Part V.

Interagency Task Force on Acid Precipitation, "National Acid Precipitation Assessment Program Annual Report to the President and Congress (1982)", June 1983, U.S. Department of Energy, Washington, D.C.

> This report summarizes the major research findings and accomplishments of the National Program's first year, with particular emphasis on their bearing on acid precipitation policy questions. It provides a brief overview of the available scientific information relevant to policy formulation. As the National Program progresses, future annual reports will provide updates and increasingly better information to assist decision-makers and the public in continuing to resolve the "acid rain" issue.

Johannes, A.H., Altwicker, E.R. and Clesceri, N.L., "Characterization of Acidic Precipitation in the Adirondack Region", Report No. EPRI EA 1826, May 1981, Electric Power Research Institute, Palo Alto, California.

> Atmospheric inputs into three lake watersheds within a 30 km radius of each other in the Adirondack Park region of New York State were quantified for the period May 1978-August 1979. This was accomplished with a wet/dry precipitation network which collected samples on an event basis. Rain and snow amounts were measured and samples were analyzed for pH, conductivity, SO_4, NO_3, Cl, NH_4, Ca, Mg, K, and Na.

Jones, K.H. and Mirra, R.R., "Estimating Future Trends in Acid Rainfall", Acid Rain: Proceedings of ASCE National Convention, American Society of Civil Engineers, 1979, pp. 120-131.

> This article discusses the cost-benefit of acid rain reduction in terms of energy and transportation needs and increases in the future. The authors approach the problem from the viewpoint of being unable to return to conditions of the 1950's, therefore, the need to decide what magnitude of impact of acid rain can be tolerated in the future.

Khan, S.M., "The Relationship Between Acid Content of Particulates and Rainfall in Bangkok", Journal of Environmental Science and Health, Vol. A15, No. 6, 1980, pp. 561-572.

> Rainfall at nine stations in Bangkok, Thailand, varied from pH 5.57 to 6.32, low values compared with cities in North America and Europe. Particulate acid content ranged from 5.38 to 10.15 micrograms per cu meter. Analysis for several ions showed that the concentration of sulfate was the controlling factor in acidity of rain. The pH was reduced by 1 unit for each 9.09 micrograms per cu meter acid content of particulates according to a relationship derived in the study; pH = 6.87-0.11 (acidity of particulates in micrograms per cu meter).

Kish, T., "Acid Precipitation: Crucial Questions Still Remain Unanswered", Journal of the Water Pollution Control Federation, Vol. 53, No. 5, May 1981, pp. 518-521.

> Information currently available concerning acid precipitation is reviewed, with specific attention given to various control measures under consideration. Precipitation found preserved in 350 yr old ice cores of glaciers and continental ice sheets has been shown to be generally above

pH 5. Ice originating as snow 150 years ago in Greenland has a pH ranging from 6 to 7.6. The earliest known pH record obtained in the U.S. was made in Brooklyn, Maine in 1939 and showed a pH value of 5.9. In the early 1950's rain was falling with pH values in the low 5's. The acidic portion in rainfall in the eastern U.S. has been analyzed to be about 62% sulfuric acid and 32% nitric acid, with 6% other acids including hydrochloric acid. Sulfur and nitrogen acids derive from the burning of coal and oil. While sulfate and nitrate emissions have increased during 1960-1970, there has been no apparent trend toward a corresponding increase in the acidity of precipitations during that time. No seasonal variations have been noted, even though sulfate and nitrate emissions are higher in the summer time. Acid rain samples have been collected in remote areas of the world. The relationship of acid precipitation to lake acidity is not well understood. The relative importance of other acid inputs is not well defined. The Canadian Ministry of Environment estimates that about 60,000 lakes are degraded or endangered due to acid precipitation. Norway estimated 20,000 of its lakes are endangered. Low pH levels can release heavy metals from soils into the lakes, further complicating the situation. Acid lakes experience changes in algal populations. While solutions for reducing acid rain problems are not clear, the desire to do so is widespread.

Knudson, D.A., "Acidic Deposition: Review of Current Knowledge", 1981, Argonne National Laboratory, Argonne, Illinois.

Events in Europe which led to the introduction of the concern for acidic deposition in eastern North America are reviewed. Some of the early conclusions regarding the existence of a trend and the extent of the phenomenon are reexamined. These reexaminations indicate that the definition of a precipitation acidity trend is questionable. Also, it was verified that any explanation of the variability of precipitation acidity must consider such factors as drought and farming and forest management activities. The chemical constituents contributing most to precipitation acidity are reviewed as an indication of primary sources. Sulfate is the primary contributor in the Northeast (60 to 70 percent). The contribution due to nitrates in this region increased from about 22 percent in the mid-1950s to about 30 percent in 1973. However, in other regions deposition acidity may be attributable to a much different combination of chemical species, and may even be dominated by an acidic constituent other than sulfate or nitrate. Several individual processes involved in the transport and transformation of precursor pollutants from source to receptor were briefly reviewed. These processes included mechanisms for the transformation of SO_2 and NO_x emissions into sulfates and nitrates, specific meteorological conditions conducive to long-range transport, major source identification, and the incorporation of this information into mathematical models to predict acidic deposition amounts and patterns. Although the information presented does not explore every facet of each process, it does address important governing features and allows for an appreciation of the complexity of quantifying the complete process through use of mathematical models.

Kramer, J.R., "Effect of Precipitation Chemistry on the Upper Great Lakes Region", Proceedings of the Second National Conference on Complete Water Reuse: Water's Interface with Energy, Air, and Soils, 1975, American Institute of Chemical Engineers, New York, New York, pp. 355-357.

Increases in emissions in industrial North America have led to large increases in loading on waters from the atmosphere. It has been shown that Cu, Ni, and Zn are loaded more heavily from the air than from all other sources in southern Lake Michigan, and estimates are that 10, 10, 25, and 70 percent of N, P, Pb, and Zn total loadings, respectively, are from the atmosphere in the Lake Ontario area. In addition, concern about acid precipitation and its effect upon land and water is increasing. Most studies suggest that large distance scales (1,000 km) are involved when source-receptor relationships from the atmosphere are considered. Commonly areal sources, rather than point sources, are important. The purpose of this paper is to outline key parameters and considerations of atmospheric loadings and to summarize estimates of total loadings obtained from a station network operated by McMaster University for the past five years.

Lakhani, K.H. and Miller, H.G., "Assessing the Contribution of Crown Leaching to the Element Content of Rainwater Beneath Trees", Effects of Acid Rain Precipitation on Terrestrial Ecosystems, 1980, Plenum Press, New York, New York, pp. 161-172.

This was a 2-year study of rain water data from Fetteresso Forest in Scotland involving sodium and potassium. The contribution of these elements from rain water and from the plant tissues themselves was studied. Equations are given and a mathematical approach was used. Crown leaching does contribute to the content of Na and K in rain water deposited on the forest floor.

Lee, Y.H. et al., "The Slope of Grant's Plot: A Use Function in the Examination of Precipitation, the Water-Soluble part of Airborne Particles and Lake Water", Water, Air, and Soil Pollution, Vol. 10, No. 4, 1978, pp. 457-469.

For some environmental samples, characteristics of acid precipitation (concentration of acids present and dissociation constants) can be determined from Grant's plot of an acid-base potentiometric titration. A sample is analyzed first for acidity. If the pH is higher than 4.5, a suitable known amount of perchloric acid is added. The sample is flushed with N_2 to remove carbon dioxide. Four milliliters of the sample are mixed with 1 ml of a 0.50 m potassium chloride solution which gives the mixture an ionic strength of 0.10 m. The titration is carried out at 25°C and pH less than with 6 with stirring, a N_2 atmosphere, and 0.01 m sodium hydroxide titer. After each titer addition, 2-3 min are waited for equilibrium adjustment; the hydrogen ion concentration is determined through measurement of EMF in the cell. The hydrogen ion concentration is plotted vs. the hydroxide concentration. The acid dissociation constant is determined from the slope of this plot. The application of this method to some rainwater samples and leaching solutions of airborne particles showed that no unknown weak acids which can affect pH in the range of 4-5 were present.

Leonard, R.L., Goldman, C.R. and Likens, G.E., "Some Measurements of the pH and Chemistry of Precipitation at Davis and Lake Tahoe, California", Water, Air, and Soil Pollution, Vol. 15, No. 2, 1981, pp. 153-167.

Chemical analyses were performed on precipitation collected at Davis, California, and Lake Tahoe, during September 1972 and water years 1973 and 1978. The water year is defined as October 1 through September 30. Prevailing winds are westerly; local emission sources at Tahoe are mostly downwind, but pollutants accumulate during inversions. At both sites the precipitation was found to be more acidic than was water in equilibrium with atmospheric carbon dioxide. Acidity at Lake Tahoe increased over the 5-yr period, as reflected by a decrease in volume-weighted average pH from 5.10 to 4.67. Sulfate was the dominant acid anion in 1972-73. Industry in the San Francisco Bay Area was probably the major sulfate and nitrate precursors to the atmosphere. Total deposition of nitrate, chloride, sulfate and calcium ions estimated in a previous study on Nevada snowpack samples was at least an order of magnitude less than the values obtained from fresh precipitation, indicating that the chemical composition of pack snow alters over time.

Li, T. and Landsbergn, H.E., "Rainwater pH Close to a Major Power Plant", Atmospheric Environment, Vol. 9, No. 1, Jan. 1975, pp. 81-89.

The distribution of pH of rain from heavy summer showers was measured in a network of rain gauges 0.5-5.0 km from a 710 MW power plant in southern Maryland. Values ranged from 3.0-5.7, with modal values from 3.6-4.0. A notable wind dependence of the acidic washout from the plume is discussed. In calm conditions during a shower, the values drop off concentrically from the source.

Likens, G.E., "Acid Precipitation", Chemical and Engineering News, Vol. 54, No. 48, Nov. 22, 1976, pp. 29-38.

The acidity of rain and snow falling on widespread areas of the world has been rising during the past two decades. While there is controversy over the causes of the phenomena and while the consequences have yet to be fully evaluated, acid precipitation clearly seems to have a far-reaching environmental impact. Acid rain has been linked to sharp declines-and-extinction-in the number of fish in many lakes and streams. Evidence also suggests that acid rain, snow, and aerosol particles may be damaging trees and other plants, that it enhances the weathering and corrosion of materials and buildings, and that it may present an added threat to human health. Current knowledge on the causes of acid precipitation and its effect on terrestrial ecosystems are summarized.

Likens, G.E. et al., "Acid Rain", Scientific American, Vol. 241, No. 4, Oct. 1979, pp. 43-52.

In recent decades the acidity of rain and snow has increased sharply over wide areas. The principal cause is the release of sulfur and nitrogen oxides by the burning of fossil fuels. A review of the general data concerning acid rain (with accompanying history of data gathering) for Europe and the U.S. was given. On an annual basis rain and snow are 5 to 30 times more acid than in unpolluted atmospheres.

Likens, G.E. and Butler, T.J., "Recent Acidification of Precipitation in North America", Atmospheric Environment, Vol. 15, No. 7, 1981, pp. 1103-1109.

Analysis of precipitation occurring prior to the industrial revolution, and preserved in glaciers or continental ice sheets, reveals a pH of greater than 5.0. Remote areas such as Greenland and Antarctica have a pH of about 5.5 for precipitation currently falling. However, rain and snow falling over certain regions of the world are currently 5 to 30-40 times more acid than the lowest value expected for unpolluted atmospheres. This phenomenon is observed to be widespread in nonurban as well as urban areas throughout most of eastern North America and western Europe. Annual pH values average about 4.0 to 4.5 in parts of the northeast U.S. and southeast Canada and northwest Europe. Individual storms have been recorded at pH 3.0. During 1959 through 1966 a sampling network was established throughout the conterminous United States. The vast majority of monthly pH readings from stations west of the Mississippi River were consistently above 5.6, with modal values in the high 6's. In contrast, monthly values east of the Mississippi were much more acidic. In the past two decades the southeastern United States has undergone the most rapid decrease in pH. This area has at the same time experienced an increase in atmospheric haze, presumably from increased urban and industrial activity, which would provide a regional source of anthropogenic emissions leading to the increased acidity. The longest continuous comprehensive record of precipitation chemistry in the U.S. was obtained for the White Mountains area of New Hampshire in the Hubbard Brook Ecosystem Study. The pH has varied between 4.03 and 4.21 since the start of record keeping in 1964. Atmospheric deposition of heavy metals and various organic pollutants also occurs, and apparently has increased concurrently with increased acidity of precipitation in eastern North America.

Likens, G.E., Edgerton, E.S. and Galloway, J.N., "The Composition and Deposition of Organic Carbon in Precipitation", Tellus, Vol. 35B, No. 1, 1983, pp. 16-24.

Wet-only precipitation collected at Hubbard Brook (HB), New Hampshire, a heavily forested site, and Ithaca, New York, from June 1976 to May 1977 was analyzed for several types of organic carbon. Annual volume weighted mean concentrations (mg/l) were: total organic carbon, 2.37 at Ithaca and 1.28 at HB; dissolved organic carbon, 1.88 at Ithaca and 1.09 at HB; and particulate organic carbon, 0.49 at Ithaca and 0.20 at Hubbard Brook. Deposition of total organic carbon was 24,230 g per ha at Ithaca and 14,031 g per ha at HB. Organic material concentrations and rainfall levels were highest in June-September. Particulate plus dissolved macromolecular (greater than 1000 molecular weight) organic matter accounted for 51 percent and 63 percent of the total organic carbon at HB and Ithaca, respectively. The remainder of the carbon, the fraction containing compounds of less than 1000 molecular weight, consisted largely of carboxylic acids, aldehydes, carbohydrates, tannin/lignin, primary amines, and phenols. Sources of the organic materials were airborne soil dust and plant material. The organic acids present in the rain at HB contributed only 2 percent of the total acidity.

Liljestrand, H.M. and Morgan, J.J., "Chemical Composition of Acid Precipitation in Pasadena, California", Environmental Science and Technology, Vol. 12, No. 12, Nov. 1978, pp. 1271-1274.

Wet-precipitation-only samplers were used to collect acid rainfall in Pasadena, California, from February 1976-September 1977. The concentrations of the cations (hydrogen+, ammonium+, potassium+, calcium2+, and magnesium2+) and the anions (chlorine-, nitrate-, and sulfate-) were measured. The chemical composition of Pasadena rainwater was compared with the composition of sea salt, soil dust, fuel oil, flyash, automobile aerosol, cement dust, and gaseous air pollutants. The results showed 35% of the total residue to be due to nitrate from nitrogen oxides, 20% due to sulfate from sulfur dioxide, 4.4% due to ammonium from ammonia, 17.2% due to soil dust, 13.6% due to sea salt aerosol, less than 7% due to fuel oil fly ash, 1.5% due to automobile aerosol, and less than 2% due to cement dust. The most important of these pollutants in Pasadena rainwater are HNO_3 and NH_4. HNO_3 is the dominant air pollutant acid in Pasadena, in contrast with the eastern U.S., where HNO_3, although increasing in importance, is still less significant than sulfuric acid.

Madsen, B.C., "Acid Rain at Kennedy Space Center, Florida: Recent Observations", Atmospheric Environment, Vol. 15, No. 5, May 1981, pp. 853-862.

Precipitation samples were collected in automated samplers designed to prevent both dry deposition and evaporation. Determinations of major cation and anion concentrations were made within 10 days after sample collection. All sampling sites were on or near the Kennedy Space Center complex in Florida. In general, summertime acidity was greater than during other times of the year. Maximum free acidity was observed during July in both 1978 and 1979. The July rainfall accounted for 21 and 10 percent of the total amounts received during the respective years. Lowest acidities were noted in November 1977 and April 1979. No pattern was noted when rain composition from a specific month during the first year of the study was compared to the identical time in the following year. The presence of sulfate ion can be attributed primarily to non-marine sources. Monthly and annual deposition of components which will be important if changes in rain composition occur due to future SRM launches are presented. Free acidity, nitrate ion and excess sulfate ion deposition account for the present acid loading, while chloride ion deposition reflects the influence of the marine environment. Monthly weighted average chloride ion concentrations ranged from 20 to 240 micromole/liter. The chloride:sodium ion ratio was slightly lower than that present in sea water.

Manzig, J.G.W. and Mulvaney, J.N., "Legal Aspects of Acidic Precipitation", International Journal of Environmental Studies, Vol. 18, No. 2, Feb. 1982, pp. 117-127.

Acid rain has been discovered to be a serious problem in Eastern Canada. The legislative competence of the Federal Parliament to deal with air pollution is not well established, but recent cases interpreting the federal general power indicate that the scope of general power has been established. Canadian common law remedies are not likely to be successful, for a variety of reasons. The Canadian Clean Air Act, especially with the 1980 amendments, lays the basis for possible adoption of emission standards for acid rain as an international pollutant. Ontario's Environmental Protection Act of 1971 may offer a basis for control of emissions resulting in acid rain through regulation of major sources and control orders of medium sized and small sources.

Martin, A., "Some Observations of Acidity and Sulfur in Rainwater from Rural Sites in Central England and Wales", Atmospheric Environment, Vol. 12, No. 6-7, 1978, pp. 1481-1487.

Weekly rainwater samples from 10 sites in rural eastern England were analyzed for acidity and ions in solution. Significant amounts of sea salts were found 80 km inland from the coast. There were nonsea deposits of chlorides and Mg in rain in the East Midlands, but not in Wales. Rainfall across the network was 450 $\pm$ 50 mm in 1976, and the most common hourly wind was from the west although the median value for the range of hourly wind direction during rain periods in 1 wk was 130°. Sodium ion rainwater values decreased rapidly away from the sea, as did Mg. Nonsea sulfate decreased rapidly away from combustion sources, but not to the same extent as acidity in rainwater. Nonsea chlorides varied from near zero in Wales and at one site on the coast to substantial values inland close to combustion sources. Sulfur dioxide air values also decreased away from sources and were similar to 1974 values. Results of other ions in rainwater--ammonium, nitrate, Ca, and K--showed no discernible spatial pattern. An inland dry deposition rate of sea sulfate was about 0.5 $mg/m^2/d$, reflecting an inland air concentration of about 1.2 ug/m^3. The measured air concentration of total SO_4 was around 8 g/m^3, so nonsea SO_4 concentration was about 6.8 g/m^3. The acidity in rainfall at a site in a shallow valley in West Wales was as much as that at sites at flat agricultural land 25 km downwind of major sources in the East Midlands. The valley rainfall acidity was accompanied by a smaller SO_2 concentration than in the East Midlands. Acidity episodes were due more to large rainfall amounts than to excessive acidities in the rain.

Martin, A., "Sulphate and Nitrate Related to Acidity in Rainwater", Water, Air, and Soil Pollution, Vol. 21, No. 1-4, Jan. 1984, pp. 271-277.

In recent years rainwater acidity and composition have been widely monitored in many networks and projects. Some authors have attempted to identify which substances and sources most influence rain acidity in practice. This paper shows that sulphate and nitrate in rainwater vary systematically with the acidities of the samples from certain rural sites and that there are major differences between strong- and weakly-acid rainwaters. Weakly acid samples can modify on standing and so special care is required in attempts to use them to identify sources. Some differences have been noted between samples from Europe and from the northeast United States.

McColl, J.G., "A Survey of Acid Precipitation in Northern California", Feb. 1980, Department of Soils and Plant Nutrition, University of California, Berkeley, California.

Wet and dry atmospheric precipitations were monitored on an event basis at the following locations during the wet season, November 1978 through May 1979; Berkeley and San Jose, Davis and Parlier, Hopland and Napa, and Tahoe City. Concentrations of 13 elements were determined. Acid rain (pH less than 5.6) was common at all eight sites. Mean pH varied from 4.42 at San Jose to 5.20 at Davis, and the lowest pH of any storm was 3.71 at San Jose. The primary cause of acidity was the NO_x air pollutants, and secondarily SO_x. Nitrate was the anion most closely correlated with H(+), and nitrogen generally occurred in greater amounts than sulfur. There were appreciable quantities of dry deposition. Although NO(-)3 concentration (microgram/1) and acidity (H(+) concentration, microgram/1) of wet precipitation were greatest in pollution source areas, total deposition (kg/ha) of NO(-)3 and H(+) were greatest in the nonurban receptor areas; this was largely a function of the greater precipitation volumes at these sites. Thus ecological effects may be expected within the general easterly "wash-out fan" of wet precipitation, as well as within pollution-source areas.

McColl, J.G., "Trace Elements in the Hydrologic Cycle of a Forest Ecosystem", Plant and Soil, Vol. 62, No. 3, 1981, pp. 337-349.

Levels of Cu, Fe, Mn, and Zn, sometimes considered toxins, in bulk atmospheric precipitation, throughfall, stemflow, and soil solutions at 10-, 15-, 25-, and 30-cm depths in a Eucalyptus globulus forest of California were measured after each main storm event during the 1974-75 wet season. Metal concentrations in litter and plant samples were also determined. The hydrologic cycles of Cu, Fe, and Zn were similar, but Mn behaved differently. Mn and Zn were found mostly on forest canopy due to dry deposition, while Cu and Fe input occurred mostly in rain. Cu, Zn and Fe levels in the soil solution fluctuated with downward movement of wetting fronts and were negatively correlated with pH. Fe levels in soil solution were approximately 10 times greater than in stemflow and throughfall. For Cu and Zn, the corresponding relative differences were much less. Mn was taken up by plants to a greater extent than Cu, Zn, or Fe. Since precipitation contained low levels of Mn, the increases in Mn levels from precipitation to throughfall and stemflow were greater than those for Fe, Cu or Zn. Mn levels in soil solution were also negatively correlated with pH. Although Mn accumulates in soil during the dry summer, it is quickly flushed out by the early rains of the wet season.

McColl, J.G. and Bush, D.S., "Precipitation and Throughfall Chemistry in the San Francisco Bay Area", Journal of Environmental Quality, Vol. 7, No. 3, July-Sept. 1978, pp. 352-358.

The chemical composition of bulk precipitation in the San Francisco Bay Area, California, is reported. Origins and changes in the primary inorganic constituents, and correlations with air pollution levels are discussed. During the wet season of 1974-75 the main ionic constituents were sulfate, chloride, bicarbonate, sodium, and calcium. The pH was 5.0. Although sulfate accounted for 50 percent of the anions in bulk precipitation, hydrogen ion concentration had the highest correlation with nitrate ion. Examples of the interaction of precipitation with forest vegetation and soil include: accumulation of impacted air pollutants on tree leaves between major rainstorms; correlation of nitrate and sulfate in bulk precipitation and leafwash with atmospheric nitrogen and sulfur; and resemblance of the ionic composition of bulk precipitation to that of surface-soil solutions in clearcut regions of the San Francisco Bay Area.

McNaughton, D.J., "Relationships Between Sulfate and Nitrate Ion Concentrations and Rainfall pH for Use in Modeling Applications", Atmospheric Environment, Vol. 15, No. 6, 1981, pp. 1075-1079.

Precipitation chemistry data for the northeastern United States were analyzed to test relationships between anthropogenic sulfate and nitrate ion contributions and rainfall pH. The data set incorporated all observations, with complete measurements of all three ion concentrations being collected from September 1976 through December 1979. A relationship was found between sulfates and nitrates and rainfall acidity expressed in terms of pH, and this relationships can be applied to deposition predictions to predict pH from regional air pollutant transport models. Variance in the pH versus ion relationships can be explained by the sulfate ion concentration alone. Thus, rainfall acidity patterns could be approximated by sulfate deposition patterns alone in cases where relative levels of nitrate ions and sulfate ions in the atmosphere are the same as those that influence precipitation. The analysis was based on the assumption that the predominant anions are anthropogenic, and this assumption could lead to an overprediction of rainfall acidity.

McQuaker, N.R., Kluckner, P.D. and Sandberg, D.K., "Chemical Analysis of Acid Precipitation: pH and Acidity Determinations", Environmental Science and Technology, Vol. 17, No. 7, July 1983, pp. 431-435.

The effects of the residual streaming potential and residual junction potential on the accuracy of pH and acidity measurements are discussed and quantified. Detailed analytical procedures minimizing these effects are described; a Gran's titration is used to assess total and strong acidity. The procedure for pH determination is shown to have a precision and accuracy of "plus or minus" 0.01 pH unit.

McWilliams, P. and Musante, L., "ARIS: Acid Rain Information System", Apr. 1982, NASA Industrial Applications Center, University of Pittsburgh, Pittsburgh, Pennsylvania.

ARIS is to provide the technical, government, and business communities with abstracted information from the world's significant technical and business literature. The subject areas covered by this acid rain data base includes: (1) the mechanism of the formation of acid rain; (2) its transport phenomena; (3) its effects on materials; (4) its effects on plants; (5) the health effects of acid rain; and (6) monitoring and analysis of acid rain. Data in ARIS comes from several government and commercial data base producers, and these include EDB DOE Energy Database, Environmental Science Index, Air Pollution Abstracts, National Technical Information Service (NTIS), and articles of regional interests from various newspapers. The types of publication source documents are: technical journals, conference proceedings, selected monographs, government reports, special studies, and newspapers. The file data is proposed to be updated quarterly and will cover selected references from 1970 with major focus on material after 1976.

Melo, O.T., "Ontario Hydro's Acid-Rain Monitoring Network", Research Review of Ontario Hydro, No. 2, May 1981, pp. 29-38.

A great deal of concern about the health effects of sulphate aerosols surfaced in the mid-70's. As one of the emitters of the sulphur dioxide implicated in the formation of sulphate, Ontario Hydro initiated a monitoring program in 1975 to establish facts about this, then, little known pollutant. The monitoring effort has since been expanded to include precipitation chemistry measurements, in response to new concerns about the ecological effects of acid rain. Monitoring results are presented and discussed in terms of influencing variables such as meteorology and sulphur dioxide source distribution, and compared with information from other networks operating in northeastern North America.

Miller, J.M., "The Collection and Chemical Analysis of Precipitation in North America", Proceedings of the Second National Conference on Complete Water Reuse: Water's Interface with Energy, Air, and Soils, 1975, American Institute of Chemical Engineers, New York, New York, pp. 331-337.

There is no question that precipitation scavenging is an important mechanism for cleansing the atmosphere of both natural and man-made materials. The complex meteorological and physical processes of rainout and washout by precipitation make direct, long-term measurement of these phenomena in the atmosphere very difficult.   To sidestep this problem, atmospheric chemists have recognized that one important way of measuring the deposition of material on the earth's surface is to collect and analyze precipitation.   Evaluating   the   chemical   nature   of precipitation not only gives us a way to investigate the atmospheric processes but also the actual chemical input into surface waters, the level of chemical constituents interacting with the biota, and other direct and indirect effects.   It is obvious that knowledge of precipitation chemistry is important to a multitude of disciplines investigating the environment.

Miller, J.M., "The Monitoring of Acid Rain", Acid Rain: Proceedings of the ASCE National Convention, 1979, American Society of Civil Engineers, New York, New York, pp. 111-119.

The monitoring of acid rain requires a long-term commitment to collect and analyze precipitation samples.   The present monitoring performed in the United States has been undertaken by a number of government and private organizations. The results have been to date very uneven. This paper reviews the methods and monitoring and puts forth a suggestion for a coordinated U.S. effort in this area.

Mohnen, V.A., "Acid Precipitation: Finding the Source", Mining Congress Journal, Vol. 68, No. 8, Aug. 1982, pp. 42-45.

This article compares the sampling and interpretation of pH determinations made on a weekly basis (such as the National Atmospheric Deposition Program-NADP) or on an event basis (such as the Multistate Atmospheric Power Production Pollution Study-MAP3S).   The discussion demonstrates the complexity of acid precipitation.

Morgan, J.J. and Liljestrand, H.M., "Measurement and Interpretation of Acid Rainfall in the Los Angeles Basin", Feb. 1980, California Institute of Technology, Pasadena, California.

The purpose of this work was to define the extent, degree and pertinent chemical characteristics of acid precipitation in the Los Angeles Basin of Southern California. Precipitation samplers were placed at nine locations: Pasadena, Azusa, Big Bear Lake, Central Los Angeles, Long Beach, Mt. Wilson, Riverside, Westwood and Wrightwood. A total of

533 individual samples were analyzed from the nine locations, and 38 different storms were sampled at one or more of the locations. Increments of precipitation collected during a storm were analyzed for pH, titration acidity, chloride, nitrate, nitrite, sulfate, fluoride, bromide, orthophosphate, total phosphate, bicarbonate, sodium, potassium, calcium, magnesium, ammonium, organic carbon and suspended solids. The mean acidity in the fall-spring 1978-79 period ranged from a high of 38.4 micro-equivalents/liter at Pasadena to a low of 2.45 micro-equivalents/liter at Big Bear Lake, with corresponding mean pH's of 4.41 at Pasadena and 5.42 at Big Bear Lake. At Pasadena, individual sample (0.25 inch increments of precipitation) acidities ranged from 1600 micro-equivalents/liter to -8.1 micro-equivalents/liter, and individual sample pH's ranged from 2.89 to 6.24. Incremental sampling during storms revealed significant changes in pH and chemical composition with time, with early stages of precipitation generally showing low pH and high nitrate and sulfate concentrations.

Moody, J., "Sharing Pollution: Quetico and BWCA", Not Man Apart-Foe, Vol. 8, No. 2, Mid-Jan.-Feb. 1978, pp. 12-14.

A power plant to be built near Quetico, Ontario, threatens to pollute the boundary waters canoe area of Minnesota with acid rain. Sulfur dioxide and other pollutants would be released from the 650 ft stack of the proposed Atikokan Generating Station. The boundary waters canoe area is only about 35 mi south of the power plant site. Environmental assessment documents focusing on this problem are accused of ignoring the fact that acid rain is especially toxic to regional flora. Other environmental concerns associated with the power plant project and recommendations for citizen actions are summarized.

Nagamoto, C.T. et al., "Acid Clouds and Precipitation in Eastern Colorado", Atmospheric Environment, Vol. 17, No. 6, 1983, pp. 1073-1082.

Rain and snow samples were collected at the eastern foothills of the Rocky Mountains and analyzed for chemical composition. Many precipitation samples had pH values considerably more acidic than the 5.6 value of pure water containing only an equilibrium amount of atmospheric CO_2. The averages of pH for rain and snow are 4.8 and 5.5, respectively. Approximately 80 percent of the snow samples show pH greater than 5.5. Clear and considerable dependencies of the acidity on seasonal synoptic scale weather patterns are demonstrated. The most acidic snow or rain in winter is prefrontal or frontal precipitation derived from deep low-pressure centers tracking across or just to the south of Colorado. The most acidic summer precipitation occurs with patterns similar to these but weaker, and with synoptic conditions conducive to isolated but relatively deep convection. Cloud water samples, collected by aircraft over eastern Colorado, also showed low pH values. The acidity of clouds was greatest near the city of Denver. Samples collected over east Denver showed pH = 3.3, indicating that the city is a major source of the acid clouds in the area. Organic anions appeared in the samples taken near the city, but not in remote samples.

National Research Council, "Atmosphere-Biosphere Interactions: Toward a Better Understanding of the Ecological Consequences of Fossil Fuel Combustion", 1981, Washington, D.C.

The NRC Committee focused its attention on the following pollutants: sulfur and nitrogen compounds, trace metals, and organic substances. The Committee noted that understanding of patterns of emission, transport, deposition, and biological effects of these pollutants is incomplete, and they provided a preliminary guideline for the sorts of integrated research needed. After discussing biogenic emissions of the substances of concern and the magnitude and form of anthropogenic emissions, the Committee described in detail atmospheric transport and deposition processes and biological accumulation. A guide was developed to predicting consequences of continued or accelerated pollution, and a case history was given of effects of acid rain. The conclusion was reached that increased scientific effort is needed in two critical areas; long-term monitoring and forecasting of future effects of these pollutants, and ecotoxicology. The Committee recommended that U.S. agencies provide funding for the needed scientific effort and in particular for the establishment of graduate training programs in ecotoxicology.

National Technical Information Service, "Acid Precipitation--June, 1976-December, 1981 (Citations from the Energy Data Base)", Dec. 1981, U.S. Department of Commerce, Springfield, Virginia.

Citations in this bibliography cover sources, causes, effects, and controls of acid rain, acid snow, and acidification of soils and fresh and salt water bodies. Coverage is mostly for the North American continent. This updated bibliography contains 380 citations, 242 of which are new entries to the previous edition.

National Technical Information Service, "Acid Precipitation--1970-December, 1982 (Citations from Pollution Abstracts)", Dec. 1982a, U.S. Department of Commerce, Springfield, Virginia.

This bibliography contains citations concerning the composition, causes, effects, sources, measurements, and controls of acid precipitation (wet and dry) and the acidification of land and water. Some attention is focused upon the geographical distribution of acid precipitation and acidification. This updated bibliography contains 289 citations, 104 of which are new entries to the previous edition.

National Technical Information Service, "Acid Precipitation--June, 1970-December, 1982 (Citations from the Engineering Index Data Base)", Dec. 1982b, U.S. Department of Commerce, Springfield, Virginia.

This bibliography contains citations concerning the causes, effects, sources, and controls of acid precipitation and acidification. Techniques and technology for measurement and analysis of acid precipitation are considered. This updated bibliography contains 156 citations, 11 of which are new entries to the previous edition.

National Technical Information Service, "Acid Precipitation--June, 1974-November, 1982 (Citations from the International Aerospace Abstracts Data Base)", Nov. 1982c, U.S. Department of Commerce, Springfield, Virginia.

This bibliography covers the measurement and analysis of the effects of acid precipitation and acidification. Causes, sources, and various controls are considered as well. This updated bibliography contains 115 citations, 20 of which are new entries to the previous edition.

National Technical Information Service, "Acid Precipitation--1977-1982 (Citations from the Selected Water Resources Abstracts Data Base)", Dec. 1982d, U.S. Department of Commerce, Springfield, Virginia.

This bibliography contains citations concerning the sources, ecological effects, and economic effects of acid precipitation. The emission of sulfur and nitrogen compounds, the role of calcium carbonate concentration on the acidification of lakes and streams, and its effects on aquatic life, are considered. The effects of acid precipitation on the soil and vegetation are also discussed. This updated bibliography contains 169 citations, 45 of which are new entries to the previous edition.

National Technical Information Service, "Acid Precipitation: Legal, Political, and Health Aspects--1976-November, 1982 (Citations from the Energy Data Base)", Nov. 1982e, U.S. Department of Commerce, Springfield, Virginia.

This bibliography contains citations concerning the political and legal aspects of acid rain pollution. Health aspects of acid rain are also addressed. Federal (U.S.) legislation, including the Clean Air Act are considered. Transfrontier aspects of acid rain pollution are excluded. This contains 110 citations fully indexed.

National Technical Information Service, "Precipitation Washout--1964-November, 1982 (A Bibliography with Abstracts)", Jan. 1983, U.S. Department of Commerce, Springfield, Virginia.

The reports cited in the bibliography cover the removal of pollutants, radioactive isotopes, and dust by rain and snow. The theory, modeling, sampling and effects of precipitation washout are presented. This updated bibliography contains 213 citations, 19 of which are new entries to the previous edition.

Newman, L., Likens, G.E. and Bormann, E.H., "Acidity in Rainwater: Has An Explanation Been Presented?", Science, Vol. 188, No. 4191, May 30, 1975, pp. 957-959.

The theory that the installation of particle-removing devices in tall smokestacks eliminates alkaline substances and permits appreciable quantities of sulfur dioxide to be converted to acid is debated. Whether or not all smokestacks and particle-removing devices alter the regional acid problem in the northeastern U.S. is an unanswered question. Included is an argument that states that the pattern and intensity of acid deposition are recent and increasing, that the root cause is most likely the combustion of fossil fuels, and that the problem may be further exacerbated by a careless rush to solve the energy crisis.

Oden, S. and Thorsten, A., "The Sulfur Budget of Sweden", Effects of Acid Precipitation on Terrestrial Ecosystems, Plenum Press, New York, New York, 1980, pp. 111-122.

Based on the discharge of sulfur by Swedish rivers (34 drainage basins) covering 78 percent of Sweden, the total atmospheric fallout of sulfur has been computed for the year 1974. The figure amounts to about 590,000 tons, which is higher than the corresponding figure based on atmospheric chemical data. The increase of the discharge of sulfur caused by increased fallout amounts to 2-5 percent, or an increase of about 20,000 tons for the whole of Sweden per year. The fallout of sulfur from different parts of Sweden during this century has been reconstructed. The total figure amounts to about 60 million tons of sulfur.

Okland, J., "Acidification in 50 Norwegian Lakes", Nordic Hydrology, Vol. 11, No. 4, 1980, pp. 25-32.

Since the 1950's significant acidification has taken place in southeastern Norway lakes. There were two observation periods, 1953-57 and 1963-70. For the 50 lakes studied, the average pH dropped from 7.0 to 6.8, a statistically significant change; total hardness showed no significant change. In the low total hardness group the average pH changed from 6.6 to 6.3 and average H+ concentration increased by 4.33×10^{-7} mol per liter, equivalent to a 233 percent change. For the high total hardness lakes there was no statistically significant change in acidity. Lakes with low total hardness appear vulnerable to acidification. Such lakes are located all over Norway and predominate in some areas.

O'Sullivan, D.A., "Norway: Victim of Other Nation's Pollution", Chemical and Engineering News, Vol. 54, No. 25, June 14, 1976, pp. 15-17.

Norway is directly in the path of southern winds that carry various pollutants from industrial centers in the U.K. and in central and eastern Europe. These airborne pollutants range from particles of tar, flyash, and such chemicals as PCB's to oxides of sulfur and of nitrogen. Over Norway's high mountain ranges, the oxides fall to the earth as acid rain and snow. A comprehensive national research project on the effects of acid precipitation on forest and fish is described. OECD's similar study on the movement of airborne pollutants is compared with the Norwegian project.

Patrick, R., Binetti, V.P. and Halterman, S.G., "Acid Lakes from Natural and Anthropogenic Causes", Science, Vol. 211, No. 4481, Jan. 1981, pp. 446-448.

Although the total amount of coal used in the U.S. has not significantly increased since 1920, changes in use patterns and pollution technology have altered the nature and quantity of stack emissions. This situation has produced acid rains and increasing acidity in many lakes in the northeastern U.S. Factors contributing to more acid emissions from fossil fuel burning are: (1) increase in the proportion of coal consumed by utilities; (2) gradual increase in use of fly ash precipitators, which remove the neutralizing effect of ash upon sulfur and nitrogen oxides; (3) the fact that sulfur scrubbers are only required on coal-burning plants built since 1975; (4) increase in the height of stacks, which introduces emissions high into the atmosphere where they can be readily transported for long distances; (5) use of hotter combustion flames and resultant increase in nitrogen oxides emitted since 1975; and (6) increase in use of petroleum in utilities and transportation where emissions are not controlled. Aquatic life in naturally acid lakes does not appear to suffer greatly from increases in acidity because the high humate content tends to chelate toxic metals released by the acid. However, lakes without this protective effect show more serious ecological disturbances.

Pellett, G.L., Bustin, R. and Harriss, R.C., "Sequential Sampling and Variability of Acid Precipitation in Hampton, Virginia", Water, Air, and Soil Pollution, Vol. 21, No. 1-4, Jan. 1984, pp. 33-49.

Rain samples were collected sequentially by amount (approximately 2.7 mm each) from individual events at a single, relatively isolated, suburban site from August 1977 to July 1980. Rain pH's for less than or equal to 3 mm samples closely fit a monomodal Gaussian distribution with a median of 4.50 and a standard deviation of 0.39. The 3-yr volume-weighted pH was 4.35 ± 0.02 for 3.16 m collected; annual pH's were 4.31, 4.37, and 4.38, and cumulative H+ deposition was 141 mg H+ m^{-2}. Event-averaged rain pH and meteorological and air quality data were correlated. Low pH was associated with low rainfall volume and rate; rain after several dry days; rains with northeast surface winds; high SO_2, NO_2, and O_3 in the ambient atmosphere; and high, strongly correlated, SO_4^{-2} and NO_3^- rainwater concentrations. The lowest 3-yr seasonal average pH (4.31) occurred during the summer; values for other seasons were approximately 4.37.

Perhac, R.M., "Acid Rain--An Overview", Mining Congress Journal, Vol. 67, No. 8, Aug. 1981, pp. 19-21.

The author discusses two claims: that acid rain, at least in the eastern United States, is increasing, both in acidity and in area being affected; and that coal burning, particularly by utility companies, is the major cause of increasing acidic precipitation.

Popp, C.J. et al., "Precipitation Analysis in New Mexico", 1980, New Mexico Technical University, Socorro, New Mexico.

A report is presented on the chemistry of precipitation collected at two sites 25 km apart in the high desert in south-central New Mexico. The mountain site at 3200 m and the valley site at 1400 m represent a large climate change within a very short distance in an area of low population and industrialization. Precipitation ranges from 50 cm annually at the mountain site to 20 cm in the valley. It is found that the standard deviations are quite large as would be expected but the average pH values of 5.22 (mountain) and 5.70 (valley) indicate that acid rain may be a problem even in areas with highly alkaline terrestrial environments and low population densities. The overall pH average at the low altitude site was 5.70 with a 5.89 average in the summer and a 5.38 average in the winter. The average pH at the mountain site was 5.22 for summer samples only. The precipitation events need to be compared on a seasonal basis because of the diverse weather patterns from summer to winter.

Poundstone, W.N., "Let's Get the Facts on Acid Rain", Mining Congress Journal, Vol. 66, No. 7, July 1980, pp. 45-47.

The author examines contentions that rain is highly acid; the acidity of rain in the eastern U.S. is increasing rapidly; it is damaging the environment; the burning of coal is a major cause; and acidity of rainfall in a given area is greatly influenced by emissions of pollutants upwind.

Pratt, G.C., Coscio, M.R. and Krupa, S.V., "Regional Rainfall Chemistry in Minnesota and West Central Wisconsin", Atmospheric Environment, Vol. 18, No. 1, 1984, pp. 173-182.

Rain samples were collected at seven sites in Minnesota and west central Wisconsin during the summer of 1981. The sites were dispersed in the study area in order to sample within the three major types of ecosystems found in the region. At the sites in northeast Minnesota (an area considered sensitive to acidic precipitation) concentrations of all major ions except hydrogen were generally lower than at the other sites. The higher rainwater acidity at the northern sites appeared to be a function of the lower concentrations of alkaline cations such as Ca^{2+} and ammonium, however, overall the best single predictor of acidity was sulfate. Analysis of the relationships between pH and concentrations of the major ions indicated that "clean" rain may be slightly more acidic than pH 5.60. Air parcel back trajectories showed that rain in air masses arriving from the NW quadrant contained the lowest concentrations of sulfate and nitrate, while the majority of rainfall sulfate and nitrate arrived from the S quadrant.

Pratt, M., "Acid Rain: Filtering Out the Facts", Agricultural Engineering, Vol. 64, No. 5, May 1983, pp. 7-9.

One problem confronting the researcher examining the problem of acid rain is the lack of essential records much before the mid-sixties. Descriptions of 19th and early 20th century industrial operations both in the United States and Europe suggest a higher level of polluted air and water than would be tolerated now. A pH of 5.6 is considered normal. Rain and snow in particularly vulnerable areas may consistently measure less than 4.5 pH. The disposition of atmospheric contaminants occurs from two related atmospheric processes often identified as rainout and washout. A plan has been proposed where Canada and the U.S. will have a permanent acid deposition measurement system with specific monitors strategically placed. With the usual systems for monitoring rain, only cations and anions can be measured, but not trace elements. The application of limestone to acidified lakes and streams required considerable knowledge both of input sources to the lake and of water turnover times. Liming tends to increase aluminum and cadmium accumulations in waters with 4-6 pH levels, thus intensifying the likelihood of fish kills. While electric utilities often are cited as the major contributor to the problem, seldom does one hear of industries contributions toward research funding to solve the problem. Without accurate long-term monitoring data, researchers have had to calculate acidity trends in presumably vulnerable areas. At times they may have had to assume that forest, crop and fish kills come from acid rain as they had no way to assess other factors such as whether some soils and waters in specific areas have ever consistently supported certain crops and certain types of aquatic life.

Rippon, J.E., "Studies of Acid Rain on Soils and Catchments", Effects of Acid Precipitation on Terrestrial Ecosystems, Plenum Press, New York, New York, 1980, pp. 499-524.

This is an excellent article describing: (1) field experiments at Snake Pass, Derbyshire, and on the Tillingbourne, Surrey, U.K.; and (2) experimental work involving lysimeter studies of leaching processes and pot-scale experiments of soils. The data presented is qualified as preliminary data. The field studies showed decreases in acidity and increases in sulphate concentrations downstream from the contamination source. Increases in rainfall appear to directly affect soil sulphur retention and hydrogen ion interactions.

Robertson, J.K., Dolzine, T.W. and Graham, R.C., "Chemistry of Precipitation from Sequentially Sampled Storms", EPA-600/4-80-004, Jan. 1980, U.S. Environmental Protection Agency, Research Triangle Park, North Carolina.

Sequential sampling techniques and applications to collect precipitation are reviewed. Chemical data for samples collected by an intensity-weighted sequential sampling device in operation at the U.S. Military Academy, West Point, New York from October 1976 to April 1978 are presented and discussed. The problem of dry deposition is explored. A newly designed intensity-weighted sequential sampler that excludes dry deposition is presented. The experiments have shown that intensity-weighted sequential sampling is a viable technique for monitoring the rapid changes in precipitation chemistry within a storm. Complete chemical data are needed from individual storms to evaluate intensity-related scavenging.

Robinson, J.W. and Ghane, H., "Continued Studies of Acid Rain and Its Effects on the Baton Rouge Area", Journal of Environmental Science and Health, Vol. A18, No. 2, 1983, pp. 165-174.

The acidity of rain water was measured in the Baton Rouge, Louisiana area from June 1981 to September 1982. Coordinated measurements were taken of the dissolved oxygen concentration in two local lakes before and after each rainfall. About 50 percent of the rainfall observed was quite acidic, with about 25 percent of the rain having a pH of 4 or less. Rain was more acidic during warm summers than in the winter weather. Rainfall during 1982 was, on a month to month comparison, more acidic than in 1981. Attempts were also made to discover any possible correlation of pH values with wind direction. The acidity of each of the two lakes increased over the time of the study. The dissolved oxygen content in each lake increased after periods of rain, probably due to a high concentration of oxygen in the rainwater. The buffering capacities of the lakes was measurable. However, it is noted that the larger lake was undergoing dredging at the time of the study and showed considerably less buffer capacity than the smaller lake. The smaller lake was far more affected by surface drainage and thus should have been more influenced by the acid rain.

Rodhe, H., Soderlund, R. and Ekstedt, J., "Deposition of Airborne Pollutants on the Baltic", Ambio, Vol. 9, No. 3-4, 1980, pp. 168-173.

The pollution of the Baltic Sea by airborne pollutants is considered with emphasis on the deposition process. Downward transport across the air-sea interface may be divided into wet deposition (fallout in precipitation) and dry deposition (gravitational setting, transport by turbulent eddies, molecular diffusion, and impaction by inertia). Deposition rates for a number of pollutants are given, suggesting that atmospheric deposition is a major source of nitrogen compounds (120,000-1.3 million tons/yr) and lead (600-5,000 tons/yr) in the contaminated Baltic Sea area. A meteorological study of the area indicates that the turnover time for water in the Baltic Sea is about 15 yr. Surface waters have a normally shorter residence time. Concentration data is provided for inorganic nonmetals, metals, radionuclides, and organic compounds in the Baltic Sea. The deposition of DDT and polychlorinated biphenyls in various areas of Sweden varies from 960 to 2,120 and 1,400 to 1.780 ng/sq m/month, respectively. Estimates of the annual flux of a number of pollutants from the atmosphere to the Baltic Sea have been calculated from the methods and data presented.

Roffman, A., "Acid Precipitation in the Pittsburgh, Pennsylvania Area", Journal of Environmental Sciences, Vol. 23, No. 2, Mar. 1980, pp. 33-37.

A survey of the pH levels of precipitation in the Pittsburgh area was conducted. Supplementary climatological data was collected. Results indicate that the pH values of samples in pollution-free areas were lower than those obtained at polluted stations. The industrial mills along the principal rivers seem to have had little or no effect on the pH values measured. The coal-burning power plants appear to have influenced the pH of precipitation. The data imply that pollution-carrying winds from the Ohio River Valley Basin contribute acidity to the stations.

Rosencranz, A., "The Problem of Transboundary Pollution", Environment, Vol. 22, No. 5, June 1980, pp. 15-20.

National control strategies, international responses, the EEC convention, limits of the law, bilateral responses, the determination of acceptable levels, and prognosis and predictions are discussed.

Rosencranz, A. and Wetstone, G., "Acid Precipitation: National and International Responses", Environment, Vol. 22, No. 5, June 1980, pp. 6-8.

The international effects of acid rain are discussed.

Rosenfeld, A., "Forecast: Poisonous Rain", Saturday Review, Vol. 5, No. 23, Sept. 2, 1978, pp. 16-19.

Pollutants, principally sulfur and nitrogen oxides, have overloaded the buffering capacity of rain, worsening the problem of poisonous, acidic rain. The antipollution measure of raising smokestacks higher has contributed to cloud pollution. Dangerously acidic rains have been found nationwide and worldwide, several U.S. authorities, including USDA, TVA, EPA, and USGS, have combined to organize a rain-monitoring network. Scandinavian countries have called upon the U.N. to establish a global network.   Acid rain poisons rivers and lakes, killing fish; accelerates erosion of metal, paint, and stone; and progressively poisons members of dependent food chains. The causes and effects of acidic rain are explored.

Rosenquist, I.T., "Alternative Sources for Acidification of River Water in Norway", Science of the Total Environment, Vol. 10, No. 1, July 1978, pp. 38-49.

The geochemistry of the acidification of rivers, including natural production of acids in the catchment areas of southern Norway, is described. These processes are compared with the airborne acids produced from human activities, and with the rate of neutralization caused by weathering of silicate minerals and by the buffering capacity and buffer level of the organic humus layer. The buffering capacity and level has changed during the last century because of such activities as forestry, agriculture, and animal husbandry. The amount of biomass and subfossil humus has also increased because of changes in these activities. Acid runoff in waterways is primarily due to ion exchange processes in the catchment areas, and only slightly due to the hydrogen ion content of precipitation.

Seip, H.M., "Acid Snow-Snowpack Chemistry and Snowmelt", Effects of Acid Precipitation on Terrestrial Ecosystems, Plenum Press, New York, New York, 1980, pp. 77-94.

This study was conducted in Norway and deals with aspects of snowmelting which may be important for acidification of soil and natural waters. The fractionation of chemical impurities during snowmelt, concentrations of impurities in snow samples, the degree and effects of contact between meltwater and the ground, and the variation in composition of river water during snowmelt are discussed quantitatively and qualitatively.

Seymour, M.D. et al., "Variations in the Acid Content of Rain Water in the Course of a Single Precipitation", Water, Air, and Soil Pollution, Vol. 10, No. 2, Aug. 1978, pp. 147-162.

The microtitrimetric method and the coulometric method are used to determine acid components of rain water collected during five different events during July and August of 1977 in Tucson, Arizona. Variations in the concentrations of a strong acid, a weak acid, and dissolved carbon dioxide are determined as a function of time. The strong acid concentrations attain maximum values in the course of a single precipitation event.

Seymour, M.D. and Stout, T., "Observations on the Chemical Composition of Rain Using Short Sampling Times During a Single Event", Atmospheric Environment, Vol. 17, No. 8, 1983, pp. 1483-1487.

Studies of changes in the chemical composition of rainfall during a single rainfall event were initiated in the summer of 1979. During that summer, 20 different events were collected and the pH of each sequentially collected sample was measured. The pH of the individual samples varied between 5.8 and 3.6. The most rapid decrease in pH occurred during the early portion of the rainfall event, while only small increases or decreases in the pH occurred during the latter portion of the event. In 1980, sequential samples of rain, each corresponding to 0.23 mm of rainfall, were collected during single precipitation events at a site in western Michigan and analyzed for specific conductance, pH, $SO_4(2-)$, $NO_3(-)$, $Na(+)$, $K(+)$, $Ca(2+)$ and $Mg(2+)$. A rapid decrease in the concentration of all the measured components except $H(+)$ was observed during the initial portion of the event. The $H(+)$ concentration of the rainwater generally increased during the initial portion of the event. This converse relationship between the H(+) concentration and the SO4(2-) and NO(3-) concentrations is attributed to the rapid washout of atmospheric particulates containing basic salts. Depending upon the type and duration of the event, this washout can cause 50 percent of the total SO4(2-) and NO3(-) deposition to occur during the first 15-20 percent of the total rainfall. While sampling of this type may not be possible for all studies of precipitation chemistry, further studies using this method may be useful in describing the effect that atmospheric particulates and rainfall amount have on the relationship between the concentrations of SO4(2-) and NO3(-) in rain and the net deposition of H(+) by a single precipitation event.

Sharp, P.G., "Acid Rain", Journal of the Air Pollution Control Association, Vol. 30, No. 6, June 1980, p. 968.

This is a brief report on the acid rain conference held in Sandefjord, Norway, March 1980. Acid rain effects and trends in Europe and North America are summarized.

Shinn, J.H. and Lynn, S., "Do Man-made Sources Affect the Sulfur Cycle of Northeastern States", Environmental Science and Technology, Vol. 13, No. 9, Sept. 1979, pp. 1062-1067.

This is an excellent article reviewing the natural and anthropogenic input sources of sulfur. Although the natural input of sulfur exceeds that of anthropogenic sources, the environment cannot adapt nor absorb sudden, high-rate anthropogenic loadings of sulfur compounds.

Spaite, P., "Acid Rain: The Impact of Local Sources", Nov. 1980, PEDCo-Environmental, Inc., Cincinnati, Ohio.

It has been assumed that acid rain is predominantly a problem of long-range transport of pollutants from large fossil fuel combustion sources, namely coal-fired utilities. However, close examination of fuel use information and source emission characteristics in the Adirondacks, Florida, and California suggests that local oil burning and automotive sources may be major contributors to the occurrence of acid rain in these areas. This report describes the possible role of local combustion sources in the production of acid rain, and discusses the implications of the findings, and their relevance to alternative control strategies for acid rain. Oil-fired boilers, especially the smaller commercial, industrial, and residential units, produce at least 3 to 10 times as much primary sulfate per unit of sulfur content as coal-fired units. Moreover, oil-fired units emit comparatively large quantities of catalytic compounds capable of rapidly converting still more sulfur oxide to sulfate in the atmosphere. Thus, in areas where large quantities of oil are burned, the direct impact from locally generated sulfates may equal or even exceed that produced by imported sulfates derived from distant coal-burning sources. Fuel consumption data show that large quantities of oil are being consumed in areas experiencing acid rain. Forty percent of the residual and 36 percent of the distillate oil burned in the United States is consumed in the eight-state area surrounding the Adirondacks. California is the next largest oil-consuming area and Florida is third. Nitric acid is responsible for about 30 percent of rainfall acidity in the Northeast and Florida, and for about 30 to 75 percent of the rainfall acidity in California.

Suleymanov, D.M. and Listengarten, V.A., "Chemical Composition of Precipitation on the Apsheronsk Peninsula", <u>Dokl. Akad. Nauk. Azerb. SSR</u>, Vol. 22, No. 12, 1966, pp. 42-44.

Analysis is based on 55 snow and rain samples taken at Baku and to the north of Novkhana. The overall mineralization of precipitation was found to be about the same in these two regions. In most cases, the occurrence of anions was in the descending order: HCO_3, SO_4, and Cl. When the total mineral residue was less than 7 mg/l, the cation content in descending order was Ca, Na plus K, and Mg; with a residue greater than 7 mg/l the order was Na plus K, Ca, and Mg. Frequently the Mg ion was lacking. Average annual values in mg/l are: Cl, 7.1; SO_4, 17.6; NO_3, 0.6; HCO_3, 63.1; Na plus K, 20.7; Mg, 1.5; and Ca, 11.9; total dry residue is 102.1. This source of salts is seen as an important factor in connection with land reclamation and the formation of ground water.

Szabo, M.F., Esposito, M.P. and Spaite, P.W., "Acid Rain: Commentary on Controversial Issues and Observations on the Role of Fuel Burning", DOE/MC/19170-1168, Mar. 1982, U.S. Department of Energy, Washington, D.C.

Even though much information has been accumulated on the subject of acid precipitation, lack of knowledge in certain technical areas precludes an adequate understanding of (1) how serious the acid precipitation problem really is and (2) what effect controlling sources of acid precipitation precursors would have in reducing acidification. It is nevertheless possible to draw some broad conclusions regarding the problem and to ascertain the direction that the required further work should take. This report presents the results of an investigation of various issues associated with acid rain. The following topics are addressed; occurrence of acid precipitation; effects of acid precipitation; sources of acid precipitation; transport, transformation, and deposition of acid pollutants; and fuel trend analysis. Recommendations for further research are included.

Tanaka, S., Darzi, M. and Winchester, J.W., "Sulfur and Associated Elements and Acidity in Continental and Marine Rain from North Florida", <u>Journal of Geophysical Research</u>, Vol. 85, No. C8, Aug. 1980, pp. 4519-4526.

The composition characteristics of north Florida rain were investigated to establish some nonurban baselines of elemental concentrations and to estimate the extent of transport of air pollutants. A new measuring technique was used, the proton-induced X-ray emission (Pixe), which could analyze a wide range of elements from S to Pb simultaneously. Elemental composition was determined for: S, K, Ca, V, Mn, Fe, Ni, Cu, Zn, Br, Sr, and Pb, from 4 ml aliquots of 38 different rain samples. Northerly air flow rain exhibited polluted and continental characteristics. Concentrations of Pb, S, and Fe were higher, and strong correlations of Br, Ca, and Fe with Pb were found. In general, southerly air flow rain had lower elemental concentrations, and element concentrations were not correlated with Pb. The average pH of northern rain was 4.4, while southern rain had an average pH of 5.3. The region of influence for atmospheric pollution from industrial regions seemed larger than is usually thought.

Tennessee Valley Authority, "The Acid Connection", Impact, Vol. 3, No. 4, July 1980, pp. 2-10.

The problems of defining and understanding the physical, chemical and biological processes involved in acid rain are discussed. The acidity of rain is measured by the pH scale; "pure" rain would have a pH of about 5.6 and the average pH measured for rains in parts of the northeastern U.S. range from 4 to 4.2, with the most extreme single value of pH 2.3. No complete, detailed historical measurement of rainfall acidity and its impacts are available in the country so a national effort is underway to coordinate all related research efforts. Air pollution is known to alter the acidity of precipitation by increasing sulfur oxide and nitrogen oxide concentrations and this in effect will increase with increasing use of coal as fuel. Fish and aquatic life in rivers and lakes with no natural "buffering" system against acid rain may experience a decline; the rivers and lakes may become completely dead in the future. The TVA is presently engaged in active research in rainfall monitoring, atmospheric movement of pollutants, and laboratory and field studies of vegetation, soils, and surface waters to define present conditions and indicate possible future trends. The long-range transport, chemical transformation, and ultimate fate of air pollutants and the effect of acid rain on soil chemistry is also being investigated. As of now, there is no clear evidence of how much acidity the environment can tolerate and how much needs to be controlled.

Trumbule, R.E. and Tedeschi, M., "Acid Rain Information: Knee Deep and Rising", Science Technology Librarian, Vol. 4, No. 2, Winter 1983, pp. 27-41.

This article briefly describes the dramatic growth in acid rain literature which has occurred in the past five years and examines various sources of that information. Included among these are: bibliographies, databases, journals, newsletters, books and monographs, congressional and Federal agency documents, industry-sponsored research, state activities, and internal sources. The article implicitly suggests that, while the sources one would use depend to a great extent on the purpose for which the information is sought, a careful selection of one bibliography, a few journals and newsletters, and a careful perusal of both congressional and executive agency documents should provide a good overview to where the best sources of knowledge on specific facets of the acid rain issue can be found.

Turk, J.T., "Evaluation of Trends in the Acidity of Precipitation and the Related Acidification of Surface Water in North America", Geological Survey Water Supply Paper No. 2249, 1983, U.S. Geological Survey, Washington, D.C.

Acidification of lakes and streams in the Northeastern United States has occurred in a time frame compatible with the hypothesis that acidification of precipitation was the cause. The acidification of surface waters appears to have occurred before the mid- to late 1960s. In Southeastern Canada, the best-documented cases of acidified lakes point to localized sources of acidic emissions as the cause. In the Southeastern United States, most data on acidification of surface waters are ambiguous, and in the West, most of the data reflect local conditions. However, recent analysis of a national network of remote stream sampling stations indicates that, since the mid- to late 1960s, sulfate concentrations have increased in the Southeast and the West, with a concurrent decrease in alkalinity.

Tyree, Jr., S.Y., "Rainwater Acidity Measurement Problems", Atmospheric Environment, Vol. 15, No. 1, Jan. 1981, pp. 57-60.

Three methods which are currently used to determine the acidity of rainwater have been studied. The first method relies on pH measurement, the second on titration, and the third assumes that any charge discrepancy in favor of anions is due to hydronium ion concentration. In this third method only strong acid is measured, and the method is used only for those solutions at or below a pH value of 5.6. The method requires the determination of concentrations of all principal ions in the rainwater sample, usually considered to be chloride, nitrate, sulfate, ammonium, sodium, potassium, calcium, and magnesium. The accuracy and reproducibility of such a method are no better than the properly combined accuracy and reproducibility of each of the principal solute ion values. The first method, the pH measurement, measures the concentration of strong acids plus some of the weak acids in rainwater. However, the accuracy and reproducibility are not sufficient. Similarly, the accuracy and reproducibility of the titration method are also unacceptable. The results of two collaborative intercomparison laboratory studies on simulated rainwater show that adequate analytical reproducibility is not being attained at present.

Ulrich, B., "Production and Consumption of Hydrogen Ions in the Ecosphere", Effects of Acid Rain Precipitation on Terrestrial Ecosystems, Plenum Press, New York, New York, 1980, pp. 255-282.

This paper presents a general scheme for estimating anthropogenic and natural sources of H ions in terrestrial ecosystems. An excellent and extensive series of chemical equations is presented with tables included. Long-term general observations are advocated for these types of studies.

U.S. Environmental Protection Agency, "Environmental Effects of Increased Coal Utilization: Ecological Effects of Gaseous Emissions from Coal Combustion", EPA 600/7-78-108, June 1978, Washington, D.C.

Ecological and environmental effects of gaseous emissions and aerosols that result from coal combustion are evaluated. Sulfur oxide emissions and nitrogen oxide emissions are projected to be higher in 1985 and 2000 than they were in 1975. Since SO_x and NO_x are major contributors to acid precipitation, substantial increases in total acid deposition can be expected in the nation as a whole. At present, acid precipitation is most abundant in the north central and northeastern states. Estimates of the nonhealth-related cost of air pollutants range from several hundred million dollars to $1.7 billion per year. These estimates include only easily measured considerations, such as known losses to cultivated crops from acute air pollution episodes. Economic losses from factors that are difficult to measure, such as long-term chronic low level exposure to crops, forests, and natural ecosystems to air pollutants, remain unmeasured nationwide.

U.S. Environmental Protection Agency, Proceedings of the Acid Rain Conference: Springfield, Virginia, April 8-9, 1980, Aug. 1980a, Research Triangle Park, North Carolina.

A transcript of the presentations heard at the conference, this document addresses the causes and effects of acid rain and proposed solutions to dealing with the problem. Many charts, diagrams and some photographs are included.

U.S. Environmental Protection Agency, "Acid Rain", July 1980b, Office of Research and Development, Washington, D.C.

Acid precipitation has become one of the major environmental problems of this decade. It is a challenge to scientists throughout the world. Researchers from such diverse disciplines as plant pathology, soil science, bacteriology, meteorology and engineering are investigating different aspects of acid precipitation. Despite some unknowns, the preponderance of the evidence to date points to the need to control the emissions of acid-forming sulfur and nitrogen compounds. Given the potential for damage from acid precipitation, prudence dictates that actions soon occur to avoid far more serious problems in the future. EPA's regulatory standards for new power plants and motor vehicles are a major step in the right direction. Further steps may be required in the near future. This background document tells what is known--and not known--about acid precipitation.

Vermeulen, A.J., "Acid Precipitation in the Netherlands", Environmental Science and Technology, Vol. 12, No. 9, Sept. 1978, pp. 1016-1021.

In 1966 the highest acid precipitation measurements in the world (on a yearly basis) were made in the Netherlands, with decreases in 1967, and expected increases in the future. In the western hemisphere acid concentrations of precipitation are 10 to 10,000 times greater than possible naturally. In 15-20 years pH values changed from 6-4. In 1966 pH values in the Netherlands fell below 4. Change to use of low content coal reduced SO_x emissions. SO_2 emission densities in Europe are given. Coal gasification produces little SO_2 emission, but currently is expensive. NO_x emissions in the Netherlands are due to traffic emissions.

Wagner, G.H. and Steele, K.F., "Nutrients and Acid in the Rain and Dry Fallout at Fayetteville, Arkansas (1980-1982)", Apr. 1983, Arkansas Water Resources Research Center, University of Arkansas, Fayetteville, Arkansas.

Wet and dry fallout at Fayetteville have been collected separately and analyzed since April, 1980. The precipitation-weighted-average pH for two yearly periods of rainfall were 4.72 and 4.75. This corresponds to a concentration of the acid ion of about 18 parts per billion (ppb). Pure water in equilbrium with the CO_2 of the air would have a pH of 5.65. The range of pH during this two year period was 3.86-7.74. Aqueous extracts of the dry fallout were always in the 6.75-7.87 pH range, i.e., neutral to slightly alkaline. Ammonium bisulfate is the major acidic chemical in the rains. Wet and dry fallout add significant amounts of nutrients to the local soils with 25-87 percent of the total flux being dry fallout. Iron and zinc were the most prevalent heavy metals in the wet fallout.

Wetstone, G.S., "The Need for a New Regulatory Approach", Environment, Vol. 22, No. 5, June 1980, pp. 9-14, 40.

Topics discussed are ambient air standards, source control, state implementation plans, motor vehicle emission controls, the common law of nuisance, and solutions to the problem of acid rain.

Wetstone, G.S., "National Recourses for International Pollution: Towards a United States, Canada Solution", Journal Air Pollution Control Association, Vol. 34, No. 2, Feb. 1984, pp. 111-118.

The most prominent transboundary concern is, of course, acid rain, caused by the long-range transport of industrial and automotive pollution. But long-range air pollution transport also results in several other serious but less publicized transboundary problems such as the movement across national boundaries of oxidant pollution, which is impairing agricultural productivity in southern Ontario and elsewhere, and the transboundary flow of toxic air pollutants, a phenomenon which may already be affecting the Great Lakes.   Over the past several years the Environmental Law Institute has conducted a series of studies of the institutional framework governing control of international pollution problems. A major conclusion of this research effort is that current national and international legal structures are poorly suited to the effective control of transboundary pollution.   Despite the recent increase in the number and severity of international environmental problems, governments continue to make energy policy, pollution control, and land use decisions without explicit consideration of transboundary impact.

Whelpdale, D.M., "The Contribution Made by Air-Borne Pollutants to the Pollution of Large Bodies of Water", Atmosphere, Vol. 10, No. 1, 1972, pp. 18-22.

The pollution of large bodies of water by airborne pollutants is discussed.   Twenty-five percent of pesticides and 50 percent of heavy metals such as lead reaching the oceans are contributed by atmospheric sources.   The chemical composition and low buffering capacity of freshwater bodies makes them susceptible to more potential pollutants such as sulfur dioxide, nitrogen, and phosphorus.   Pollutants enter the water directly from precipitation scavenging and dry deposition or indirectly from surface run-off and ground water.   The scarcity of direct measurements of atmospheric pollutants contributions to the pollution of large bodies makes it difficult to establish pollutant budgets for specific substances and specific bodies of water.   The Canadian Atmospheric Environment Service is investigating airborne pollution of water including the measurement of vertical fluxes of ozone, carbon dioxide, and sulfur dioxide of a lake.

Wilson, J., Mohnen, V. and Kadlecek, J., "Wet Deposition in the Northeastern United States", Dec. 1980, U.S. Department of Energy, Washington, D.C.

Attempts are made to examine concentration and wet deposition of pollutant material at selected stations within the northeastern United States and to characterize as many events as possible with respect to air mass origin. Further attempts are made to develop a regional pattern for the deposition of dominant ion species.   MAP3S (US Multistate

Atmospheric Power Production Pollution Study) data for 1977 to 1979 are used to determine concentration and deposition on an event basis from which monthly, seasonal, annual, and cumulative averages are developed. The ARL-ATAD trajectory model is used to characterize individual events as to air mass origin. Case studies are examined to illustrate variability in the chemical composition of precipitation originating from distinctly different air mass trajectories. A difference in concentration of pollution-related ions in precipitation is noted between Midwest/Ohio Valley and Great Lakes/Canadian air mass origins for carefully selected cases. Total deposition of the major ions is examined in an effort to develop a regional pattern for deposition over a period of at least one year. For that purpose, total deposition is normalized to remove the variability in precipitation amounts for inter-station comparison. No marked gradient is noted in the normalized deposition totals within the northeast of the United States. The Adirondack region exhibited the lowest normalized ion deposition value, while the Illinois station showed the highest of the MAP3S network. The data analysis suggest that the acid rain phenomena covers the entire northeast. The concept of large scale mixing emerges to account for the lack of a significant gradient in the normalized deposition.

Wisniewski, J., "The Potential Acidity Associated with Dews, Frosts, and Fogs", Water, Air, and Soil Pollution, Vol. 17, No. 4, 1982, pp. 361-377.

In general special events of dews, frosts, and fogs are the result of water vapor condensation or sublimation in the layers of the atmosphere closest to the ground. The meteorology of these special events is briefly reviewed, including mechanisms of their formation as well as their frequency of occurrence. While research into the chemistry of the water involved in these special events is sparse, there is direct and indirect evidence available which indicates a potential acidity. The chemistry of this water is primarily dependent on two factors. The first is the amount and type of material absorbed by dew and fog directly from the atmosphere. The second controlling factor is the presence of substances, on the surface of which the moisture is deposited. The source of most of these materials is dry deposition of natural and anthropogenic substances from the atmosphere, including soil particles, fly ash, pollen, microbes, trace metals, aerosols, and adsorbed gases. Values of pH associated with these events have been observed to be below 3.0 for fog. Theoretical calculations indicate that synergistic acidity from dew combined with previous acidic dry deposition may result in pH values of less than 2.0. Neutralization due to leaching of plant metabolites or surface reactions on contact could raise the pH, but either of these two processes would damage the plant or material surface in question.

Wisniewski, J. and Keitz, E.L., "Acid Rain Deposition Patterns in the Continental United States", Water, Air, and Soil Pollution, Vol. 19, No. 4, May 1983, pp. 327-339.

Maps of both pH and H(+) deposition in precipitation have been developed for the continental United States by analyzing laboratory pH data from nine precipitation chemistry networks and two single stations spread across the continental United States and southern Canada during the late 1970's. Average laboratory pH values were obtained or calculated for approximately 100 stations, and isopleths of weighted mean pH and mean annual H(+) deposition in precipitation were drawn. In spite of a wide variety of collection methods and sampling intervals, there is remarkable uniformity in the average pH among the various stations. Precipitation remains most acidic in the heart of the northeastern section of the United States. The next highest level of precipitation acidity occurs in the surrounding area, which includes the Ohio Valley and southern Ontario regions. Evidence of precipitation acidity is now appearing in Florida, the Colorado Rockies and in population centers along the west coast, where pH values at or below 5 are now recorded. Huge data gaps are apparent in the western portion of the country. Even though most western areas are not experiencing acidities at this time, the need for additional monitoring is important due to projected western energy development.

Wisniewski, J. and Kinsman, J.D., "An Overview of Acid Rain Monitoring Activities in North America", Bulletin of the American Meteorological Society, Vol. 63, No. 6, June 1982, pp. 598-618.

Acid rain is known to acidify natural waters, resulting in damage to fish and other components of the aquatic ecosystem, degradation of drinking water supplies, deterioration of man-made structures, erosion of soils and damage to forests and crops. Acidic components may be deposited from the atmosphere by wet deposition (rain and snow), dry deposition (particles and gases), and special events (dews, frosts, and fogs). Most current monitoring activities focus on wet deposition or acid rain. Recent monitoring devices and 71 studies conducted or on-going in North America are surveyed. Tables are presented that describe the name or title of the study, the organization or agency that funds each study, the chemical parameters monitored, the geographic extent and location of the study, the time period of operation, the types of samples used, where samples are analyzed, and a contact for further information. The Aerochem Metrics wet-dry collector is the most widely used instrument for collection of wet deposition and appears to be reliable in collecting precipitation samples for chemical analysis. Much of the wet deposition monitoring focuses on the between-year differences in precipitation acidity. No simple method for monitoring dry deposition is available on an experimental or commercial basis. The frequency of special events needs to be analyzed using existing climatological data.

Wright, R.F. and Dovland, H., "Regional Surveys of the Chemistry of the Snowpack in Norway", Atmospheric Environment, Vol. 12, No. 8, 1978, pp. 1755-1759.

Some major conclusions of four regional snowpack surveys in Norway are summarized. The usefulness of such surveys in estimating regional patterns in precipitation chemistry is assessed. While the snowpack gives a rather poor measure of the deposition of chemicals in precipitation over the winter, concentrations of major ions provide valuable and relatively easily gained information on precipitation chemistry. Portions of northernmost Norway apparently get precipitation with pH levels well below 5.0.

APPENDIX B

ATMOSPHERIC REACTIONS

Bolin, B., Aspling, B. and Persson, C., "Residence Time of Atmospheric Pollutants as Dependent on Source Characteristics, Atmospheric Diffusion Processes and Sink Mechanisms", Report No. AC-25, May 1973, Stockholm University - Institute of Meteorology, Stockholm, Sweden.

A brief discussion of the sink mechanisms of atmospheric pollutants is given and particular emphasis is given to the importance of rain-out (wash-out) and dry deposition. The turbulent transfer processes that determine the latter are analyzed in the light of the classical theory for the atmospheric surface boundary layer. The transfer rates to the free atmosphere (where rain-out and wash-out are the sink mechanisms) and to the earth's surface (when dry deposition occurs) are computed as dependent on height of emission, roughness of the earth's surface, deposition velocity at the earth's surface, and the intensity of the turbulent processes as dependent on wind velocity. On the basis of such computations the residence time of atmospheric pollutants can be computed. The relative importance of rain-out (wash-out) and dry deposition in this regard is also discussed.

Calvert, J.G. and Stockwell, W.R., "Mechanism and Rates of the Gas Phase Oxidations of Sulfur Dioxide and the Nitrogen Oxides in the Atmosphere", EPA-600/D-83-079, July 1983, U.S. Environmental Protection Agency, Research Triangle Park, North Carolina.

As knowledge of the atmospheric chemistry of the SO_2, NO, and NO_2 continues to grow, it becomes increasingly clear that many different chemical reactions contribute to the oxidation of these oxides in the atmosphere. Solution phase and gas phase chemistry are both important to "acid rain" development, and gas-solid, gas-liquid, liquid-solid as well as simple gaseous molecule interactions are seemingly important in some circumstances as well. The evaluation of the mechanism and rates of solution phase and heterogeneous pathways for SO_2 and NOx oxidation within the troposphere, the significance of surface removal processes, and the transport and diffusion processes are much less amenable to simple laboratory studies and quantification. Although a significant wealth of important and necessary information related to SO_2 and NOx chemistry has been defined in recent years and a reasonable depth of knowledge of the many fundamental gas phase processes exists today, many uncertainties still remain in this, the "simplest" of the research areas related to "acid rain" development.

Chang, T.Y., "Rain and Snow Scavenging of HNO_3 Vapor in the Atmosphere", Atmospheric Environment, Vol. 18, No. 1, 1984, pp. 191-197.

The coefficients for the wet removal of HNO_3 vapor from the atmosphere by rain (in-cloud and below-cloud) and snow have been derived under a number of approximations. The wet removal coefficients are parameterized in terms of precipitation rate. These coefficients are intended to represent the average scavenging from large precipitation bands or frontal systems where there is widespread weakly ascending air motion. Consequently, the derived coefficients would be appropriate, on an interim basis, for inclusion into regional or mesoscale models which include the wet removal of HNO_3 vapor. These coefficients, when combined with altitude-dependent precipitation rates and vertical profiles of HNO_3 concentrations, are also useful to estimate the flux of HNO_3 to the earth's surface. Both rain and snow scavenging coefficients are consistent with recent observations of ionic composition in winter precipitation samples which indicated that snow removes HNO_3 vapor more efficiently than rain.

Davenport, M.H. and Peters, L.K., "Field Studies of Atmospheric Particulate Concentration Changes During Precipitation", Atmospheric Environment, Vol. 12, No. 5, 1978, pp. 997-1009.

The effects of rain on changes in the atmospheric particulate concentrations were evaluated. Measurements were compared with theoretical calculations of particle washout coefficients. Theories underlying the processes of particle capture by raindrops are examined. Equations defining such concepts are discussed. Experimentally determined washout coefficients vary with the raindrop size distribution and rain intensity. The collection of atmospheric particles during rain by the mechanisms of inertial impaction, interception, brownian diffusion, thermophoresis, and electrostatic charge effects cannot adequately explain the decrease in the particle number density observed during field experiments.

Dittenhoefer, A.C. and Dethier, B.F., "The Precipitation Chemistry of Western New York States: A Meteorological Interpretation", Report No. GBAI7674, Apr. 1976, Cornell University, Ithaca, New York.

Precipitation chemistry data for western New York were studied through the use of 12- and 24-hour isentropic trajectories. The distant source regions and paths of long-range transport were determined. Meteorological variables such as surface wind speed and direction, height of the mixed layer, mean relative humidity below cloud base, and the past history of precipitation were monitored to assess their importance in controlling rainwater concentrations. Ionic constituents such as sulfate owe much of their presence in rainwater to wet removal from the rain-producing layers (1-5 km), where their concentration is largely influenced by long-range transport. Constituents derived from the soil, such as potassium, calcium, ammonium, and magnesium, seem to be directly related to trajectory wind speed and atmospheric instability. Results indicate that air mass precipitation is characterized by markedly higher ionic content compared to frontal rainfall, and that rainfall pH is significantly lower in air mass precipitation.

Dupoux, N., "Variations in Atmospheric Aerosols as a Function of the Properties of the Medium", Report No. NSA2909, July 1973, Centre d'Etudes Nucleaires de Saclay, 91-Gif-sur-Yuette, France.

A study of variations in the concentration of atmospheric aerosols and the average radius as a function of certain properties of the medium was conducted. Some meteorological parameters affect the behavior of the atmospheric aerosol: rain "washes" the atmosphere; fog adsorbs the SO_2 in the droplets, and in its presence the aerosol is made up of large particles.

Durham, J.L., Overton, J.H. and Aneja, V.P., "Influence of Gaseous Nitric Acid on Sulfate Production and Acidity in Rain", Atmospheric Environment, Vol. 15, No. 6, 1981, pp. 1059-1068.

A physico-chemical subcloud rain model was used to study the pollutant chemistry leading to rain acidification. In the model, drops fall through a polluted region containing trace gases CO_2, O_3, SO_2, NHO_3, NO, and NO_2. The concentration of each chemical species within a drop was calculated. Subcloud scavenging of HNO_3 may control acidification during the beginning of a rain event and may have more control over the final pH than SO_2. In fact, gaseous HNO_3 inhibits the production of sulfate in rain by lowering the pH. Acidification was not caused by the oxides of nitrogen, NO and NO_2. In-cloud scavenging can have greater control over final acidity than subcloud scavenging. Furthermore, gaseous HNO_3 is removed from the atmosphere more rapidly than SO_2 or O_3. In this simulation, cumulative NO_3 decreased by 42 percent in one hour, while SO_4 and pH slightly increased. This theoretical study did not consider the formation of rain in the polluted zone, and the effect of in-cloud processes are ignored.

Engelmann, R.J., "Scavenging Prediction Using Ratios of Concentrations in Air and Precipitation", Journal of Applied Meteorology, Vol. 10, June 1971, pp. 493-497.

The ratios of the concentrations of SO_2 and of water vapor in precipitation to those in air on a mass basis are in the range 19 to 500. For particulates, the ratio usually falls in the range 290 to 2700. The ratio varies inversely with mixing ratio and precipitation rate. Stronger updrafts and rapid condensation produce lower nuclei concentrations because coalescence of cloud particles contributes less water to the final precipitation particle. The relative contribution to precipitation water of coalescence as compared to condensation is expected to be greater at the beginning and end of storms and at the edges of convective cells, and this partially explains the time variation of pollutants in precipitation. Ratios should be determined by

$$\frac{K}{X} = \frac{pn}{qE_2} + \frac{(1-n)pa}{q} + \frac{HA}{R}$$

where K and X are the concentrations in rain and air, respectively, p the density of water, q the absolute humidity, E_2 the efficiency of the cloud at removing water vapor as precipitation, n the fraction of the pollutant which nucleates and is subsequently scavenged, a a dimensionless reactivity factor for gases with water, H the height of the cloud base, and A the washout coefficient for the precipitation rate R. Measurements of washout ratios will, therefore, provide information on cloud processes.

Enger, L. and Hogstrom, U., "Dispersion and Wet Deposition of Sulfur from a Power Plant Plume", Atmospheric Environment, Vol. 13, No. 6, 1979, pp. 797-811.

Transformation rates near an oil-fired 1000 MW power plant in Sweden were studied. Five out of seven wet deposition tests revealed that 66 percent of the emitted sulfates were deposited within 80-120 km of the source. SO_2, sulfate, and an inert tracer were measured in ground-based cross plume traverses and in flight. Two of the dispersion tests indicated that SO_2-to-SO_4 transformation rates vary with relative humidity. Analysis of ammonium deposition and back trajectories indicated that rapid transfer background concentrations, the rate of turbulent flux into the plume, is sufficient to account for the observed NH_4 fallout rate.

Fowler, D., "Wet and Dry Deposition of Sulfur and Nitrogen Compounds from the Atmosphere", Effects of Acid Precipitation on Terrestrial Ecosystems, Plenum Press, New York, New York, 1980, pp. 9-27.

The transport, chemistry, and deposition mechanisms for sulfur and nitrogen compounds are reviewed. The dry deposition of gases and particulates and subsequent uptake by surface waters and vegetation are discussed. Rainout and washout processes involved in wet deposition are summarized by process for hours of lifetime and concentration in rain.

Fuquay, J.J., "Scavenging in Perspective", Report No. CONF-700601-2, July 1970, Pacific Northwest Laboratory, Battelle-Northwest, Richland, Washington.

Mechanisms of precipitation scavenging in the atmosphere are reviewed. It is noted that precipitation scavenging is usually thought of as being in major divisions including transport of the material to the scavenging site, in-cloud scavenging and precipitation, and below-cloud scavenging by the precipitation. Discussions are included concerning various metallic and nonmetallic pollutants.

Gatz, D.F., "Estimates of Wet and Dry Deposition of Chicago and Northwest Indiana Aerosols into Southern Lake Michigan", ERDA 2nd Federal Conference on the Great Lakes, Mar. 25-27, 1975, Argonne National Laboratory, Argonne, Illinois, pp. 277-290.

Pollutant aerosol input into Lake Michigan is calculated using measured wet and dry deposition rates. The calculation yields only a rough estimate, and better estimates must await: better predictions of emissions of individual elements; better characterization of elemental size distribution and solubilities; measurements of velocity for water surfaces, especially as a function of atmospheric stability; and further measurements of washout ratios for various precipitation types. Deposition estimates suggest that annual wet and dry deposition are approximately equal, that from 3-15 percent of elemental emissions from Chicago and northwestern Indiana enter the Lake, and that the fraction of emissions to be deposited in the Lake increases with particle size.

Gatz, D.F., "Urban Influence on Deposition of Sulfate and Soluble Metals in Summer Rains", 1979, Illinois State Water Survey, Urbana, Illinois.

Recent observations of abnormally acidic precipitation have raised questions regarding the distribution of acidic precipitation, its time trend, its sources of acidity, and the relevant physical and chemical processes involved in its formation. An attempt was made to answer some of these questions based on the content of sulfate and other materials in the atmosphere and in samples of summer convective rainfall from mesoscale sampling networks near St. Louis. These and other observations in the literature agree that rainfall deposits locally emitted sulfur at short distances downwind of cities. This causes enhanced deposition and concentration of sulfur in local rainfall and increases the local variability of these parameters relative to that of rain and crustally derived materials, on both daily and seasonal scales. Airborne sulfate concentrations vary by a factor of at least 2 to 4 from urban to rural areas on individual days. This is similar to the observed variation of sulfate deposition or concentrations in rain. Thus, there may be no need to invoke extensive SO_2 scavenging in rain systems to explain the observed enhancements; nucleation scavenging of atmospheric sulfate appears adequate.

Georgii, H.W. and Beilke, S., "Atmospheric Aerosol- and Trace-Gas-Washout", Report No. USGRDR6616, Mar. 1966, Institut Fuer Meteorologie Und Geophysik, Frankfurt University, Frankfurt, West Germany.

This report summarizes the results of detailed investigations in the laboratory on washout and rainout of SO_2 by droplets of known size distribution and concentration. The results show clearly the effect of drop size, intensity and of the chemical composition (pH-value) of rain and fog on the scavenging efficiency. The results of the experiments were used as the basis of a model calculation of the effect of washout and rainout by natural precipitation at a given vertical distribution of SO_2. The report demonstrated under which circumstances which mechanisms become predominant for the chemical composition of rainwater at the ground.

Granat, L., "On the Relation Between pH and the Chemical Composition in Atmospheric Precipitation", Tellus, Vol. 24, 1972, pp. 550-560.

Based on chemical analyses of some 2,000 rainwater samples, the relation between the pH and the total amount of acid or of base, respectively, is studied. Considering the carbon dioxide-water system the theoretical relation between pH and total amount of acid or bicarbonate is calculated. A regular deviation is observed between the empirically found and theoretically calculated relation. Starting from chemical analyses of the most important compounds in atmospheric precipitation, a quantitative hypothesis is formulated of how these compounds originally were associated with acids or bases. It is further assumed that these have neutralized each other and that the system has come to an equilibrium with the carbon dioxide pressure of the atmosphere. Good agreement is obtained between the predicted amount of acid or base and the measured one, and this is taken as a clear indication of the validity of the model. This model on the stoichiometric relation between acids and bases turns out to be a useful tool both for an interpretation of the amount of acid found in precipitation and in estimating future deposition of acid by precipitation.

Gravenhorst, G. et al., "Sulfur Dioxide Absorbed in Rain Water", Effects of Acid Precipitation on Terrestrial Ecosystems, Plenum Press, New York, New York, 1980, pp. 41-55.

The mechanisms of the incorporation of sulfur dioxide into rain drops is discussed as well as the contribution to free acidity in rain. Equations and formulas are given. Meteorological mechanisms (local and global) are reviewed.

Hales, J.M., "Pollutant Transformation and Removal Measurements at METROMEX", Report No. CONF-770210-8, Jan. 1977, Battelle Pacific Northwest Laboratories, Richland, Washington.

Tracer studies of convective storm scavenging at METROMEX represent the most comprehensive effort to date in the field of multielement tracer applications for scavenging research. These began in the summer of 1971 with rather modest attempts to inject tracers into storm inflow regions from stacks and ground level positions, and developed into comparatively sophisticated, coordinated aircraft release schemes during subsequent years. While the surface-release experiments provided some indication of the efficiency of scavenging in specific cases, the aircraft experiments were much more valuable as indicators of the pertinent mechanisms of storm behavior. Some rather qualitative aspects of the tracer experiments, including experiment design and pertinent conclusions which have arisen from the studies, are discussed.

Hales, J.W., "How the Air Cleans Itself", Proceedings of the SCI Sulfur Symposium, Feb. 1979, Battelle Pacific Northwest Laboratories, Richland, Washington.

The various mechanisms for atmospheric recovery are discussed, and means by which these pathways can be treated mathematically to formulate models of air quality and pollutant behavior are described. Many of the essentials of atmospheric recovery processes are outlined. The important problems currently at the forefront of wet- and dry-deposition research are also discussed.

Harrison, R.M. and Pio, C.A., "Comparative Study of the Ionic Composition of Rainwater and Atmospheric Aerosols: Implications for the Mechanism of Acidification of Rainwater", Atmospheric Environment, Vol. 17, No. 12, 1983, pp. 2539-2543.

Measurements of the chemical composition of rainwater and suspended particles collected in parallel at a rural site in Northwest England have shown that sulfate, nitrate, chloride, sodium, magnesium and potassium exists in similar proportions in both media. However, rainwater shows a marked enhancement of hydrogen ions, and a corresponding decrease in ammonia ions relative to aerosols. It is concluded that the major contribution to rainwater acidity at this site is due to sulfuric acid incorporated at cloud level. The scavenging of nitric or hydrochloric acids, or incorporation of acid sulfates close to ground level can account for only a minor contribution to acidity in this locality.

Hegg, D.A., "Sources of Sulfate in Precipitation--Part 1--Parameterization Scheme and Physical Sensitivities", Journal of Geophysical Research, Vol. 88, No. C2, Feb. 1983, pp. 1369-1374.

The previously derived parameterization scheme for sulfate washout is modified to allow estimation of the relative contributions of nucleation scavenging, in-solution production, and below-cloud scavenging to the sulfate content of precipitation. The parameterization scheme predicts changes in precipitation sulfate concentration by up to a factor of 6 for large variations in key cloud physics parameters. The results suggest in-solution sulfate production can be a substantial contributor to the sulfate content of precipitation.

Hidy, G.M. et al., "Precipitation-Scavenging Chemistry for Sulfate and Nitrate from the SURE and Related Data", EPRI-EA-1914-V.2, Feb. 1983, Environmental Research and Technology, Inc., Westlake Village, California.

   Parallel ground-based observations of ambient aerosol chemistry and precipitation water chemistry from samples taken at rural sites in the northeastern United States offer opportunities for elucidation of the wet deposition process. Data taken between 1978 and 1979 were selected and analyzed to estimate Junge's rainout efficiencies for sulfur oxides and nitrogen oxides, which are directly related to washout ratios. Calculations indicate that the mean rainout efficiency for particulate sulfate based on 500 samples is approximately 0.6 to 0.8. This is a necessary, but not a sufficient condition that particulate matter scavenging can account for most of the sulfate in precipitation water. Additional chemical data were used to infer that significant acid production may take place in precipitation by SO_2 absorption and oxidation at some locations, particularly in Ohio and northeastward. In contrast with sulfate, the calculations indicate that nitrate is dominated by scavenging of gases, such as NHO_3, NO_2 or $N_2 O_5$, rather than particles. The rainout efficiencies tend to be smaller in winter, and are largely independent of the phase of precipitation elements, and storm conditions. There is also a tendency for the rainout efficiencies to decrease with precipitation intensity. Supplemental results are included which illustrate the informational value in combined aerosol and precipitation water data which include both cation and anion chemistry. These data are useful in testing hypotheses about the enrichment mechanisms of hydrogen ions, sulfate and nitrate in precipitation relative to the ambient air. Examples show that the precipitation acidity is dominated by scavenging of sulfuric acid or aqueous SO_2 absorption followed by oxidation rather than HNO_3 scavenging.

Melo, O.T., "Ontario Hydro Studies of Atmospheric Chemistry", Research Review Ontario Hydro, No. 2, May 1981, pp. 21-28.

   Since the mid 1970s, emissions of sulphur and nitrogen oxides from power plants have been implicated in the formation of sulphate aerosols and acid rain and with suspected health and ecological effects far from the point of emission. Several groups have performed many studies of the chemical conversions taking place in power-plant plumes, by use of airborne and ground-based measurement platforms. Work at Ontario Hydro indicates that most of the nitrogen oxides are emitted as nitric oxide, about 5 percent of the NO is quickly converted to nitrogen dioxide through reaction with oxygen and that the rest is converted to NO_2 by ambient ozone in terms which are typically 1 to 2 hours. Further conversion of NO_2 to nitric acid proceeds at appreciable rates only after this 1 to 2 hour induction period. The reaction of sulphur dioxide to sulphate aerosol is generally quite slow except on a few occasions when it reaches values of up to 5 percent per hour. A recent extensive review of the literature revealed results generally consistent with Ontario Hydro's.

Molenkamp, C.R., "Numerical Modeling of Precipitation Scavenging by Convective Clouds", Report No. UCRL-75896, Sept. 1974, California University - Lawrence Livermore Laboratory, Livermore, California.

Very good discussion of the mechanisms of nuclei condensation. Most action occurs just above the base of a cloud. Input/output of pollutants in a cloud are discussed.

Overton, J.H. and Durham, J.L., "Acidification of Rain in the Presence of SO_2, H_2O_2, O_3, and HNO_3", EPA-600/D-82-150, Jan. 1982, U.S. Environmental Protection Agency, Research Triangle Park, North Carolina.

The production of acid sulfate and the accumulation of acid nitrate are calculated for falling raindrops using a physico-chemical model that accounts for the mass transfer of SO_2, H_2O_2, O_3, HNO_3, and CO_2. The acidification is postulated to occur through the absorption of free gaseous HNO_3 and the absorption and reaction of SO_2, H_2O_2, and O_3 to yield H_2SO_4. Investigated are the relative effectiveness of $H_2O_2(aq)$ and $O_3(aq)$ for oxidizing $SO_2(aq)$ to yield $H(+1)$ and $SO-2(-2)$, and the role of $HNO_3(aq)$ in acidifying raindrops and influencing $SO_4(-2)$ formation. Results indicate: that H_2O_2 is more effective than O_3, HNO_3 inhibits $SO_4(-2)$ formation due to O_3 oxidation, and in all cases, HNO_3 is important in acidifying rain.

Overton, Jr., J.H., Aneja, V.P. and Durham, J.L., "Production of Sulfate in Rain and Raindrops in Polluted Atmospheres", Atmospheric Environment, Vol. 13, 1979, pp. 355-367.

A physico-chemical model for the accumulation of sulfur species in raindrops is developed in which account is taken of the mass transfer of SO_2, O_3, NH_3, and CO_2 into ideal raindrops containing the catalyst $Fe(III)$. The sulfur accumulation was calculated for the droplets as a function of fall distance. The model predicts the formation of sulfate due to the oxidation of dissolved SO_2 by O_3 and catalytic ions in the presence of NH_3 and CO_2. The initial pH of a drop was taken as 5.56. The final value depended on drop size, fall distance, and ambient concentrations, but in all cases was between 4.2 and 6.6. Sulfate values also depended on the same conditions and ranged from 2 to 2,000 umol L^{-1} for individual drops. For a precipitation rate of 10 mm h^{-1} and a fall distance of 2,000 m, the raindrop pH varied from 4.5 to 6.5 and the average sulfate concentration from 3 to 87 umol L^{-1} depending on ambient pollutant concentrations. These results conform to the experimentally measured values of the rain pH (3-9) and sulfate concentration (20-150 umol L^{-1}), and corresponding air SO_2 concentration (1-10 ppb) as reported in the literature.

Pack, D.H., editor, Proceedings: Advisory Workshop to Identify Research Needs on the Formation of Acid Precipitation, May 1979, Sigma Research, Inc., Richland, Washington.

Results of an advisory workshop sponsored by the Electric Power Research Institute to identify research needs on the formation of acid precipitation are presented. The state of knowledge on instrumentation and sampling, atmospheric chemistry, cloud processes, meteorological transport, and mathematical modeling related to acid precipitation is described. Thirty-eight research projects spanning five areas--instrumentation, field studies, modeling, laboratory studies, and data analysis and interpretation--are described, along with a background discussion of rationale and resources required. The total program described encompasses 210 professional man-years of effort and an estimated cost of $20 million.

Penkett, S.A. et al., "The Importance of Atmospheric Ozone and Hydrogen Peroxide in Oxidizing Sulfur Dioxide in Clouds and Rainwater", Atmospheric Environment, Vol. 13, No. 1, 1979, pp. 123-128.

The measured rates of sulfur dioxide oxidation by oxygen, ozone, and hydrogen peroxide across a range of pH and temperature are compared to evaluate their relative importance in the formation of sulfate in atmospheric droplets. Oxidation will occur readily in cloud or fog droplets. If pH is above 6, ozone will be primarily responsible. If pH is below about 5.5, hydrogen peroxide is the favored oxidant.

Rogowski, R.S. et al., "Carbon-Catalyzed Oxidation of SO_2 by NO_2 and Air", Apr. 1982, National Aeronautics and Space Administration, Hampton, Virginia.

A series of experiments was performed using carbon particles (commercial furnace black) as a surrogate for soot particles. Carbon particles were suspended in water, and gas mixtures were bubbled into the suspensions to observe the effect of carbon particles on the oxidation of SO_2 by air and NO_2. Identical gas mixtures were bubbled into a blank containing only pure water. After exposure each solution was analyzed for pH and sulfate. It was found that NO_2 greatly enhances the oxidation of SO_2 to sulfate in the presence of carbon particles. The amount of sulfate found in the blanks was significantly less. Under the conditions of these experiments no saturation of the reaction was observed and SO_2 was converted to sulfate even in a highly acid medium.

Semonin, R.G. et al., "Study of Atmospheric Pollution Scavenging--Seventeenth Progress Report", July 1979, Illinois State Water Survey, Urbana, Illinois.

The continued study of historical precipitation chemistry, air quality, and emissions data shows that the nitrate concentrations in precipitation have increased over the past 20 years with little change in sulfate concentrations. An analysis of aerosol concentrations at MAP3S precipitation collection sites shows consistently higher elemental concentrations at urban Champaign, Illinois, than sites at Whiteface Mountain, New York, and rural Champaign. Scavenging ratios at Whiteface exceeded those obtained previously at St. Louis, Missouri, possibly due to mid-tropospheric long-range transport and differing synoptic situations. Factor analysis of 6 storms during METROMEX shows that different deposition patterns were found for the soluble and insoluble concentrations of the same element. This suggests different scavenging processes may be active for these fractions. Using METROMEX data, no correlation was found between pollutant source strength and the urban-related precipitation anomaly. Four different Nuclepore air filter setups were used to compare air concentrations of sulfate, nitrate, and ammonium. The sulfate comparison was good, but the large variability for nitrate and ammonium raise serious questions on the credibility of measurements using Nuclepore as the collection medium. The details of a case study of wet deposition from SCORE-78 are presented and show the pollutant concentrations are more variable than the rainfall. A brief description of the progress on the SCORE-79 project is presented. The progress on acid rainfall studies shows that the high pH values in the

Midwest in the mid-1950's were due in part to elevated concentrations of calcium and magnesium. A variety of model calculations are presented to show the effects of adjusting past data to currently observed values. Recent results of continuing research on ionic stability of precipitation samples are given.

Sullivan, J.L., Wen, Y.P. and Frantisak, F., "Smelter Stack Plume Kinetic Studies in Northern Ontario", Proceedings of the Second National Conference on Complete Water Reuse: Water's Interface with Energy, Air, and Soils, 1975, American Institute of Chemical Engineers, New York, New York, pp. 311-314.

The potential for lakes acidification by sulfur oxides emissions exists and has been the subject for multifaceted studies funded by the Ontario Ministry of the Environment in the past few years. Of major interest in the study, indeed in all such studies, is the elucidation of possible air-water transfer mechanisms for acid sulfur gases or salts. Several potential routes of water uptake, such as direct solution of sulfur dioxide, deposition of particulates containing sulfur oxides or sulfuric acid, and deposition of acid sulfates exist. In various specific circumstances relating to atmospheric conditions or source distance, one or the other of these mechanisms will be more important.

APPENDIX C

ATMOSPHERIC MODELS

Adamowicz, R.F., "A Model for the Reversible Washout of Sulfur Dioxide, Ammonia, and Carbon Dioxide from a Polluted Atmosphere and the Production of Sulfates in Raindrops", Atmospheric Environment, Vol. 13, No. 1, 1979, pp. 105-112.

> The washout of trace atmospheric gases and the production of acids in rain are modeled. The washout model is modular and can be coupled with meteorological transport and dispersion models of pollutants. The model is applied to washout of sulfur dioxide, carbon dioxide, and ammonia. When simultaneous mass transfer of SO_2 and ammonia is involved, the effective driving force for the transport of both gases can be reduced, resulting in long relaxation times for the absorption process.

Adamowicz, R.F. and Hill, F.B., "Model for the Reversible Washout of Sulfur Dioxide, Ammonia and Carbon Dioxide from a Polluted Atmosphere and the Production of Sulfates in Raindrops", Report No. CONF-77/102-12, 1977, Brookhaven National Laboratory, Upton, New York.

> A model has been developed to describe the washout of trace atmospheric gases and the production of acids in rain. The model is applied to washout of sulfur dioxide, carbon dioxide and ammonia and incorporates reversible mass transfer of the trace gases, all possible ionic equilibria of the compounds in solution and catalyzed oxidation of the dissolved sulfur species to sulfates. The significance of ammonia and carbon dioxide on the raindrops capacity for sulfur and on sulfate production based solely on bisulfite oxidation are explored in detail. The influence of raindrop size, rainfall intensity, cloud-base height, the presence of oxidation catalyzing compounds in the atmosphere and the initial composition of the raindrops as they enter the polluted atmospheric layer on the detailed chemical composition of rain at ground level and the time scale for gaseous sulfur dioxide removal are evaluated.

Barrie, L.A., "The Prediction of Rain Acidity and SO_2 Scavenging in Eastern North America", Atmospheric Environment, Vol. 15, No. 1, 1981, pp. 31-41.

> The role of sulfur dioxide as a rainwater acidifying agent is examined. A quantitative relationship between the sulfur (IV) content of rain, rainwater pH and temperature is derived which can be used in long-range transport models to simulate the sulfate scavenging process mathematically. Methods of calculating precipitation acidity and thus sulfur dioxide washout are demonstrated for various rainwater chemistry data sets. Based on the equilibrium solution chemistry of sulfur dioxide, it is concluded that, for the purposes of long-range transport models, sulfur dioxide scavenging by rain can be calculated using a washout ratio approach, provided that the temperature and acidity of rain are taken into account. It is also concluded that dissolved sulfur dioxide can contribute a significant fraction of the total hydrogen ions deposited in rainfall, particularly at near zero degrees C, in air containing aerosols of low acidity such as ammonia neutralized sulfuric acid particles. If long term trends in the acidity of rain are to be explained, dissolved sulfur dioxide should be measured in precipitation. The sulfur dioxide scavenging process must also be realistically simulated in models. To accomplish this one must be able to predict precipitation acidity. The strong relation between the hydrogen ion and sulfate ion contents of rain in acid sensitive areas of eastern North America enables one to predict precipitation

acidity from models of the sulfur cycle. Nitrogen oxides are also important, but their cycles are not so well known. More study is also needed in order to parameterize sulfate aerosol washout.

Bhumralkar, C.M. et al., "ENAMAP-1 Long-Term Air Pollution Model: Adaptation and Application to Eastern North America", EPA-600/4-80-039, July 1980, U.S. Environmental Protection Agency, Research Triangle Park, North Carolina.

The long-term EURMAP-1 model, a trajectory-type regional air pollution model extensively evaluated and applied in Europe in conjunction with studies of interregional sulfur transport and deposition, was adapted, tested, and applied to eastern North America. The adapted version, ENAMAP-1, was used to calculate monthly, seasonal, and annual distributions of sulfur dioxide and sulfate (SO_2 and $SO_4(-2)$) concentrations and wet and dry depositions over the eastern United States, as well as over the southern portions of the Canadian provinces of Quebec and Ontario. This geographical area was partitioned into 13 different regions and interregional sulfur exchanges calculated. Model calculations were based on emission data that included both the specialized data prepared for the Sulfate Regional Experiment (SURE) and the U.S. Environmental Protection Agency's National Emissions Data System (NEDS). Model results using emissions and meteorological data for the year 1977 are presented. Results include tables showing the calculated interregional exchanges of sulfur deposition between 13 regions of eastern North America. Comparisons were made between calculated and measured SO_2 and $SO_4(-2)$ concentrations. Calculated and measured values show reasonable agreement and indicate that improvements in the calculated values may be achieved by refinements in the modeling of mixing heights and stability. Results are also shown for an application of the model using projected 1985 emissions in conjunction with meteorological data for 1977.

Bolin, B. and Persson, C., "Regional Dispersion and Deposition of Atmospheric Pollutants with Particular Application to Sulfur Pollution Over Western Europe", Report No. AC-28, May 1974, Stockholm University - Institute of Meteorology, Stockholm, Sweden.

The basic transfer equation interrelating sources and sinks for atmospheric pollutants through the turbulent processes in the atmosphere was reformulated in a statistical manner. Since on the regional scale the synoptic disturbances are the essential turbulent elements, their statistics were considered using geostrophic trajectory computations. The various scales of atmospheric turbulence were analyzed with regard to their importance for vertical transfer. The sink mechanisms of rain-out and wash-out (interior sinks) and direct deposition at the lower boundary (boundary transfer) are also described statistically. The skeleton of a regional dispersion model is developed and its advantages and limitations are discussed. As an example, a simplified study of the emission, dispersion, and removal of atmospheric sulfur is presented.

Carmichael, G.R., Yang, D.K. and Lin, C., "Numerical Technique for the Investigation of the Transport and Dry Deposition of Chemically Reactive Pollutants", Atmospheric Environment, Vol. 14, No. 12, 1980, pp. 1433-1438.

A numerical technique to approximate solutions to the atmospheric advection-diffusion equation, which applies the concept of Locally-One-Dimensional methods, and utilizes the Crank-Nicolson Galerkin method for the diffusion-dominated vertical transport and the Egan and Mahoney method for the advection-dominated horizontal transport, is discussed. This technique is used to simulate the dynamic response of an elevated plume to the growth and dissipation of the daytime mixing layer.

Chen, C.W. et al., "Acid Rain Model: Hydrologic Module", Journal of the Environmental Engineering Division, American Society of Civil Engineers, Vol. 108, No. EE3, June 1982, pp. 455-472.

An Integrated Lake-Watershed Acidification Study (ILWAS) model is presented and findings based on simulation applications of the model to date are evaluated. The ILWAS model includes hydrologic, canopy chemistry, snowmelt chemistry, soil chemistry, and stream and lake water quality modules. For modeling purposes a drainage basin is divided into terrestrial subcatchments, stream segments, and a lake. Vertically, each subcatchment is further segmented into canopy, snow pack, and soil layers. The lake is also vertically layered. All these physical compartments are coupled to form a network that allows water to be routed through the system to the lake outlet. The physical compartments are described by geometric input parameters such as area, aspect, slope, and soil layer thickness. Subcatchments are allowed to have different percentage distributions of deciduous and coniferous trees and open areas. The hydrologic module of the ILWAS model realistically simulated the total outflow from a lake basin and the hydrology of the subcatchments. The flowpath calculations are supported in part by observations of the soil solution and lake water pH values. The flowpaths of rain and snowmelt water through the terrestrial system are important factors in determining the susceptibility of lakes to acidification by atmospheric deposition. Several mechanisms can force rain and snowmelt water to become lateral flow in the upper soil horizons instead of percolating into the inorganic horizons. Low hydraulic conductivity in any soil layer may result in lateral flows in overlying layers.

Corbett, J.O., "Validity of Source-Depletion and Alternative Approximation Methods for a Gaussian Plume Subject to Dry Deposition", Atmospheric Environment, Vol. 15, No. 7, 1981, pp. 1207-1213.

The source-depletion method, which is used to account for dry deposition in gaussian plume models of the atmospheric dispersion of pollutants, is investigated mathematically on the basis of a diffusivity model. The results are used to derive simple criteria for the validity of the method, which can be applied within the framework of the gaussian model without explicit recourse to diffusivity theory. A method for estimating long-range transport is similarly assessed.

Dana, M.T. et al., "Natural Precipitation Washout of Sulfur Compounds from Plumes", Project No. BNW-389/B46621, June 1973, Battelle-Pacific Northwest Laboratories, Richland, Washington.

This report describes field measurement and modeling of the washout of SO_2 and sulfate from plumes. Field measurements of precipitation washout were conducted in conjunction with both controlled test sources and actual power plant plumes. A primary achievement of this work has been the formulation of an SO_2 washout model, which predicts rain-borne SO_2 concentrations that agree favorably with those observed. An approximate theoretical analysis of sulfate washout in conjunction with field observations indicates that sulfate formation and scavenging exhibit a strong inverse dependence on acidity levels in the background rain.

Dana, M.T. et al., "Precipitation Scavenging of Inorganic Pollutants from Metropolitan Sources", Report No. EPA 650/3-74-005, June 1974, Battelle Pacific Northwest Laboratories, Richland, Washington.

This report describes the initial results of a program to measure and model the precipitation scavenging of urban pollutants in the St. Louis area. The primary field measurements of the program are of concentrations of trace inorganics in rainwater collected at specific locations in the area. A review of possible field experimental designs in the context of the modeling objectives of this study indicates that the concept of a regional pollution material balance is an appropriate approach. The scavenging term in the balance is the scavenging rate, the mass of a given pollutant removed per unit distance along the storm path per unit time. These rates were computed from the concentrations measured during 5 convective storms in August 1972. For one storm where the scavenging rates were determined at 3 distances from the city, the derived downstream removal rates for SO_4 and NO_3 were comparable in magnitude to estimates of the urban area emission rates.

Davis, W.E., Eadie, W.J. and Powell, D.C., "Users Guide to REGIONAL-1: A Regional Assessment Model", Sept. 1979, U.S. Department of Energy, Washington, D.C.

A guide was prepared to allow a user to run the PNL long-range transport model, REGIONAL 1. REGIONAL 1 is a computer model set up to run atmospheric assessments on a regional basis. The model has the capability of being run in three modes for a single time period. The three modes are: (1) no deposition, (2) dry deposition, and (3) wet and dry deposition. The guide provides the physical and mathematical basis used in the model for calculating transport, diffusion, and deposition for all three modes. Also the guide includes a program listing with an explanation of the listings and an example in the form of a short-term assessment for 48 hours. The purpose of the example is to allow a person who has past experience with programming and meteorology to operate the assessment model and compare his results with the guide results. This comparison will assure the user that the program is operating in a proper fashion.

Draxler, R.R. and Heffter, J.L., "Workbook for Estimating the Climatology of Regional-Continental Scale Atmospheric Dispersion and Deposition Over the United States", NOAA-TM-ERL-ARL-96, Feb. 1981, U.S. National Oceanic and Atmospheric Administration, Silver Spring, Maryland.

A version of the Air Resources Laboratories' (ARL) regional trajectory model was developed to permit simultaneous calculation of trajectories from 70 hypothetical sources for very long periods. Five years of data were used to develop a climatology of atmospheric dispersion. Air concentrations were calculated at receptors coincident with the 70 source locations. Each source and receptor was treated independently to develop a source/receptor matrix that can be used to produce air concentration patterns over the U.S. for any combination of sources or to evaluate the impact of different sources on any receptor. Five-year averages of seasonal and annual air concentrations are given for various combinations of wet and dry deposition. Example calculations are shown.

Golomb, D. et al., "Sensitivity Analysis of the Kinetics of Acid Rain Models", Atmospheric Environment, Vol. 17, No. 3, 1983, pp. 645-653.

The great number of variables in acid rain modeling makes it difficult to pinpoint those parameters to which the model output (rain acidity) is most sensitive. The approach taken here separates the kinetic and dynamic parts, and analyzes the sensitivity of the kinetic model alone. Further simplifications and linearizations are introduced; however, the essential steps of the transformation processes are believed to be preserved. The major conclusions are: (a) rain acidity is most sensitive to both the oxidation rates of SO_2 to $SO_4(-2)$ and NO_x to $NO_3(-)$; (b) dry deposition of the emitted gases, but not the formed anions, is important in determining the wet-deposition fraction; (c) wet deposition is much faster than oxidation, and acidic matter is removed very rapidly from the air, but it is the oxidation rate that determines the total amount of acidity in rain; and (d) for similar initial concentrations of SO_2 and NO_x, nitrate ions may be the predominant species in wet deposition due to the faster oxidation and slower dry deposition of NO_x compared to SO_2.

Hales, J.M. and Lee, R.N., "Precipitation Scavenging of Organic Contaminants", July 1975, Battelle Pacific Northwest Laboratories, Richland, Washington.

Mathematical expressions were formulated to describe the observed solubility behavior, and these were incorporated into the EPAEC scavenging model to predict concentrations of these materials in rain for comparison with experimental measurements. Agreement between experiment and theory was found to improve over previous estimates which were based upon less accurate solubility data. Diethylamine results showed generally good agreement, while those for ethyl acetoacetate exhibited considerable deviation. Disagreement in this latter case was attributed primarily to the neglect of aqueous-phase mixing effects in the model calculations.

Hill, F.B. and Adamowicz, R.F., "Model for Rain Compostion and the Washout of Sulfur Dioxide", Contract No. E(30-1)-16, July 1976, Brookhaven National Laboratories, Upton, New York.

A continuous model for the washout of sulfur dioxide from the atmosphere by rain was developed in which account was taken of mass transfer of SO_2 into well-mixed drops, ionic equilibrium of sulfur compounds in solution, oxidation of dissolved species to sulfate ion, and the presence in the rain of background strong acid or base. Expressions were developed to predict the composition of raindrops as a function of fall distance, the time scale of atmospheric SO_2 removal, and ground level composition transients during a rain event. Illustrative calculations were made for single drop sizes and for the full spectrum of a model drop-size distribution, with the following results. In the absence of bisulfite oxidation and with an acidic background pH (pH = 4) the composition of rain was in equilibrium with SO_2 in the atmosphere after falling 100 to 200 m through a mixed layer of pollutant. In contrast, the equilibrium composition was not reached within a fall distance of 200 m for a basic background pH (pH = 10). Furthermore, a unimodal distribution of pH with drop size was found in initially neutral or acidic rain whereas a bimodal distribution appeared with a strongly basic background pH. Introduction of bisulfite oxidation led to enhanced SO_2 uptake at pH = 4, and diminished uptake at pH = 10. Also, rejection of SO_2 from raindrops of small size was found in the presence of oxidation at pH = 10. Half-lives for SO_2 removal from the atmosphere at a 1 mm/hr rainfall rate and a mixed layer height of 1 Km ranged from hours to days depending on background pH and initial SO_2 atmospheric concentration.

Hill, F.B. and Adamowitz, R.F., "A Model for Rain Composition and the Washout of Sulfur Dioxide", Atmospheric Environment, Vol. 11, No. 10, 1977, pp. 917-928.

A model for rain composition and the washout of sulfur dioxide from the atmosphere by rain is discussed. Two models were combined by modifying the mass transfer relation for an inert gas to incorporate the ionic equilibria of the chemical model. The resulting model was used to calculate the composition of rain as a function of fall distance and to obtain the time scale for SO_2 removal from the atmosphere. Illustrative calculations were made for single drop sizes and for the full spectrum of a model drop size distribution.

Horst, T.W., "Review of Gaussian Diffusion-Deposition Models", 1979, Battelle Pacific Northwest Laboratories, Richland, Washington.

The assumptions and predictions of several Gaussian diffusion-deposition models are compared. A simple correction to the Chamberlain source depletion model is shown to predict ground-level airborne concentrations and dry deposition fluxes in close agreement with the exact solution of Horst.

Huff, D. et al., "TEHM: A Terrestrial Ecosystem Hydrology Model", EDFB/IBP-76'8, Apr. 1977, Oak Ridge National Laboratory, Oak Ridge, Tennessee.

The terrestrial ecosystem hydrology model (TEHM) combines mechanistic models for climatic and hydrologic processes with vegetation properties to explicitly simulate interception and throughfall; infiltration; root zone evaporation, transpiration, and drainage; plant and soil water potential; unsaturated and saturated subsurface flow; surface runoff; and open channel flow. It is also possible to use the TEHM earth models for forest stand biomass dynamics and chemistry and exchange of heavy metals to study the transport and fate of trace contaminants at a watershed scale.

Walker Branch Watershed has been used as an example to illustrate development of the required input parameters and variables that are necessary to execute the TEHM. In all cases, emphasis has been placed on objective, physically based methods. When simulations of interception loss, soil moisture content, and base flow and storm flow are compared with observation, the overall adequacy of the model may be assessed.

For user convenience, the documentation includes a complete discussion of input formats, example data input sets, output summaries, and a microfische listing of complete source deck and program output. As presented, the TEHM provides an operational tool and a model structure and data management capabilities that will be useful for future hydrologic simulation work.

Junod, A., "Prediction Methods of Cooling Tower Plumes", International Center for Heat and Mass Transfer Conference on Heat Disposal from Power Generation, Aug. 23-28, 1976, Dubrovnik.

The possible meteorological effects of cooling tower operation can be predicted by: production of a visual plume; stimulation of precipitation; increases in ambient humidity; rise of ambient temperatures; perturbation of air flow near the ground; and stimulation of convective cloud formation. The possible synergistic interaction of the plume with airborne pollutants, such as sulfur dioxide, could also form acid droplets and cause subsequent precipitation. Two conceptual mathematical models of a cooling tower plume are presented, the Sauna model and the Walkure model. The combined use of several modeling approaches is also discussed, and necessary meteorological information is reviewed.

Lee, Y., "Numerical Models for Precipitation Scavenging", Report No. C00-1407-52, Apr. 1974, Department of Atmospheric and Oceanic Science, University of Michigan, Ann Arbor, Michigan.

Two models are developed. Model 1 is for precipitation scavenging in stratiform clouds. Model 2 addresses the problem of air cleaning by a convective rain-generating system. The article describes these two types of models and has many references concerning precipitation scavenging.

Lehmann, E.J., "Atmospheric Modeling of Air Pollution (A Bibliography with Abstracts)", Report No. GLAT7515, June 1975, National Technical Information Service, U.S. Department of Commerce, Springfield, Virginia.

All aspects of lower atmospheric modeling of air pollution from both mobile and stationary sources are covered in this bibliography of Federally funded research reports. Included are models concerning local diffusion, climatology, and smog. Stratospheric modeling concerning supersonic aircraft are excluded.

Lewellen, W.S. and Sheng, Y.P., "Modeling of Dry Deposition of SO_2 and Sulfate Aerosols", July 1980, Aeronautical Research Associates of Princeton, Inc., Princeton, New Jersey.

A model for dry deposition of SO_2 and sulfate aerosol is formulated, and calculations are made for four flow regimes; the outer boundary layer, the constant-flux region, within the vegetative canopy, and the viscous sublayer next to a smooth surface. The results of calculations are presented to demonstrate sensitivity of deposition to atmospheric stability, surface resistance, plant area density and structural shape, aerosol particle size, and Rossby number and canopy Reynolds number. A simple parameterized model for the SURE regional model is constructed.

Liu, M.K., Stewart, D.A. and Henderson, D., "Mathematical Model for the Analysis of Acid Deposition", Journal of Applied Meteorology, Vol. 21, No. 6, June 1982, pp. 859-873.

The use of a regional-scale air quality model as a diagnostic tool for analyzing problems associated with acid rain is described. The model, which is hybrid in nature, consists of a puff module and a grid module. The puff module computes the evolution of individual puffs, such as the horizontal and vertical standard deviations of the puff spreads and the location of the center of mass, emitted continuously from each major point source. It also determines the location at which the puff will be released to the grid module and the amount of oxidation and deposition along the trajectory. The grid module then follows the transport, diffusion, and chemical reactions of these aged puffs, as well as emissions from a variety of diffuse sources. On the basis of model calculations, atmospheric budgets for SO_2 and sulfate over the modeling region have been estimated.

Mayerhofer, P.M. et al., "ENAMAP-1A Long-Term SO_2 and Sulfate Air Pollution Model: Refinement of Transformation and Deposition Mechanisms", EPA-600/3-82-063, May 1982, U.S. Environmental Protection Agency, Research Triangle Park, North Carolina.

The ENAMAP-1 model for long-range air pollution transport has been modified in several ways to produce a newer version, ENAMAP-1A. The geographic region of the modeled domain has been increased to include southeastern Canada, and the meteorological and emission data for this area have been added to the U.S. data base. The transformation rate for SO_2 to $SO_4(-2)$ and the deposition rates of SO_2 and $SO_4(-2)$ have been updated. In ENAMAP-1 these rates are all constants; in ENAMAP-1A they are variable in space and time. The transformation rate has been made dependent on the amount of sunshine (i.e., a function of latitude and season) and is about twice as large as the previous rate. In ENAMAP-1A the dry deposition rate has been made dependent on the type of underlying terrain and vegetation, on thermal stability in the boundary layer, and on time of day. Wet deposition is treated as a function of rainfall rate and cloud type (convective, warm process, or Bergeron process). Boundary exchanges of SO_2 and $SO_4(-2)$ have been computed for each of 41 states (or provinces) and also for 12 especially sensitive areas of special interest such as parks. The computations show the history of pollution from emission to deposition and are documented in the form of maps and tables. In contrast to ENAMAP-1 computations, ENAMAP-1A computations for January and August 1977 have much larger amounts of $SO_4(-2)$ in the form of airborne concentration and deposition, while the amount of SO_2 deposition is decreased.

McNaughton, D.J., "Emission Source Specification in a Regional Pollutant Transport Model", June 1979, Battelle Pacific Northwest Laboratories, Richland, Washington.

A series of sensitivity and verification tests are examined to evaluate various means of emission source specification in a regional model. The regional air quality model incorporates pollutant transport, dispersion, chemical transformation, dry deposition, and precipitation scavenging of primary and secondary pollutants. It is currently under development and evaluation for sulfur oxide predictions as part of the Multistate Atmospheric Power Production Pollution Study (MAP3S). The effects of spatial averaging of emissions as would be applicable to models using gridded emissions as input are discussed. The analysis makes use of observed ambient air concentration data for sulfate and sulfur dioxide and sulfate concentrations in precipitation. Cases for study were selected from the intensive study periods of the Sulfate Regional Experiment of the Electric Power Research Institute.

Michael, P. and Raynor, G.S., "Modification of Trajectory Models Needed for Pollutant Source-Receptor Analysis", BNL-29924, 1981, Brookhaven National Laboratory, Upton, New York.

The purpose of this paper is to discuss some of the difficulties encountered when using the usual trajectory models to calculate source-receptor relationships for the incorporation of pollutants into precipitation. Section II discusses sources of difficulty; Section III gives the results of trajectory calculations that were performed to test the sensitivity of source-receptor relationships to transport layer assumptions; and Section IV presents some suggestions for future developments.

Mills, M.T. and Reeves, M., "A Multi-Source Atmospheric Transport Model for Deposition of Trace Contaminants", ORNL-NSF-EATC-2, 1973, Oak Ridge National Laboratory, Oak Ridge, Tennessee.

An Atmospheric Transport Model (ATM) was developed from a Gaussian plume model to calculate the toxic material deposition rate at any point within a watershed, given the location of various air pollution sources. Sources included the point source (stack), area source (city), line source (road or railbed), and windblown source (dry tailings pond). These rates were then used by the Wisconsin Hydrological Transport Model (WHTM) to determine the subsequent transfer of the toxic materials to stream and ground water. The result was a Unified Transport Model (UTM) which couples the air, land, and water systems. Sample calculations and computer program input and output are given for a test area.

Moore, R.E., "Calculation of the Depletion of Airborne Pollutant Plumes Through Dry Deposition Processes", Environment International, Vol. 3, No. 1, 1980, pp. 3-10.

The depletion of airborne pollutant plumes resulting from dry deposition on ground surfaces should be taken into account when estimating pollutant concentrations in air at distances downwind from their sources. This can be done by using a reduced release rate instead of the actual release rate for the pollutant in atmospheric dispersion equations. The reduced release rate is the actual rate of release multiplied by a depletion fraction. Depletion fractions for use with the Gaussian atmospheric dispersion equation of Pasquill and Gifford were calculated by numerical integration for downwind distances ranging from 35 to 90,000 m for release heights from 1 to 400 m, and they are tabulated for convenient interpolation for intermediate values of distance and release height. Pasquill atmospheric stability categories A-G are included.

Moroz, W.J., "Prediction of Deposition of Acid Precursors", Research Review of Ontario Hydro, No. 2, May 1981, pp. 75-80.

Acidic deposition originates from sources ranging from exceptionally large--almost single-point emitters--to small, scattered, diffuse emitters distributed over the area of entire cities or regions. In these circumstances, numerical models are needed to quantify the deposition rates over receptors and to identify the significance of both individual and collective sources. Modeled calculations of this type indicate that Ontario Hydro's contribution to total acidic deposition in sensitive receptor areas in Ontario is relatively small.

Patterson, M.R., Mankin, J.B. and Brooks, P.A., "Overview of Unified Transport Model", Proceedings of First Annual National Science Foundation Trace Contaminant Conference, 1973, National Science Foundation, Washington, D.C., pp. 12-23.

A computer model is being developed which couples atmospheric deposition of trace contaminants to a hydrologic transport mode. This Unified Transport Model consists of two major submodels: the Atmospheric Transport Model and the Wisconsin Hydrologic Transport Model. These models are described and discussed. Physical transport, chemical exchange processes, and coupling of toxicant transport to biological uptake and food chain processes are actively being modeled. A brief summary is given of the programming considerations, coupling of the submodels, and the capabilities of the unified model. A summary of model validation studies is given along with a projection of the modeling effort during the next year. The Wisconsin Hydrologic Transport Model has been translated from Fortran V to Fortran IV and was implemented on the IBM 360 computer. The Air Transport Model was coupled to the Wisconsin Model to obtain washout coefficients, provide deposition values, and form the nucleus of the United Transport Model. The soil erosion process in the unified model is being reformulated, and the suspended particulates will be coupled to the Texas model for sediment exchange in the channel system. Sediment transport in channels is being studied. Soil moisture and ground water transport is actively being developed in a finite element formulation.

Powell, D.C. et al., "Variable Trajectory Model for Regional Assessments of Air Pollution from Sulfur Compounds", Contract EY-76-C-06-1830, Feb. 1979, Battelle Pacific Northwest Laboratories, Richland, Washington.

This report describes a sulfur oxides atmospheric pollution model that calculates trajectories using single-layer historical wind data as well as chemical transformation and deposition following discrete contaminant air masses. Vertical diffusion under constraints is calculated, but all horizontal dispersion is a function of trajectory variation. The ground-level air concentrations and deposition are calculated in a rectangular area comprising the northeastern United States and southeastern Canada. Calculations for a 29-day assessment period in April 1974 are presented along with a limited verification. Results for the studies were calculated using a source inventory comprising 61 percent of the anthropogenic SO_2 emissions. Using current model parameterization levels, predicted concentration values are most sensitive to variations in dry deposition of SO_2, wet deposition of sulfate, and transformation of SO_2 to sulfate. Replacing the variable mixed-layer depth and variable stability features of the model with constant definitions of each results in increased ground-level concentration prediction for SO_2 and particularly for sulfate.

Ragland, K.W. and Wilkening, K.E., "Intermediate-Range Grid Model for Atmospheric Sulfur Dioxide and Sulfate Concentrations and Depositions", Atmospheric Environment, Vol. 17, No. 5, 1983, pp. 935-947.

A three-dimensional time-dependent grid type model for two chemically reacting species which undergo atmospheric transport, diffusion and wet and dry deposition over a region of several hundred km is presented. Accuracy and sensitivity of the model are discussed. The model is applied to the Rainy Lake Watershed using the available emissions and meteorological data. The model calculations are compared to National Atmospheric Deposition Program wet deposition data and snow core data.

Ritchie, I.M., Bowman, J.D. and Burnett, G.B., "Mesoscale Atmospheric Dispersion Model for Predicting Ambient Air Concentration and Deposition Patterns for Single and Multiple Sources", Atmospheric Environment, Vol. 17, No. 7, 1983, pp. 1215-1223.

A mesoscale modified Gaussian model (MGM) that incorporates chemical oxidation of sulfur dioxide, dry deposition and precipitation scavenging for single and multiple sources is described. The model was verified in the range 5-250 km using 1.5 years of monthly deposition data, 1 year of 24-h ambient air SO_2 data, and by comparison to two dispersion models. The calculated deposition rates which were compared to data collected around the Sudbury, Ontario, Canada, smelter complex were found to be within a factor of two to three of measured values depending on the averaging time. The comparison with two existing dispersion models gave results to within a factor of two. Sensitivity testing on the input parameters is also discussed.

Ritchie, L.T., Brown, W.D. and Wayland, J.R., "Effects of Rainstorms and Runoff on Consequences of Atmospheric Releases from Nuclear Reactor Accidents", Nuclear Safety, Vol. 19, No. 2, Mar.-Apr. 1978, pp. 220-239.

The preliminary model describing the effects of washout and runoff on the consequences of a nuclear reactor accident is presented. This new model is compared with the consequence model--consequence of reactor accident code (CRAC)--developed for the NRC Reactor Safety Study (Wash-1400). Wash-1400 used a simplified washout model in which there was no runoff. The new rain model differs from CRAC in two main ways: the average rain rate in the new model is determined from rain-gauge data rather than being fixed at 0.5 mm/hr; and the spatial structure of rain is manifested in the new model by four levels of rain activity. In the original model, the spatial structure of the rain is uniform. Calculations resulting from use of the models indicate that runoff and the spatial and temporal structure of rainstorms can have large effects on the consequences of a nuclear reactor accident.

Schmidt, J.A., "Models of Particle Dry Deposition to a Lake Surface and Some Effects of Surface Microlayers", Journal of Great Lakes Research, Vol. 8, No. 2, 1982, pp. 271-280.

A review of the theoretical and empirical information that has been used in modeling dry deposition over water reveals uncertainty up to the order of magnitude level for particles in the important 0.5 to 5 "mu" m range of diameter. Much of this is attributable to uncertainty in identifying the mechanisms controlling this transport. The results do indicate that dry deposition velocities are a function of particle size as well as windspeed, surface roughness, reference height, stability, and possibly relative humidity. Surface microlayers may affect roughness, but they are unlikely to exert significant control over deposition rates except on a very local level. Estimates of potential lake-derived concentrations in air indicate that bubble ejection may prove important in evaluating net atmospheric loading rates for some metallic species and perhaps organic substances as well.

Schnoor, J.L., Carmichael, G.R. and van Shepen, F.A., "Integrated Approach to Acid Rainfall Assessments", 1980, University of Iowa, Iowa City, Iowa.

A tiered approach to the mathematical modeling of acid precipitation, its fate and effects, might include: (1) a steady state susceptibility model to assess long-term effects from average annual atmospheric loadings on surface and ground waters; (2) a dynamic, regional model to assess the role of long range versus point source pollutant loadings for regulatory purposes; and (3) a three-dimensional dynamic event model to assess the maximum environmental insult to a specific ecosystem during critical conditions such as a snow-melt event. An integrated approach is needed which considers the fate of pollutants from their origin through terrestrial and aquatic ecosystems to the ultimate water resource, ground water. In this study the first step, the steady state model, is developed and applied to the northern Wisconsin, Minnesota, and Michigan water resource area.

Sehmel, G.A., "Model Predictions and a Summary of Dry Deposition Velocity Data", Aug. 1979, Battelle Pacific Northwest Laboratories, Richland, Washington.

Literature values of independent measurements of dry deposition velocities are summarized as a function of particle diameter and gas speciation. In most of the experiments reported in the literature, there are uncertainties that have hindered the development of general predictive deposition velocity models. However, one model (Sehmel and Hodgson, 1978) offers a more useful approach for predicting particle dry deposition velocities as a function of particle diameter, friction velocity, aerodynamic surface roughness, and particle density.

Sehmel, G.A. and Hodgson, W.H., "Model for Predicting Dry Deposition of Particles and Gases to Environmental Surfaces", American Institute of Chemical Engineers Symposium Series, Vol. 76, No. 196, 1980, pp. 218-230.

A predictive model is demonstrated for correlating particle and gas removal rates from the atmosphere by dry deposition. Predicted deposition rates can vary over several orders of magnitude and are complex functions of pollutant, air, and surface variables. The validity of many published field deposition measurements is discussed in terms of these complexities.

Shannon, J.D., "Advanced Statistical Trajectory Regional Air Pollution Model", 1978, Argonne National Laboratory, Argonne, Illinois.

The statistical trajectory technique has been used in the Statistical Trajectory Regional Air Pollution (STRAP) model. There are no temporal variations in quantities such as dry deposition velocities of sulfur dioxide and sulfate, rate of transformation of SO_2 to sulfate, and stability or vertical eddy diffusivity. In order to utilize state-of-the-art knowledge about typical variations of the above parameters, the Advanced Statistical Trajectory Regional Air Pollution (ASTRAP) model has been developed. ASTRAP contains hourly variations of the parameters, specified according to best estimates of typical values on the regional scale, yet still maintains a computational simplicity nearly equivalent to that of STRAP. ASTRAP consists of subprograms for calculation of horizontally integrated vertical profiles, horizontal dispersion statistics, and resulting concentrations.

Shannon, J.D., "Model of Regional Long-Term Average Sulfur Atmospheric Pollution, Surface Removal, and Net Horizontal Flux", Atmospheric Environment, Vol. 15, No. 5, 1981, pp. 689-701.

The Advanced Statistical Trajectory Regional Air Pollution model presented combines efficient calculation of long-term regional-scale concentrations and fluxes of pollutant sulfur with improved parameterizations of boundary-layer processes. The parameterizations include diurnal and seasonal variations of dry deposition velocities for SO_2 and sulfate, rate of transformation from SO_2 to sulfate, vertical structure of the planetary boundary layer, and emission rates.

Tsai, Y.J. and Johnson, D.H., "Cooling Tower Drift Model", Fifth Pittsburgh Conference Proceedings, 1974, Pittsburgh University, Pittsburgh, Pennsylvania, pp. 143.

A computer model has been developed to predict the distribution of drift from brackish or salt-water cooling towers for power plant condensate cooling. Input into the model consists of the cooling tower characteristics and field weather measurements. The model calculates the drift transport for each set of meteorological data at three hour intervals. For each droplet class size, the initial rise in the plume is computed. Droplet fall, evaporation, and downwind transport are calculated at 10 sec intervals until ground level is reached. The model has four groups of output; annual deposition of dissolved solids, annual deposition of water, maximum near-ground air concentration, and annual near-ground air concentration. This data is used to evaluate the impact on the environment of the cooling tower, including the determination of the effects on plant foliage and growth, and ground water quality.

Williams, R.M., "Model for the Dry Deposition of Particles to Natural Water Surfaces", Atmospheric Environment, Vol. 16, No. 8, 1982, pp. 1933-1938.

More realistic estimates of the deposition velocity, for aerosol particle deposition on natural water surfaces, are obtained by modifying earlier resistance models to include the effects of wave breaking and spray formation in high winds and particle growth in the humid regions near the air/water interface. Both processes act to enhance deposition by shunting the high transfer resistance to a smooth water surface.

APPENDIX D

LONG RANGE TRANSPORT OF AIR POLLUTANTS

Barnes, R.A., "The Long Range Transport of Air Pollution--A Review of European Experience", Journal of American Pollution Control Association, Vol. 29, No. 12, Dec. 1979, pp. 1219-1235.

This is a general review of the acid rain problem. It includes long range transport, emissions, atmospheric chemistry, LRTAP climatology, effects, costs of damage and abatement options and costs.

Chung, Y.S., "The Distribution of Atmospheric Sulfates in Canada and Its Relationship to Long-Range Transport of Air Pollutants", Atmospheric Environment, Vol. 12, No. 6-7, 1978, pp. 1471-1480.

Analyses were carried out using data obtained in April-May 1975. Atmospheric sulfate loadings were relatively low (less than 5 ug/m^3) in the Western Provinces while high values (less than 15 ug/m^3) were frequently observed in eastern Canada. The natural background level of ambient sulfates appeared to be less than 1.5 ug/m^3 in Canada. Maximum values (greater than 20 ug/m^3) occurred in southern Ontario, however. Generally, values recorded in the Atlantic Provinces were lower than the values observed in Ontario. The observed levels of sulfates could be associated with 3 types of typical weather situations--extratropical cyclones or the forward sides of anticyclones with cold northerly flows (low sulfates); the rear sides of anticyclones (high sulfates); and quasi-stationary fronts (mixed). Air-parcel trajectory analyses of the low-level atmosphere show that high sulfate levels were often associated with S-SW airflows on the rear side of a warm, moist air mass. The results suggest the long-range transport of airborne S pollutants, mainly from several industrial areas in the U.S.

Galvin, P.J. et al., "Transport of Sulfate to New York State", Environmental Science and Technology, Vol. 12, No. 5, Aug. 1979, pp. 580-584.

Trajectory analysis provides evidence that high sulfate concentrations observed during summer high-pressure systems at three rural sites in New York State are products of sulfur dioxide emissions to the south and southwest of New York State. The highest concentrations occur in air that first stagnates over the area surrounding the Ohio River Valley and then is advected into New York State as the high-pressure system begins to move eastward. The amount of oxides of nitrogen that emissions data indicate should accompany these high sulfate concentrations is not present in the particulate samples as inorganic nitrate.

Kerr, R.A., "Global Pollution: Is the Arctic Haze Actually Industrial Smog?", Science, Vol. 205, No. 4403, July 20, 1979, pp. 290-293.

Heavy haze in the Arctic that can reduce visibility from more than 100 km to less than 10 km may have its origins 10,000 km away in the same polluted air that produces acid rain over the U.S. and Europe. Analyses of samples collected by the Arctic sampling network tend to support the theory of extremely long-range transport of pollutants into the Arctic. An alternative explanation for the dirty samples of Arctic air would be that the pollution could originate from a local Arctic source. Although data on the chemistry and meteorology of the Arctic haze continue to be collected, the possible environmental effects of the haze are still unknown.

Lewis, D.H. and Ball, R.H., "Long-Range Transport of Air Pollution", DOE/EP-0037, Jan. 1982, U.S. Department of Energy, Washington, D.C.

Gaseous pollutants such as SO_2, NO_x, and ozone are transported over distances on the order of 100 km. These pollutants may cause significant deterioration in areas which meet air quality standards or, when combined with locally generated pollutants, they may result in the nonattainment (violation) of air quality standards. Pollutants transported longer distances (several hundred to several thousand kms) tend to be various forms of fine particulates, including sulfates and nitrates formed from emitted SO_2 and NO_x, emitted particles (fly ash, heavy metals, minerals), and organic products of photochemical processes. Fine particulates are responsible for much of the known and suspected adverse impacts of long-range transport. Current regulatory activity is limited to situations where sources in one state may be contributing to violations of ambient standards for particulates, SO_2, NO_x, and ozone in another state, and to EPA proposed rules to mitigate visibility impairment due to plumes from single sources. Most of the long-range transport effects (e.g., acid precipitation and regional visibility impairment) which occur great distances (hundreds to thousands of kms) from sources have not yet been regulated using the existing Clean Air Act provisions for ambient standards and source standards of performance, primarily because the sulfates and nitrates which cause these effects are secondary pollutants, rather than primary (or emitted) pollutants the Clean Air Act was designed to control. Models of sulfur transport indicate that sulfur dioxide (SO_2) is carried a distance of about 250 km on the average, and sulfates (SO_4) about 1000 km, before being deposited, but values vary greatly with atmospheric conditions; ambient violations of SO_2 standards ordinarily will not be caused by pollutants transported over such distances unless reinforced by local emissions.

Martin, H.L., "Long-Range Transport of Airborne Pollutants and Acid Rain Conference", Water, Air, and Soil Pollution, Vol. 18, No. 1-3, July/Oct. 1982.

Proceedings include 30 papers presented at seven technical sessions that deal with the problems of environmental protection and examine the interfaces between various disciplines such as soil-atmosphere interactions, atmosphere-vegetation interactions, etc. Available models of all components of the problem including atmospheric delivery, geochemistry and hydrology, and aquatic effects are also examined and discussed. Technical and professional papers from this conference are indexed with the conference code no. 01339 in the Ei Engineering Meetings (TM) databases produced by Engineering Information, Inc.

McMillan, A.C., "Long-Range Transport of Atmospheric Pollutants", Research Review of Ontario Hydro, No. 2, May 1981, pp. 11-20.

The meteorological factors affecting long-range transport of pollutants in the atmosphere are discussed. The literature describing the relation of meteorology to precipitation acidity is briefly surveyed and an example of an acid rain incident in eastern North America is presented. It is concluded that both large-scale flows and mesoscale features must be considered if the meteorological conditions leading to long-range transport are to be understood.

Miller, J.M., "Acidity of Hawaiian Precipitation as Evidence of Long-Range Transport of Pollutants", 1979, National Oceanic and Atmospheric Association, Silver Spring, Maryland.

Precipitation has been collected at five sites on the island of Hawaii from sea level to 3400 m for a period of four years. Samples were taken on a daily or biweekly basis and analyzed for pH, conductivity, and certain ions such as sulfate and nitrate. The data show that the most acidic rains fall above the 2500 m level. Since the precipitation forming clouds must scavenge material above this level, it is suggested that acid-forming pollutants reach mid-troposphere levels and are transported long distances until being scavenged in the Hawaiian rains.

National Technical Information Service, "Acid Precipitation: Transfrontier Transport of Air Pollutants--1976-November, 1982 (Citations from the Energy Data Base)", Nov. 1982, U.S. Department of Commerce, Springfield, Virginia.

This bibliography contains citations concerning transfrontier transport of air pollutions implicated in acid rain production. The legal and political ramifications of transfrontier pollution are discussed. International remedies including international cooperation in enforcement of pollution abatement regulations and treaty agreements are considered. U.S./Canadian pollution problems are highlighted. This contains 69 citations fully indexed and includes a title list.

OECD, "The OECD Programme on Long Range Transport of Air Pollutants", 1977, Office of Economic Cooperation and Development, Paris, France.

A study was initiated to determine the relative importance of local and distant sources of sulfur compounds to the air pollution over a region. Special attention was focused on acidity in atmospheric precipitation. The research program developed covered northwestern Europe and comprised three essential elements: a survey of sulfur dioxide emissions all over the region; the measurement of sulfur compounds at a network of ground sampling stations and by aircraft; and the development and testing of mathematical air dispersion models to relate emission data with concentration and deposition data. Results indicate that sulfur compounds do travel long distances in the atmosphere and that the air quality in any one European country is measurably affected by emissions from other European nations.

Ostergaard, K., "The Problem of Regional Transportation of Air Pollutants in Northern Europe", Clean Air-Australia, Vol. 8, No. 2, May 1974, pp. 30-33.

An OECD report has indicated that sulfur dioxide emissions from sanitary sources will double from 1968-80 if no special action is taken to control these sources. This is a matter of concern in Norway and Sweden where acidified precipitation, possibly transported from industrial areas in central Europe and the U.K., damages fish populations and forests. Evidence for long-range transportation of air pollutants to northern Europe is reviewed.

Ottar, B., "International Agreement Needed to Reduce Long-Range Transport of Air Pollutants in Europe", Ambio, Vol. 6, No. 5, 1977, pp. 262-270.

A recently completed OECD study on long-range transport of air pollutants shows that all European countries receive considerable amounts of pollutants from outside. Scandinavia, Switzerland, and Austria receive considerably larger contributions from abroad than from their own sources. Acid precipitation in Scandinavia has a pronounced episodic character. A monitoring program has been established within the U.N. Economic Commission for Europe, but the only efficient way to deal with the problem would be an international agreement to reduce emissions.

Sandusky, W.F., Eadie, W.J. and Drewes, D.R., "Long-Range Transport of Pollutants in the Pacific Northwest", Jan. 1979a, Battellle Pacific Northwest Laboratories, Richland, Washington.

Air quality impacts associated with future utility and industrial siting as defined by the business as usual scenario have been analyzed. This analysis is based on assumptions regarding emission rates, implementation of Best Available Control Technology (BACT), siting data generated by the ORNL regional studies program, and a regional scale transport, transformation, and removal model for SO_2, sulfates, and particulates. Results of this assessment show that industrial siting in the Portland-Seattle area may be constrained due to regulations for Prevention of Significant Deterioration (PSD) in terms of both incremental SO_2 and particulate concentrations at sites near Class I areas. Topography influences the concentration patterns of the pollutants. Generally, these patterns reflect the wind flow characteristics but are modified by dry deposition processes. Maximum predicted ground-level concentrations of SO_2, sulfate, and particulates occur within 63 km of the source. Over 80 percent of the sulfur emissions in the Pacific Northwest U.S. will ultimately be deposited within the region. The percentage deposited for industrial sources is slightly higher than utility sources due to the lower effective release height.

Sandusky, W.F., Eadie, W.J. and Drewes, D.R., "Long Range Transport of Sulfur in the Western United States", Contract: EY-76-C-06-1830, Jan. 1979b, Battelle Pacific Northwest Laboratories, Richland, Washington.

Pollutants, such as SO_2 and sulfate, emitted from both utility and industrial coal burning processes have long residence times in the atmosphere. Therefore, the long-range atmospheric transport and diffusion of these pollutants must be considered in any environmental assessment of proposed plant operation. The most useful tool in predicting the long-range transport of pollutants is a computer simulation technique for the Gaussian diffusion equation. Information produced by the model includes: SO_2 and sulfate ground-level air concentrations, the amount of SO_2 and sulfate deposited on the ground surface, the minimum pH value in the rainfall, and the budget of SO_2 and pH value in the rainfall, and the budget of SO_2 and sulfate material over the diffusion grid. Information on siting and emissions is also required. For this study, siting was based on projected coal use in 1985 and 1990 based on a two-thirds increase in coal production. Results of the modeling for the western United States indicate that the maximum incremental ground-level air concentrations for SO_2 are 8.4 and 14 mu g/m^3 for utility and industrial sources, respectively. Maximum predicted incremental ground-level sulfate concentrations for utility and industrial sources are 0.8 and 1.2 mu g/m^3 respectively. The minimum calculated pH values for both utility and industrial sulfur emissions were 5.3. Maximum SO$_2$ deposition amounts range from 0.5 to 0.8 gm/m^2 for both the utility and industrial coal use scenarios. The largest sulfate deposition amounts range from a factor of 55 to 24 smaller than SO$_2$ deposition amounts.

Scriven, R.A. and Fisher, B.E.A., "The Long-Range Transport of Airborne Material and Its Removal by Deposition and Washout - II. The Effect of Turbulent Diffusion", Atmospheric Environment, Vol. 9, No. 1, Jan. 1975, pp. 59-69.

A diffusion approximation is used to estimate long distance transport in the mixing layer to assess the accuracy of a simple model and to investigate the buffering effect of diminishing atmospheric turbulence as the absorbing ground or sea is approached from above. For sulfur dioxide in the absence of rain and chemical reactions, mean travel distances of the order of 700 km may occur when vertical diffusion is limited by an inversion at a height of about 1 km. Deposition rate for long distance travel is largely governed by the diffusive resistance of the low turbulence region adjacent to the surface.

Shaw, R.W., "Acid Precipitation in Atlantic Canada", Environmental Science and Technology, Vol. 13, No. 4, Apr. 1979, pp. 406-412.

Long-range transport of sulfur oxides and nitrogen oxides may pose a serious pollution problem to forest areas of Canada's eastern coast. Precipitation monitoring data show that sulfur compounds are becoming more evident in the area. An August 1976 study found aerosol sulfate concentrations in Nova Scotia were as high as 20 mg/cu m. Monitoring conducted by the World Meteorological Organization is discussed. Aerosol sulfate concentrations in forest areas of Canada's Atlantic provinces increase when air masses have passed through populated and industrialized areas. Current research and development on possible effects of pollutant transport on the environment in eastern Canada are reviewed.

Slinn, W.G., "Estimates for the Long-Range Transport of Air Pollutants", May 1981, Battelle Pacific Northwest Laboratories, Richland, Washington.

Different atmospheric, source, and surface conditions can result in substantially different ranges of atmospheric transport of air pollutants; for example, even for anthropogenic sulfur and nitrogen, the ranges can vary from about 10 exp 1 to 10 exp 5 km. In this report, the emphasis is on indicating some of the reasons for the great variability of these ranges. Thus, some of the complexities of dry deposition, atmospheric chemistry, and precipitation scavenging are described, and it is demonstrated how synoptic-scale meteorologic conditions can control both dry and wet deposition. On the other hand, it is suggested that the mean, tropospheric-residence time of particulate sulfur and nitrogen from fossil-fuel combustion in temperate latitudes is probably about a week, but the amount of this material remaining airborne after a week can be large, since the amount is expected to have a log-normal distribution over an ensemble of realizations. Applications of the results to the U.S./Canadian acid-rain issue, to episodic-deposition events, and to global-scale atmospheric pollution are indicated.

Tollan, A. and Hagerhall, B., "Deterioration of Water Quality Due to Long-range Transport of Air Pollution", Proceedings of the United Nations Water Conference, (mar del Plata, Argentina), 1978, United Nations, New York, New York, pp. 2059-2070.

Increased acid precipitation is the main cause of extensive losses of salmonid fish stocks and other fish populations in southern Scandinavia, northeastern United States, and southeastern Canada. In addition, airborne transport of sulfur compounds are a threat to forest ecosystems and productivity, and to human health. Extensive surveys conducted in Sweden and Norway have documented adverse effects of the transport of pollutants through the atmosphere, and research has indicated that the major source of acid precipitation in Scandinavia is the combustion of fossil fuels in the heavily industrialized parts of Europe. Highly acid precipitation in a central area of Europe as early as the 1960s had a pH of 3-4, and the area is growing steadily larger. All governments should reduce emissions of pollutants to the air. The Governments of Norway and Sweden have prohibited use of heavy oil containing over 1 percent sulfur. World efforts to monitor and evaluate long-range transmissions of air pollutants across national frontiers must be intensified, and data collection should be integrated in the United Nation's Global Environmental Monitoring System. The present policy in some countries of dispersing harmful pollutants by means of tall stacks is no longer acceptable. International cooperation in reducing air pollutant emissions is needed.

World Meteorological Organization, "Papers Presented at the WMO Symposium on the Long-Range Transport of Pollutants and Its Relation to General Circulation Including Stratospheric/Tropospheric Exchange Processes", WMO No. 538, 1979, Geneva, Switzerland.

This volume contains 57 papers presented at the conference, 39 of which are indexed separately. Among the subjects covered are emission source specification, airborne air pollution monitoring, meteorological factors affecting pollutant concentrations, dry deposition of gaseous pollutants, sulfur aerosols in power plant plumes, transformation of pollutants during transport, wet removal of sulfur by rainfall, plume lateral spread, trajectory models, Lagrangian characteristics of turbulence and diffusion in the planetary boundary layer, transport and distribution of pollutants in the troposphere, chemical exchange between the stratosphere and troposphere through turbulent mixing, and others.

APPENDIX E

EFFECTS OF ACID PRECIPITATION

Abrahamsen, G. et al., "Impact of Acid Precipitation on Forest and Freshwater Ecosystems in Norway", Report 6/76, Mar. 1976, Agricultural Research Council of Norway, Oslo, Norway.

The phenomenon of acid precipitation and its effects on natural ecosystems in Norway are discussed. The geographical distribution of wet deposition of sulfate and acid in Norway, and the long-term changes in precipitation quality are described. Estimates of the dry deposition based on dispersion model calculations are included. The effects of acid precipitation on coniferous forests, vegetation canopies, and soil are surveyed. The geographic distribution and extent of effects of acid precipitation on the chemistry of Norwegian lakes are examined. Some pertinent research data on the effects of high acidity on freshwater organisms, with emphasis on fish, are summarized.

Abrahamsen, G., Houland, J. and Haguar, S., "Effects of Artificial Acid Rain and Liming on Soil Organisms and the Decomposition of Organic Matter", Effects of Acid Rain Precipitation on Terrestrial Ecosystems, Plenum Press, New York, New York, 1980, pp. 341-362.

This paper presents results of Norwegian studies concerning the effects of artificial acid rain on decomposer organisms and decomposition effects in coniferous forests. Descriptions of the experimental conditions are given. Many of the invertebrates studied appeared to tolerate and even thrive in increased acidic conditions but showed reductions in numbers under liming conditions. Decomposition studies showed initial plant decomposition influence by acid rain to be minimal while the decomposition of humus material was found to be pH dependent. Since these effects were observed under "artificial" conditions, the authors advocate that the results not be applied directly to an actual acid rain situation.

Alexander, M., "Effects of Acidity on Microorganisms and Microbial Processes in Soil", Effects of Acid Rain Precipitation on Terrestrial Ecosystems, Plenum Press, New York, New York, 1980, pp. 363-374.

This article is a review of the factors and mechanisms and inter-relationships among acidity, soils, and microorganisms found in soil with emphasis on the reactions of nitrogen and phosphorus.

Allen, J.D., "Laboratory Simulation of Acid Rain Effects on Freshwater Microcrustaceans", OWRT-A-054-MD(1), 1980, Water Resources Research Center, University of Maryland, College Park, Maryland.

The principal objective of this study was to investigate the sensitivity of freshwater microcrustaceans to stress associated with acid rain. All experiments were performed in large (50-gallon) aquaria in order to have satisfactory pH control, to be able to use sufficient numbers of replicates for each pH treatment, and to be sufficient to make small additions of sulfuric acid since the test water had low alkalinity. The potential effect of acid rain on Dephnia pulex (Crustacea: Cladocera) was studied in laboratory populations subjected to exposures for 1, 12, 24, 48, and 96 hours, and at pH levels ranging from 4.0 to ambient (6.6).

Amthor, J.S. and Bormann, F.H., "Productivity of Perennial Ryegrass as a Function of Precipitation Acidity", Environmental Pollution Series A, Vol. 32, No. 2, 1983, pp. 137-145.

Glasshouse-grown perennial ryegrass plants were treated with simulated precipitation of pH 5.6, 4.0 or 3.0 twice a week for 14 weeks during which the forage was sequentially harvested five times to determine the effects of precipitation acidity upon forage dry mass accumulation. Forage dry mass present during the first, second, third and fourth harvests was not affected by precipitation acidity. During the last (fourth) regrowth period, dry mass accumulation (fifth harvest) was inhibited by increased precipitation acidity. Regression analysis indicated that the relationship between precipitation acidity and forage production during the fourth regrowth period was significant if acidity was expressed as hydrogen ion concentration ($p = 0.099$) or the logarithmic transformation to pH. It is concluded that short-term responses may not reflect long-term growth patterns by plants subjected to acid precipitation.

Andersson, F., Fagerstrom, T. and Nilsson, S.I., "Forest Ecosystem Responses to Acid Deposition--Hydrogen Ion Budget and Nitrogen/Tree Growth Model Approaches", Effects of Acid Rain Precipitation on Terrestrial Ecosystems, Plenum Press, New York, New York, 1980, pp. 319-334.

Diagramatic models of soil and forest hydrogen ion budgets are given with discussions of input-output, root uptake, mineralization, and weathering effects. The nitrogen dependent tree growth model approach was used for this study on data from Sweden.

Arnold, D.E., Light, R.W. and Dymond, V.J., "Probable Effects of Acid Precipitation on Pennsylvania Waters", EPA-600/3-80-012, Jan. 1980, U.S. Environmental Protection Agency, Washington, D.C.

In order to identify any trends in water chemistry and fish communities in Pennsylvania waters which would indicate that acid precipitation was affecting them adversely, five existing data bases were examined for the existence of water analyses from the same or nearby locations, separated by at least one year. Of 983 analysis reports, there were 314 cases with two or more such points. Of these 107 or 34 percent showed a decrease in pH, alkalinity, or both. The average decrease in pH was 0.4 units, with a maximum of 1.3 units. The average decrease in alkalinity was 15.1 mg/l (as calcium carbonate) with a maximum of 105 mg/l. The average time span between the earliest and latest sample was 8.5 yr. When the data was separated by physiographic provinces, it became apparent that although the majority of the decreases occurred in the streams on the relatively insoluble rocks of the Allegheny Plateau, there were also many cases in the ridge-and-valley province and other regions. Many of these decreases are to pH levels considered marginal for growth and reproduction of trout and other fishes. Seventy-one of the 107 analyses showing decreased pH or alkalinity included fish collection data. In 40 of these cases (58 percent), the number of fish species present decreased as well.

Arthur, M.F. and Wagner, C.K., "Response of Agricultural Soils to Acid Deposition", July 1982, Battelle Columbus Laboratories, Columbus, Ohio.

Proceedings of the workshop, Response of Agricultural Soils to Acid Deposition, which was held May 12-13, 1981, in Columbus, Ohio, and which evaluated the potential beneficial and harmful impacts of atmospheric acid deposition on agricultural soils are presented. Those issues requiring further research are also identified. Five working papers and a literature review prepared by soils specialists are included as is a summary of conclusions reached by the participants.

Baath, E. et al., "Soil Organisms and Litter Decomposition in a Scots Pine Forest--Effects of Experimental Acidification", Effects of Acid Rain Precipitation on Terrestrial Ecosystems, Plenum Press, New York, New York, 1980, pp. 375-380.

This paper presents preliminary results of experiments (Sweden) of 3 block units of Scots pine using "artificial" acid rain. Initial results indicate increased leaching in soils, lower decomposition rates of pine needles and root litter after acidification, and decreases in numbers of bacteria and fungus.

Baath, E., Lundgren, B. and Soderstrom, B., "Effects of Artificial Acid Rain on Microbial Activity and Biomass", Bulletin of Environmental Contamination and Toxicology, Vol. 23, 1979, pp. 737-740.

Respiration, FDA-active fungal length and FDA-active bacteria are all parameters related to microbial activity in the soil. Total fungal length and bacterial numbers did not change significantly in response to artificial acid rain, indicating that the microbial immobilization of mineral elements was little affected in this experiment. Bacterial cell size was smaller in the most acidified plots. Mean cell size in this treatment was 0.10 m^3 compared to 0.13 m^3 in the control and the pH 3 treated plot. The reduced cell size could be due to changes in the bacterial population, or it could reflect a lower bacterial growth rate after treatment with pH 2 water. It seems most likely that the decreased microbial activity, directly or indirectly, is a hydrogen ion effect. However, since pH was adjusted with sulphuric acid, it cannot be excluded that the added sulphate influenced the microorganisms negatively.

Babich, H. and Stotzky, G., "Atmospheric Sulfur Compounds and Microbes", Environmental Research, Vol. 15, No. 3, June 1978, pp. 513-552.

The effects of atmospheric sulfur compounds on microbes are assessed. The following topics are discussed: the possible contributions of volatized organic sulfur-containing compounds to the cycling of sulfur; the involvement of microorganisms in both production and removal of volatile organics from the atmosphere; and the effects of volatile organics on microbial activities in natural environments. Anthropogenic sources of sulfur are discussed. Natural abiotic and biotic sources of sulfur are examined. Atmospheric reactions involving sulfur are described. Ecological implications of inorganic sulfur-containing gases are explored. Additional studies are needed to evaluate both overt and covert effects of volatile organic sulfur-containing compounds on microorganisms.

Bache, B.W., "The Acidification of Soils", Effects of Acid Rain Precipitation on Terrestrial Ecosystems, Plenum Press, New York, 1980a, pp. 183-202.

This paper discusses the principles of soil acidification and the factors affecting it, especially concerning soil properties which modify these acidifying effects. Soil acidification by polluted precipitation at unacceptably high rates was studied and recommendations for additional research were given.

Bache, B.W., "The Sensitivity of Soils to Acidification", Effects of Acid Rain Precipitation on Terrestrial Ecosystems, Plenum Press, New York, New York, 1980b, pp. 569-572.

This paper very briefly discusses factors influencing soil sensitivity to acidification such as difference in lime capacity, the soil profile buffer capacity, and the proportion of water-soil reaction.

Baker, J.P., "Aluminum Toxicity to Fish as Related to Acid Precipitation and Adirondack Surface Water Quality", Ph.D. Dissertation, Jan. 1981, Cornell University, Ithaca, New York.

The Adirondack Region of New York State contains over 2000 lakes. Many of these lakes are experiencing acidification and declines or loss of fish populations. The objective of this investigation was to improve the understanding of effects of acidification of surface waters on Adirondack fish populations. It was found that on the average, aluminum complexed with organic ligand was the dominant aluminum form in the dilute acidified Adirondack surface waters studied. In laboratory bioassays, speciation of aluminum was shown to have a substantial effect on aluminum toxicity to early life history stages of fish. Based upon available field and laboratory data, concentrations of inorganic aluminum and hydrogen ions appear to be important factors determining fish survival in Adirondack surface waters affected by acidification. For the range of pH and aluminum levels, and fish species and life history stages studied, sensitivity to low pH levels decreased with increasing age while sensitivity to elevated inorganic aluminum levels increased with increasing age (through the post-swim-up fry stage). Water quality in Adirondack lakes and streams sampled was temporally highly variable. In general, in streams affected by acidification pH level was at a minimum (and conversely inorganic aluminum level at a maximum) during early phases of snowmelt in the spring and also during long periods of heavy rainfall, often concentrated in the fall season. Temporal water quality fluctuations in lakes were dampened by the larger water volume and longer water retention times within the system.

Beamish, R.J., "Growth and Survival of White Suckers (Catostomus Commersoni) in an Acidified Lake", Journal of the Fisheries Research Board of Canada, Vol. 31, No. 1, 1974, pp. 49-54.

White suckers (Catostomus commersoni) in the acidic Lumsden Lake in 1967 and 1968 exhibited reduced annual growth followed by death. The reduced growth and death appeared directly related to the low pH and not to a shortage of food caused by the decreasing pH. Examination of heavy metals in the lake from precipitation fallout showed that only zinc levels (24-33 micrograms/l) were sufficiently high to be potentially hazardous to fish. The heavy metal, although not of sufficient magnitude to be considered the principle stressing agent, may have acted synergistically with the acid, influencing its concentration.

Bell, J.N., "Acid Precipitation--A New Study from Norway", Nature, Vol. 292, No. 5820, July 1981, pp. 199-200.

An accelerating decline in fish stocks which resulted in the extinction of fish populations in an area of 13,000 square kilometers and severe problems over another 20,000 square kilometers led to the establishment in 1972 of a major Norwegian research program (the SNSF project) into the biological effects of acid precipitation. The final report of this project, published in 1980, demonstrates that precipitation in southern Scandinavia has become more acidic as a result of long-distance transport of air pollution. Nitrogen oxides are contributing an increasing proportion of the acidity. While routine analysis of precipitation samples provides estimates of the inputs of hydrogen ions, sulfur oxides, and nitrogen oxides, the importance of the dry deposition of gaseous and particulate sulfur and nitrogen compounds directly onto vegetation and soils is less well defined. Lake waters have become significantly more acidic in areas subject to acid precipitation. Acid precipitation also mobilizes aluminum in soils, which is a major factor in the destruction of fish populations. In southern Norway, where much of the precipitation falls as snow, particularly severe problems result from the flush of acid water into lakes during snow melt. Intensive studies failed to reveal any consistent effects of acid precipitation on forest productivity. Also, the effects of changes in land use on lake acidification are not fully understood. The SNSF project predicted that a reduction of at least 70 percent in hydrogen ions and sulfate concentrations in precipitation is necessary to restore lakes to their previous conditions.

Botkin, D.B. and Aber, J.D., "Some Potential Effects of Acid Rain on Forest Ecosystems: Implications of a Computer Simulation", BNL-50889, Apr. 1979, Brookhaven National Laboratory, Upton, New York.

Potential effects of acid precipitation on forest productivity and species composition were considered through the use of the JABOWA forest growth simulator. Physiological studies suggest that current levels of acid rain would result in a mortality of 5.5 percent or less of tree leaf tissue depending on the species. Such reductions produce nonsignificant changes in the total productivity of the simulated forest. Significant changes in total productivity and relative species composition do occur at much higher levels of effects and are discussed. The simulations reported here consider only effects due to species interactions and direct impact on leaves; no changes in soil chemistry are assumed.

Boylen, C.W. et al., "Microbiological Survey of Adirondack Lakes with Various pH Values", Applied and Environmental Microbiology, Vol. 45, No. 5, May 1983, pp. 1538-1544.

Atmospheric acidic deposition has not been well described in terms of its effects on aquatic microbial populations. Many of the lakes in the Adirondack region of New York State are particularly sensitive to high H+ inputs from acidic deposition. It is important that these ecosystems be examined so that future changes can be appraised. Nine high-altitude oligotrophic Adirondack lakes having water of pH 4.3 to 7.0 were surveyed for total bacterial numbers and possible adaptation of the microbial communities to environmental pH. The number of heterotrophic bacteria from water samples recoverable on standard plate count agar were low

(10 to 1000 per ml) for most of the lakes. Acridine orange direct counts were approximately two orders of magnitude higher than plate counts for each lake. Sediment aerobic heterotrophs recovered on standard plate count agar ranged from 14,000 to 1,300,000 per g of sediment. Direct epifluorescence counts of bacteria in sediment samples ranged from 3,000,000 to 1.4×10^7 per g. Low density values were consistent with the oligotrophic nature of all the lakes surveyed. There were no apparent differences in numbers of bacteria originally isolated at pH 5.0 and pH 7.0 between circumneutral lakes (pH greater than 6.0) and acidic lakes (pH less than 5.0). Approximately 1,200 isolates were recultured over a range of pH from 3.0 to 7.0. Regardless of the original isolation pH (pH 5.0 or pH 7.0), less than 10 percent of the isolates grew at pH less than 5.0. Those originally isolated at pH 5.0 also grew at pH 6.0 and 7.0. Those originally isolated at pH 7.0 preferred pH 7.0, with 98 percent able to grow at pH 6.0 and 44 percent able to grow at pH 5.0. A chi-square contingency test clearly showed (P less than 0.005) that two distinct heterotrophic populations had been originally isolated at pH 5.0 and pH 7.0, although there is undoubtedly some overlap between the two populations.

Bradford, G.R., Page, A.L. and Straughan, I.R., "Are Sierra Lakes Becoming Acid", California Agriculture, Vol. 35, No. 5/6, May/June 1981, pp. 6-7.

Sierra lakes, most of which have little buffering capacity, showed little change in acidity during the last 15 years. Sampling of 170 lakes in 1965 resulted in an average pH of 6.0 (range 4.7 to 7.3). The mean 1980 pH's measured in 114 randomly selected Sierra lakes in July and October were 6.1 and 6.5, respectively. The higher fall pH probably reflects the seasonal effect of increased residence time of the water in the lake and associated buffering by lake sediments. Ten additional lakes where fish kills had occurred had a mean pH of 6.6.

Brown, D.J. and Sadler, K., "The Chemistry and Fishery Status of Acid Lakes in Norway and Their Relationship to European Sulphur Emissions", Journal of Applied Ecology, Vol. 18, No. 2, Aug. 1981, pp. 433-441.

The effect of reducing sulfur emissions on the fish populations in southern Norway lakes (below 63 degrees North latitude) is estimated by analyzing the data of Wright and Snekvik (1978). This includes full chemical analyses of the water and fishery status of 471 lakes above 200 m elevation. In this group of lakes 48% have no fish, 43% are sparsely populated, and 9% have good populations. Regression analysis indicated that the pH and excess (nonsea) sulfate concentration were correlated, with a slope of 0.225. Assuming that the excess sulfate is largely derived from the combustion of fossil fuels, a 50% reduction in European sulfur emissions would increase pH by 0.2 units. This would improve the fishery status in lakes with pH above 4.9 (9% of all lakes). No improvement would take place in the 67% of lakes in the pH range 4.3-4.9, where pH and fish status are independent. Lakes north of the severely impacted region would not be expected to improve in fishery status following a reduction in sulfur emissions.

Canfield, Jr., D.E., "Sensitivity of Florida Lakes to Acidic Precipitation", Water Resources Research, Vol. 19, No. 3, June 1983, pp. 833-839.

To assess the potential vulnerability of Florida lakes to damage by acidic precipitation, data from a survey of 165 lakes located in the major physiographic and geologic regions of Florida were used to determine pH, total alkalinity, calcium hardness, and calcite saturation index values. Mean lake pH ranged from 4.1 to 8.9. Mean total alkalinity concentrations ranged from 0 to 4100 ueq/l and mean calcium hardness concentrations ranged from 20 to 4300 ueq/l. Total alkalinity averaged below 200 ueq/l in 49 percent of the sampled lakes. Calcite saturation index values were greater than 4 in 36 percent of the sampled lakes. Based on these data, Florida has a large number of lakes that are vulnerable to reductions in pH and alkalinity by acidic precipitation. Data on chlorophyll-a concentrations, zooplankton abundance, and recent fisheries data, however, suggest that Florida lakes may not be as biologically sensitive as the alkalinity and calcite saturation indices might suggest. Low phosphorus and nitrogen concentrations are primarily responsible for reduced animal and plant populations in acidic Florida lakes. Consequently, predictions of future impacts or trends based on current sensitivity indices should be regarded with caution.

Chang, F.H. and Alexander, M., "Effect of Simulated Acid Precipitation on Algal Fixation of Nitrogen and Carbon Dioxide in Forest Soils", Environmental Science and Technology, Vol. 17, No. 1, Jan. 1983, pp. 11-13.

Indigenous algae growing in soil samples from the Panther, Woods, and Sagamore Lake watersheds of the Adirondacks regions of New York were exposed to acid precipitation of pH 3.5 or 5.6 at rates of 50, 100, 200, and 300 cm. Rates of nitrogen fixation as measured by acetylene reduction were significantly less for soils watered with pH 3.5 water compared with the pH 5.6 water for all rainfall intensities, for light and dark, and for all soils. As rain volume increased, acetylene reduction decreased for the 3.5 pH water and increased for the 5.6 pH water. The carbon dioxide fixation rate was significantly less for all soils exposed to the more acid rain, and the extent of inhibition became more pronounced with increasing volumes of rain applied.

Chen, C.W. et al., "Acid Rain Model: Canopy Module", Journal of Environmental Engineering, Vol. 109, No. 3, June 1983, pp. 585-603.

As a part of the Integrated Lake-Watershed Acidification Study, an acid rain model is being developed to provide quantitative linkage between acid deposition and lake acidification. A canopy model has been developed to calculate throughfall characteristics based on canopy properties, ambient air quality, and precipitation quantity and quality. The processes considered include wet and dry deposition, leaf exudation, nitrification, and oxidation of SO_2 and NO_x. The canopy model requires only standard input data that can be obtained by use of rain gauges, thermometers, and dry and wet collectors. The model has been calibrated with data collected at Woods Lake in the Adirondack Mountains of New York. The model has accurately simulated throughfall volume and the concentration of 15 throughfall chemical constituents. Ammonium, NH_4^+ accumulated on the canopy is nitrified rapidly, resulting in an increase in acidity and nitrate fluxes. The dominant process occurring in the coniferous canopy is dry deposition. Calculations suggest that dry deposition measured by bulk collectors grossly underestimates the total amount of dry deposition to the forested watershed. The enrichment of acidity in coniferous throughfall is derived primarily from the accumulation of acidic air particulates. The dominant process occurring in the deciduous canopy is exudation. This partially neutralizes the acidic deposition.

Ciolkosz, E.J. and Levine, E.R., "Evaluation of Acid Rain Sensitivity of Pennsylvania Soils", OWRT A-058-PA(1), Sept. 1983, Institute for Research on Land and Water Resources, Pennsylvania State University, University Park, Pennsylvania.

Pennsylvania shares with its neighboring states the most acidic deposition that falls in the United States. The effect of acidity in the deposition on soils has not been fully evaluated. A computer simulation model was developed to determine the impact of various inputs of acid deposition on Pennsylvania soils. The model simulates the changes that occur in the solid phase of soils in humid, temperate climates that are undergoing acidification and cation leaching. To obtain predictions about soil properties which closely matched actual field conditions, the state was divided into 10 regions based on physiography, county borders, and proximity to 1 of the 10 precipitation monitoring stations established by the Pennsylvania Department of Environmental Resources. Information from the Pennsylvania State University Soils Data Base for the major soil associations within each of these 10 regions, and the accompanying precipitation data, were used as input to the model. Simulations creating the "worst case scenario" were designed using data for uncultivated and unlimed soils, and without a forest canopy or litter layers. Soils were classified into sensitivity classes based on the amount of time required for the soil to reach a threshold value of pH less than or equal to 4.0, aluminum concentration in the soil solution of 1.0 mg/l in the upper horizon, or aluminum concentration in the soil solution of 0.1 mg/l in the lower horizon. Very sensitive soils reached a threshold value within 30 years, sensitive soils attained this value within 60 years, slightly sensitive soils reached it within 90 years; nonsensitive soils did not attain critical levels within 90 years. Using the model, simulation results showed the Pennsylvania soils were either very sensitive or nonsensitive. Soils at the threshold values at the beginning of the simulation were very sensitive, while those not at the critical levels from the start of the simulation were classified as nonsensitive. Thus, the nonsensitive soils contain sufficient buffer capacity to withstand acid deposition inputs at present rates for at least 90 years. Correlation analysis of these results indicates that base saturation, exchangeable cations, exchangeable acidity, and pH are most closely associated with the sensitivity of Pennsylvania soils to acid deposition. Cation exchange capacity and sulfate adsorption showed very little association with soil sensitivity class predicted by the model.

Coffin, D.L. and Knelson, J.H., "Acid Precipitation: Effects of Sulfur Dioxide and Sulfate Aerosol Particles on Human Health", Ambio, Vol. 5, No. 5-6, 1976, pp. 239-243.

While human health impairment has been attributed to pollution by sulfur dioxide, data from inhalation studies in animals show that its oxidation products are more irritating. Population surveys in which suspended sulfate was a covariant suggest that certain health parameters are associated more strongly with sulfate than with SO_2. Recent work with biological models indicate that sulfates and sulfuric acid act on the lung through the release of histamine, and the degree of release is related to the specific cation present. Attention should be given to sulfates and specific cations in atmospheric monitoring.

Cole, C.V. and Stewart, J.W., "Impact of Acid Deposition and P Cycling", Environmental and Experimental Botany, Vol. 23, No. 3, 1983, pp. 235-241.

The productivity of agricultural soils requires an active cycling of soil P pools through plants and microbial populations. Phosphorus cycling rates are thus closely linked to transformations of C, N, S and other mineral nutrients. A systems approach to an understanding of factors controlling chemical and biological processes in ecosystems is necessary for an evaluation of the impacts of acid deposition on agricultural soils. Acid deposition effects on P cycling will be most pronounced on soils subject to intense weathering and erosional processes. Management practices developed to relieve these stresses will be most useful in mitigating impacts of acid deposition.

Cole, D.W. and Johnson, D.W., "Atmospheric Sulfate Additions and Cation Leaching in a Douglas Fir Ecosystem", Water Resources Research, Vol. 13, No. 2, Apr. 1977, pp. 313-318.

The effects of present levels of atmospheric sulfuric acid input on cation leaching at the Thompson Research Center, Washington, Douglas Fir Ecosystem Cedar River watershed are described. Precipitation sulfuric acid input is small compared with sulfate and cation transfers within the ecosystem. Most of the annual incoming hydrogen ions are removed in the forest canopy, presumably through an exchange process. Canopy leaching appears to be most strongly affected by acid rainfall, whereas soil leaching appears to be dominated by mechanisms internal to the ecosystem, such as carbonic acid leaching and sulfur release from decomposing litter.

Coleman, D.C., "Impacts of Acid Deposition on Soil Biota and C Cycling", Environmental and Experimental Botany, Vol. 23, No. 2, 1983, pp. 225-233.

Abiotic and biotic factors in agricultural soils, including range and cropland, are examined in the context of soil biota and carbon cycling. This review looks at a variety of man-managed ecosystems over a range of from minor to major man-managed perturbations. Major emphasis is placed on the results from experimental manipulations of acidity in grasslands and forests and on developing knowledge in zero-till vs. conventional tillage agriculture. In terms of both carbon and energy flow, a majority of the action follows the primary production detritus/decomposer pathway. Certain litter-dwelling organisms and attendant processes, such as litter decomposition, are seriously impacted by decreasing pH. A conceptual model was developed for comparing microbial faunal and plant growth dynamics in conventional vs. no-till situations. The flows indicate breakdown mediated by microbial activity and controlling faunal populations either by direct microbial grazing or predators and top carnivores feeding on the grazers, returning N, P, and S to the soil solution, which is then taken up by plant roots. There are possible inhibiting effects of acid deposition, particularly if the target organisms are concentrated near the soil surface. Further studies in both theoretical and applied ecology are required to answer important questions of stress or effects, such as acid deposition in agro-ecosystems.

Conway, H.L. and Hendrey, G.R., "Ecological Effects of Acid Precipitation on Primary Producers", 1981, Brookhaven National Laboratory, Upton, New York.

Nonacidic, oligotrophic lakes are typically dominated by golden-brown algae, diatoms and green algae. With increasing acidity, the number of species decrease and the species composition changes to dinoflagellates and golden-brown algae, with blue-green algae dominating in some cases. For macrophytic plants, dense stands of Sphagnum and Utricularia are found in some acidic lakes which may reduce nutrient availability and benthic regeneration. Hydrogen ion concentration does not appear to be as important as inorganic phosphorus in controlling primary production and biomass in acidic lakes. In acidic, oligotrophic lakes, benthic plants may have a competitive advantage over pelagic algae because of the high concentrations of inorganic carbon and phosphorus available to them in the sediment.

Cowling, E.B. and Davey, C.B., "Acid Precipitation: Basic Principles and Ecological Consequences", Pulp and Paper, Vol. 55, No. 8, Aug. 1981, pp. 182-185.

The pulp and paper industry is involved with both the cause and effects of acid precipitation. It is responsible for small, but locally important, emissions, and it owns a significant share of forest land impacted by this pollution. Although significant quantities of desirable plant nutrients (nitrogen and sulfur) are added to the forest system by precipitation, the acidity and its detrimental effects may outweigh the benefits. Damage to the ecosystem is most likely to occur when major inputs of acid precipitation coincide with sensitive stages of a life form (such as fish eggs and larvae), and in poorly buffered, noncalcareous soils and rocks. In the laboratory and field, many biological effects of acid precipitation have been demonstrated--necrotic lesions on foliage, nutrient loss from foliar organs, reduced resistance to pathogens, accelerated erosion of waxes on leaf surfaces, reduced rates of decomposition of leaf litter, inhibited formation of terminal buds, increased seedling mortality, and heavy metal accumulation. Soil microbiological processes such as nitrogen fixation, mineralization of forest litter, and nitrification of ammonium compounds are inhibited, the degree depending on the degree of cultivation and soil buffering capacity. Water quality is impacted by contact with vegetation, soil, and bedrock. Acid precipitation mobilizes cations, especially the toxic Al, Mn, and Zn, and nutrients, K, Ca, and Mg.

Cowling, E.B. and Dochinger, L.S., "Effects of Acid Rain on Crops and Trees", Acid Rain: Proceedings of ASCE National Convention, 1979, American Society of Civil Engineers, New York, New York, pp. 21-54.

Acid rain is becoming a dominant feature of anthropogenic changes in the physical and chemical climate of the earth. But acidity in rain is only part of the changing composition of precipitation, dry particulate matter, aerosols, and gases. Recent research has demonstrated that: (1) atmospheric deposition contains both beneficial nutrients and injurious substances; (2) plants and ecosystems vary greatly in susceptibility, tolerance, and adaptability to changes in atmospheric deposition; (3) injury is most likely when rapid deposition coincides with a vulnerable life form or life stage; and (4) simulated acid rain has caused leaching of nutrients from soil and both direct and indirect injury to terrestrial vegetation; with rare exceptions, however, economic damage to crops by naturally occurring acid rains has not been demonstrated to date. These results support the establishment of a National Atmospheric Deposition Program to determine the amount, chemical form, and geographical distribution of atmospheric deposition and its effects on vegetation, soils, and aquatic ecosystems.

Crisman, T.L. and Brezonik, P.L., "Acid Rain: Threat to Sensitive Aquatic Ecosystems", June 1980, University of Florida, Gainesville, Florida.

The effects of acid precipitation on sensitive (softwater) temperate lakes are reviewed, and comparative results are presented on 20 warm temperate/subtropical lakes in Florida.

Cronan, C.S., "Effects of Acid Precipitation on Cation Transport in New Hampshire Forest Soils", DOE/EV/04498-1, July 1981, U.S. Department of Energy, Washington, D.C.

This report describes the results of an investigation of the effects of regional acid precipitation on forest soils and watershed biogeochemistry in New England. The report provides descriptions of the following research findings: (1) acid precipitation may cause increased aluminum mobilization and leaching from soils to sensitive aquatic systems; (2) acid deposition may shift the historic carbonic acid/organic acid leaching regime in forest soils to one dominated by atmospheric H_2SO_4; (3) acid precipitation may accelerate nutrient cation leaching from forest soils and may pose a particular threat to the potassium resources of northeastern forested ecosystems; (4) while acid rain may pass through some coniferous canopies without being neutralized, similar inputs of acid rainfall to hardwood canopies may be neutralized significantly by Bronsted base leaching and by leaf surface ion exchange mechanisms; and (5) progressive acid dissolution of soils in the laboratory may provide an important tool for predicting the patterns of aluminum leaching from soils exposed to acid deposition.

Dailey, N.S. and Winslow, S.G., "Health and Environmental Effects of Acid Rain--An Abstracted Literature Collection--1966-1979", Mar. 1980, Toxicology Information Response Center, Oak Ridge National Laboratory, Oak Ridge, Tennessee.

The long distance transport and ultimate deposition of acidic contaminants along air sheds has made acid precipitation both a regional problem and an international concern. Acid precipitation has been reported in Sweden, Norway, Japan, the United States, and Canada. This abstracted bibliography contains 961 citations on the health and environmental effects of acid precipitation. Contaminants of primary interest are sulfuric and nitric acids, sulfates and nitrates, and ammonium. The research directory, which can be found at the end of the bibliography, contains information concerning current research projects including the title of the project, organization conducting the research, principal investigator, sponsor, and supplementary material with a brief project description.

Davis, M.J., "Effects of Atmospheric Deposition of Energy-Related Pollutants on Water Quality: A Review and Assessment", ANL/AA-26, May 1981, Argonne National Laboratory, Argonne, Illinois.

The effects on surface water quality of atmospheric pollutants that are generated during energy production are reviewed and evaluated. Atmospheric inputs from such sources to the aquatic environment may include trace elements, organic compounds, radionuclides, and acids. Combustion is the largest energy-related source of trace-element emissions to the atmosphere. This report reviews the nature of these emissions from coal-fired power plants and discusses their terrestrial and aquatic effects following deposition. Several simple models for lakes and streams are developed and are applied to assess the potential for adverse effects on surface-water quality of trace-element emissions from coal combustion. The probability of acute impacts on the aquatic environment appears to be low; however, more subtle, chronic effects are possible. The character of acid precipitation is reviewed, with emphasis on aquatic effects, and the nature of existing or potential effects on water quality, aquatic biota, and water supply is considered. The response of the aquatic environment to acid precipitation depends on the type of soils and bedrock in a watershed and the chemical characteristics of the water bodies in question. Methods for identifying regions sensitive to acid inputs are reviewed. The observed impact of acid precipitation ranges from no effects to elimination of fish populations. Coal-fired power plants and various stages of the nuclear fuel cycle release radionuclides to the atmosphere. Radioactive releases to the atmosphere from these sources and the possible aquatic effects of such releases are examined. For the nuclear fuel cycle, the major releases are from reactors and reprocessing. Although aquatic effects of atmospheric releases have not been fully quantified, there seems little reason for concern for man or aquatic biota.

Derrick, M.R. et al., "Aerosol and Precipitation Chemistry Relationships at Big Bend National Park", Water, Air, and Soil Pollution, Vol. 21, No. 1-4, Jan. 1984, pp. 171-181.

Aerosol chemistry, precipitation and visibility parameters are currently being measured at Big Bend National Park in Texas. This is part of a large-scale air resource evaluation program which the National Park Service is sponsoring in several southwestern national parks and monuments to determine the potential impact of local and distant pollutant sources on the environmental quality within these areas. Analysis of aerosol samples collected at six sites in the Southwest indicates that soil-derived components, organic materials and the acid-base ions of sulfate, nitrate, and ammonium are the major constituents of suspended airborne particulate matter in the remote areas of the arid region. Comparison of particulate matter chemistry and precipitation chemistry data at Big Bend National Park shows consistent features which indicate that the airborne alkaline soil material and NH_3 largely neutralize the atmospheric acidic species of H_2SO_4 and HNO_3.

Dochinger, L.S. and Seliga, T.A., "Acid Precipitation and the Forest Ecosystem", Journal of the Air Pollution Control Association, Vol. 25, No. 18, Nov. 1975, pp. 1103-1106.

The role of global acidification, its nature and extent, and disposition as a means for defining current and future impacts of wet and dry acidic deposition on the forest ecosystem were topics considered at the First International Symposium on Acid Precipitation and the Forest Ecosystem sponsored by USDA and Ohio State University. The problem of acidity; Norway's acid precipitation research project; transport, chemistry, and precipitation; acid precipitation and aquatic ecosystems; acid precipitation and forest soils; and acid precipitation and forest vegetation are discussed.

Dochinger, L.S. and Seliga, T.A., Proceedings of the International Symposium on Acid Precipitation and the Forest Ecosystem, Aug. 1976, Ohio State University - Atmospheric Sciences Program, Columbus, Ohio.

The increasing acidity of precipitation in northwestern Europe, the northeastern United States, and eastern Canada, caused by emissions of sulfur and other acid-forming compounds into the air, has had serious effects, and scientists are concerned about subtle long-term changes not yet understood.   About 300 scientists from 12 countries attended a symposium at Columbus, Ohio, May 12 to 15, 1975, to assess what is known about atmospheric chemistry, transport, and precipitation, and the effects of acid precipitation on aquatic ecosystems, forest soils, and forest vegetation.   The Proceedings contain 85 papers presented at the symposium.

Dynamac Corporation, "Bibliography on Air Pollution and Acid Rain Effects on Fish, Wildlife, and Their Habitats", Mar. 1982, Rockville, Maryland.

This bibliography is the result of the development of a series of nine reports synthesizing information from scientific research related to the effects of air pollution and acid deposition on fish and wildlife resources. The reports include an Introduction, Deserts and Steppes, Forests, Grasslands, Lakes, Tundra and Alpine Meadows, Rivers and Streams, Urban Ecosystems, and Critical Habitats of Threatened and Endangered Species.

Evans, L.S., "Generation of Dose-Response Relationships to Assess the Effects of Acidity in Precipitation on Growth and Productivity of Vegetation", 1981, Brookhaven National Laboratory, Upton, New York.

Experiments were performed with several plant species in natural environments as well as in a greenhouse and/or tissue culture facilities to establish dose-response functions of plant responses to simulated acidic rain in order to determine environmental risk assessments to ambient levels of acidic rain.   Response functions of foliar injury, biomass of leaves and seed of soybean and pinto beans, root yields of radishes and garden beets, and reproduction of bracken fern are considered. The dose-response function of soybean seed yields with the hydrogen ion concentration of simulated acidic rainfalls was expressed by the equation $y = 21.06-1.01 \log x$ where y = seed yield in grams per plant and x = the hydrogen concentration in mu eq 1 exp -1. The correlation coefficient of this relationship was -0.90. A similar dose-response function was generated for percent fertilization of ferns in a forest understory. When percent fertilization is plotted on logarithmic scale with hydrogen ion concentration of the simulated rain solution, the Y intercept is 51.18, slope -0.041 with a correlation coefficient of -0.98. Other dose-response functions were generated that assist in a general knowledge as to which plant species and which physiological processes are most impacted by acidic precipitation. Some responses did not produce convenient dose-response relationships. In such cases the responses may be altered by other environmental factors or there may be no differences among treatment means.

Evans, L.S. et al., "Comparison of Experimental Designs to Determine Effects of Acidic Precipitation on Field-Grown Soybeans", BNL-32176, 1982, Brookhaven National Laboratory, Upton, New York.

Two experiments were performed to determine changes in seed yields of soybeans grown under standard agronomic practices exposed to simulated acidic rain during the summer of 1981. Seed yields of soybeans exposed twice weekly to simulated rainfalls of pH 4.1, 3.3, and 2.7 were decreased 10.7, 16.8, and 22.9 percent, respectively, compared with plants exposed to simulated rainfalls of pH 5.6. A treatment-response function of seed yield versus rainfall pH was $y = 7.40 + 1.025$ x and had a correlation coefficient of 0.997. In a second experiment, soybean plants were not shielded from ambient rainfalls (weighted mean hydrogen ion concentration equal to pH 4.04) and received only small volumes of simulated rainfalls three times weekly. Plants exposed to simulated rainfalls of pH 4.1, 3.3, and 2.7 exhibited yield reductions of 2.7, 7.0, and 7.6 percent, respectively, below yields of plants exposed to simulated rainfalls of pH 5.6. By best fit analyses, the equation that fits this latter relationship is expressed by $y = 9.68 + 0.318$ x where y is seed mass per plant and x is the pH of the simulated rain. The correlation coefficient for this latter relationship was 0.97. The decrease in seed yield observed in both experiments was due to a decrease in number of pods per plant.

Evans, L.S., Conway, C.A. and Lewin, K.F., "Yield Responses of Field-Grown Soybeans Exposed to Simulated Acid Rain", Mar. 1980, Brookhaven National Laboratory, Upton, New York.

An important area of interest is to determine the effects of acid precipitation on the yield of agronomic crops under field conditions. Experiments described herein were performed with field-grown soybeans at Brookhaven National Laboratory during the summer of 1979. A preliminary experiment was performed the preceding year at the same site to estimate the most appropriate plot design and statistical analyses. Soybeans were seeded to provide six Latin Squares. Five treatments (no rain, simulated rainfalls of pH levels of 4.0, 3.1, 2.7, and 2.3) replicated five times in each Latin Square were used to produce a total of 30 plots per treatment. These results show that additions of small amounts of simulated acid rain to soybeans decreased the number of pods per plant. This decrease in the number of pods per plant produced a small but significant decrease in seed mass. The decreases (3, 5 and 8 percent at simulated rain pH levels of 4.0, 3.1, and 2.7) were present in soybeans already exposed to rainfalls at Brookhaven National Laboratory with an average of about pH 4.0 over the period of this experiment.

Evans, L.S., Francis, A.J. and Raynor, G.S., "Acid Rain Research Program--Annual Progress Report--July 1976-September 1977", Contract No. EV-76-C-02-0016, Dec. 1977, Brookhaven National Laboratory, Upton, New York.

Experiments were carried out and chemical aspects of ambient precipitation were determined using a sequential precipitation collector for the period July 1976 through September 1977. A related report provides experimental details. In experiments with plants, experiments were aimed to document: the foliar response of six clones of hybrid poplar to simulated acid rain; effects of buffered solutions and various anions on vegetative and sexual development of gametophytes of the fern (Pteridium aquilinum) and the acid-sensitive steps of symbiotic nitrogen fixation of the garden pea (Pisum sativum). After five 6 min daily exposures to simulated rain of pH 2.7, up to 10 percent of the leaf area of some poplar clones was injured. Lesions developed mostly near stomata and vascular tissue as shown with other plant species. Acidic solutions have a marked effect on sperm motility and fertilization (sexual reproduction) of bracken fern. Since sexual reproduction of ferns is very sensitive to mildly acidic conditions under laboratory conditions, experiments are planned to view the response of sexual stages of other plant species. Nodulation and symbiotic nitrogen fixation in Pisum is very sensitive to nutrient solution acidity. In experiments to determine the effects of excess acidity on soil microbiological processes, the rate of denitrification may be slowed so drastically that increases in N_2O in the atmosphere may result with a subsequent reduction in soil nitrogen levels.

Evans, L.S., Gmur, N.F. and Da Costa, F., "Foliar Response of Six Clones of Hybrid Poplar", Phytopathology, Vol. 78, No. 6, June 1968, pp. 847-857.

Experiments conducted to determine the effects of simulated acid rain on six clones of hybrid poplars were discussed. The experiments included data on: percent of leaves injured; percent of leaf area with leaf lesions; lesion development at low magnifications and with use of scanning electron microscopy; and a histological description of lesion development. Results showed that percent leaf area with lesions and percent leaves injured were similar among all six clones at all pH levels tested. Lesions developed mainly near stomata and vascular tissues and occurred most frequently on leaves just prior to maximum leaf enlargement. Very young and older leaves were less affected. Results support the hypothesis that the adaxial leaf surface is the most affected after exposure to simulated acid rain.

Evans, L.S., Gmur, N.F. and Mancini, D., "Effects of Simulated Acid Rain on Yields of Raphanus Sativus, Lactuca Sativa, Triticum Aestivum and Medicago Sativa", Environmental and Experimental Botany, Vol. 22, No. 4, 1982, pp. 445-453.

The effects of simulated acidic rain were determined on several crops grown under greenhouse conditions. Plants were radishes (Raphanus sativus), lettuce (Lactuca sativa), wheat (Triticum aestivum), and alfalfa (Medicago sativa). Simulated rainfalls of pH 5.6, 4.6, 4.2, 3.4, 3.0, and 2.6 decreased root yields of radishes by 26, 42, 37, 41, 66, and 73 percent, respectively, compared with controls. Similar reductions were seen in radish shoot fresh mass, leaf area, and root diameter. The efficiency of radish foliage in increasing root mass decreased with increased acidity. Lettuce yields (fresh mass) were 11, 10, and 14 percent less than controls for plants exposed to rainfall of pH 4.0, 3.1, and 2.7, respectively. Rain water of pH 5.7 produced some foliar injury on the outer leaves of lettuce but did not affect marketable quality or yield. Yields of wheat exposed to 46 rainfalls with pH as low as 2.7 during anthesis and caryopsis development were unaffected. Alfalfa showed no differences in fresh mass among treatments even after 57 simulated rainfalls of pH 2.7 water over 105 days.

Evans, L.S. and Hendrey, G.R., Proceedings of the International Workshop on the Effects of Acid Precipitation on Vegetation, Soils, and Terrestrial Ecosystems, Brookhaven National Laboratory, June 12-14, 1979, 1979, Brookhaven National Laboratory, Upton, New York.

The objectives of this workshop were to determine the levels of current knowledge of the effects of acid precipitation on vegetation, soils, and terrestrial ecosystems; research needed in these areas to understand the environmental impacts of acid rain; and to help coordinate research groups to avoid excessive duplication of research. The workshop was designed so that researchers in the areas of effects of acid precipitation on vegetation, soils, and whole ecosystem approaches could communicate effectively. There was a general consensus that acid rain at extreme ambient levels, or in artificial systems that simulate extreme ambient levels, causes injury to plant tissues. A major area of concern of acid rain injury was thought to be plant reproduction. The overall levels of significance of plant injury among various plant species remains unknown. The most important priorities in the area of effects of acid rain on crops were an evaluation of effects on crop yields and interaction of acid rain in combination with pollutants on various plants. Few participants thought that ambient acid rain loadings have altered soils to such a degree that plants are affected at present, but many thought that acid rain could cause some alterations in soils. The most important research priorities were in the areas of the effects of acid rain on increased leaching of exchangeable plant nutrients and alterations in phosphorous availability. All participants agreed that there are alterations in terrestrial ecosystems from acid precipitation. However, no demonstrated harmful effects were presented from natural ecosystems. Further research on the effects of acid rain on terrestrial ecosystems should be directed mostly toward the interaction of acid rain with toxic elements such as Al, Fe, and Mn and on the effects of nutrient cycling, especially that of nitrogen.

Evans, L.S. and Lewin, K.F., "Effects of Simulated Acid Rain on Growth and Yield of Soybeans and Pinto Beans", 1979, Laboratory for Plant Morphogenesis, Manhattan College, Bronx, New York.

In order to assess the degree of damage that acid rain has or might have on plants, experiments were performed to determine the change in seed yield of two agronomic crops, soybeans and pinto beans, after exposure to simulated rain of pH 5.7, 3.1, 2.9, 2.7, and 2.5. Moreover, the effects of simulated acid rain were determined on a variety of other experimental parameters to understand further how plants respond to this environmental stress. Simulated acid rain of pH 3.1 and below decreased the dry mass of seeds, leaves, and stems of pinto beans. On a percentage mass basis the decrease in seed yield was comparable to reductions in biomass of leaves and stems. The decrease in yield of pinto beans by simulated acid rain was attributed to both a (1) decrease in the number of pods per plant and (2) a decrease in the number of seeds per pod. In soybeans, simulated acid rain decreased the dry mass of both stems and leaves. Seed yield also decreased after treatment with rain of pH 2.5. However an increase in seed yield occurred when plants were exposed to rain of pH 3.1. A larger dry mass per seed was responsible for the larger dry mass of seed per plant.

Faust, S.D. and McIntosh, A., "Buffer Capacities of Fresh Water Lakes Sensitive to Acid Rain Deposition", Journal of Environmental Science and Health, Part A, Vol. 18, No. 1, 1983, pp. 155-161.

The Van Slyke definition of buffer capacity, the increment of a strong base or strong acid that causes an incremental change in the pH value of water, is better than total alkalinity for defining a water's resistance to acid rain. This Van Slyke value, designated by beta, shows a peak at pH 6.3 for the bicarbonate-carbonate pair, indicating that the effect of acid rain on the pH and alkalinity of natural waters is not deleterious until this peak is traversed. A beta value of zero indicates a dead water with no capacity to neutralize acid. The beta values, pH and total alkalinity of lakes, reservoirs, and streams in New Jersey are given. Data clearly show that pH and alkalinity alone cannot determine buffer capacity. For example, Fairview Lake (pH of 5.5 and alkalinity of 10.2 mg/l) has a beta value 11 times that of Clyde Potts Reservoir (pH of 7.3, alkalinity of 8.1 mg/l).

Ferenbaugh, R.W., "Acid Rain: Biological Effects and Implications", Environmental Affairs, Vol. 4, No. 4, Fall 1975, pp. 745-759.

In Scandinavian countries and in the northeastern portion of the U.S., rainwater consistently has a pH of 3.0-4.0. An obvious consequence of acid rain is its corrosive effect on exposed stone and metal structures. The biological effects of acid rain on terrestrial mammalian animals, aquatic systems, soil systems, and plants are described. Possible methods of removing sulfur compounds from stack gases before the gases are emitted into the atmosphere are discussed. The use of tall stacks is also discussed.

Flinn, D.R. et al., "Acidic Deposition and the Corrosion and Deterioration of Materials in the Atmosphere: A Bibliography, 1880-1982", EPA-600/3-83-059, July 1983, U.S. Environmental Protection Agency, Research Triangle Park, North Carolina.

This bibliography contains more than 1300 article citations and abstracts on the effects of acidic deposition, air pollutants, and biological and meteorological factors on the corrosion and deterioration of materials in the atmosphere. The listing includes citations for the years 1950 to 1982, with selected citations for the years 1880 to 1949. The citations are catalogued by year in six sections for metallic materials--ferrous material, aluminum, copper, nickel, zinc and galvanized steel, and other metals--and six sections for nonmetallic materials--masonry, stone and ceramics, elastomers, fabrics, paints, plastics, and other nonmetals. An author index and an index of chemical, biological, and meteorological variables are provided.

Francis, A.J., Olson, D. and Bernatsky, R., "Effect of Acidity on Microbial Processes in a Forest Soil", Mar. 1980, Brookhaven National Laboratory, Upton, New York.

The objective of this study is to obtain comprehensive baseline information on the microbial activity in a forest soil in order to begin an evaluation of the impact of pollutants resulting from energy related activities. Since most of the soils on Long Island, New York, are poorly buffered and may respond to acid precipitation more quickly, it was decided to study the effect of acidity on various microbial processes. A Riverhead sandy loam soil collected from an oak-pine forest with a pH of 4.6, organic matter content of 3.9 percent, and 0.1 percent total nitrogen was used in this study.

Francis, A.J., Olson, D. and Bernatsky, R., "Microbial Activity in Acid and Acidified Forest Soils", Mar. 1981, Brookhaven National Laboratory, Upton, New York.

Effects of soil acidity on microbial decomposition of organic matter, transformation of nitrogen, and soil chemical and biological properties of an acid forest soil were investigated. The rates of organic matter decomposition by natural acid soil and by pH-adjusted acid and neutral soils which were preincubated for 14 and 150 days were determined by monitoring CO_2 evolution. In the control (unamended) pH-adjusted acid soil, reductions in CO_2 production of 14 percent by 14-day preincubated samples and of 52 percent by 150-day samples were observed. In the oak-leaf-amended acidified soils, the CO_2 production of 14- and 150-day preincubated samples decreased by about 6 and 37 percent, respectively. Ammonification and nitrification in the natural acid and pH-adjusted acid soils were determined. In addition, the contribution of nitrate by heterotrophic microbial communities in natural acid and pH-adjusted acid soils was examined. Ammonia formation in the pH-adjusted acid soil was 50 percent less than in the natural acid soil. An increase in nitrate concentration in the pH-adjusted acid soil over that in the natural acid soil was observed. Increased rates of ammonification and nitrification were observed in the pH-adjusted neutral soil. Little autotrophic nitrifying activity was detected in natural acid and acidified forest soils and no detectable heterotrophic nitrification was observed. The data indicate that ammonification and nitrification are affected by an increase in soil acidity. These results suggest that further acidification of acid forest soils by addition of sulfuric acid or by acid precipitation may lead to significant reductions in the leaf litter decomposition, ammonification, and nitrification, and thus affect the nutrient recycling in the forest ecosystem.

Fritz, E.S., "Potential Impacts of Low pH on Fish and Fish Populations", FWS/OBS-80/40.2, Oct. 1980, U.S. Environmental Protection Agency, Washington, D.C.

This report provides up-to-date information on the effects of low pH on freshwater fish and shellfish. The discussion of direct mortality includes information obtained from laboratory experiments and field investigations. Several aspects of indirect mortality are discussed: susceptibility to disease, destruction or modification of osmoregulatory and ionregulatory tissues, endocrine imbalance and genetic damage, changes in predator-prey relationships, habitat degradation, and changes in the availability of toxic substances.

Frizzola, J.A. and Baier, J.H., "Contaminants in Rainwater and Their Relation to Water Quality, Part I", Water and Sewage Works, Vol. 122, No. 8, Aug. 1975, pp. 72-76.

Data collected to examine the effect of rainwater quality on ground water in Long Island, New York, yielded information on major ion concentrations in precipitation. The data indicate that rainfall cannot be considered as the pure dilution elixir for ground water. The quality of precipitation that falls on Long Island is discussed. Concentrations of sulfate, sodium and chloride, ammonia and nitrate, acidity, chloride and sodium ion, and nitrate are described.

Galloway, J.N. et al., "A National Program for Assessing the Problem of Atmospheric Deposition (Acid Rain)", Dec. 1978, North Carolina Agricultural Experiment Station, North Carolina State University, Raleigh, North Carolina.

The objectives of this report are to assess the present scope of available knowledge about atmospheric deposition and its effects on agricultural lands, forests, ranges, parks, surface waters, and aquatic life in the United States and to recommend a coordinated program of research and monitoring necessary to intelligent management of atmospheric emissions and for the amelioration of the adverse effects of atmospheric deposition on plant and animal life.

Galloway, J.N., Norton, S.A. and Church, M.R., "Freshwater Acidification from Atmospheric Deposition of Sulfuric Acid: A Conceptual Model", Environmental Science and Technology, Vol. 17, No. 11, Nov. 1983, pp. 541A-545A.

A simple conceptual model is presented that incorporates only a few key processes which are felt to determine how aquatic ecosystems respond to acid deposition with respect to time. The model is based on the assumption that if the rate of acid deposition changes, a variety of constituents in both the terrestrial and aquatic systems will respond. The model has seven stages: preacidification stage--steady state prior to significant emissions or anthropogenic sulfur; undersaturated sulfate adsorption capacity; saturated sulfate adsorption capacity in soil; steady-state period of lake acidification; supersaturated sulfate adsorption capacity; recovery of the percent base saturation; and the stable period of lake recovery. The model can be used by defining the pertinent characteristics of each stage and identifying geographical regions that have those characteristics.

Gauri, K.L., "Effects of Acid Rain on Structures", Acid Rain: Proceedings of ASCE National Convention, 1979, American Society of Civil Engineers, New York, New York, pp. 70-91.

The effect of acid rain on structures has been illustrated by the example of the deterioration of the exterior of the Field Museum of

Natural History, Chicago. The field survey revealed that the marble cornices, etc., protected from the direct impact of rain has developed black crust. The unprotected areas, however, are clean due to perpetual dissociation of calcite grains. Fractures in marble blocks were found along dark banks which terminated at joints with polymeric caulking. Atomic absorption spectrophotometry, scanning electron microscopy, and x-ray diffraction of samples from the museum revealed that the black crust consisted of gypsum which had formed due to SO_2 attack on calcite. The gypsum was absent at the naturally cleaned surface. By comparing these findings with the deterioration of marble at other antique structures it has been concluded that the protected regions shall, eventually, suffer greater deterioration.   The conservation technology is as yet in its incipient state.

Gjessing, E., "Effect of Polluted Precipitation on Water Quality--the Situation in Norway", Aqua, No. 7, 1980, pp. 139-140.

Chemical analysis of precipitation and surface water in Norway shows the direct and indirect effects of atmospheric pollution.   All samples in this study were taken in areas unaffected by agriculture, forestry, or human settlement. Heavy metals and organic micropollutants have been found in addition to the long-established sulfate ion. The composition of a winter's snowfall (1972) in a southern Norway community, Langtjern, contained the following constituents (in mg per sq meter): sulfate, 870; nitrate, 561; Cu, 9.71; Zn, 16.75; Pb, 4.53; Cd, 0.70; organic C, 807; and organic matter, 1793. At Birkenes, near the southern tip of the country, 84 organic components (in nanograms per liter) were identified in rain water; alkanes, C21-C27 and C29-C33, 50-60; alkanes, C23-C28, 50-60; polyaromatic hydrocarbons, benzofluoranthene, 50-300 and benzopyrene, 50-300; fatty acid ethyl esters, C13, 50-150, C15, 50-200, and C17, 50-500; polychlorinated biphenyls, 2-5; and organic chlorine, 219. In "unpolluted" lakes and brooks 10-27 ng per liter polychlorinated biphenyls and 148-247 ng per liter polyaromatic hydrocarbons were found.

Glass, N.R., "Environmental Effects of Increased Coal Utilization: Ecological Effects of Gaseous Emissions from Coal Combustion", EPA-600/7-78-108, June 1978, Corvallis Environmental Research Laboratory, U.S. Environmental Protection Agency, Corvallis, Oregon.

This report is provided for the "Health and Environmental Effects of Coal Utilization" Committee which was created by the request of the DOE in response to the President's Environmental Message. It evaluates ecological environmental effects of gaseous emissions and aerosols of various type which result from coal combustion. The report deals with NO_x, SO_x fine particulate, photochemical oxidants and acid precipitation as these pollutants affect natural and managed resources and ecosystems. The economic implications of ecological effects are identified within acceptable limits. In addition, the reliability of the data base upon which conclusions or estimates are made is evaluated to the degree possible. Aquatic and terrestrial effects are distinguished where the pollutants in question are clearly problems in both media. Sulfur oxide (SO_x) emissions and nitrogen oxide (NO_x) emissions are projected to be higher in 1985 and 2000 than in 1975. Since SO_x and NO_x are major contributors to acid precipitation, substantial increases in total acid deposition can be expected in the nation as a whole. At present, acid precipitation is most abundant in the North Central and Northeastern States. Estimates of the nonhealth-related cost of air pollutants range from several hundred million dollars per year to 1.7 billion dollars per year. In general, these estimates include only those relatively easily measured considerations such as crop losses resulting from acute pollution episodes or cost of frequent repainting as a result of air pollution.

Glass, N.R., "U.S. Federal Program on Effects of Acid Rain", Acid Rain: Proceedings of ASCE National Convention, 1979, American Society of Civil Engineers, New York, New York, pp. 92-110.

Recent evidence points strongly to acid rain as a growing environmental problem of far reaching consequences and increasing geographical extent. Acid rain is but one aspect of the broader problem of atmospheric deposition and is also meant to include snow and dry deposition of acidic material. First noticed and studied in the Scandinavian countries, acid precipitation has now been well documented in the United States, first in the northeast and now more recently throughout much of the U.S. east of the Mississippi. Lakes and streams in regions with poorly buffered soils have become depleted of fish and other aquatic life, and are changing toward conditions leading to low aquatic productivity. Evidence also indicates that acid precipitation may cause damage to forest growth and crop production. Man-made materials such as metals, paints, and statuary are being damaged by the deposition of acidic chemical species. Drinking water supplies in some areas of the country may be adversely affected as the acidity in surface waters causes chemical composition changes. Present government programs relating to acid precipitation have been expanded in fiscal year FY-79. The major categories of effort consist of monitoring and quality assurance, atmospheric chemistry and transport, source emissions inventory, environmental effects, and historical trend analysis.

Glass, N.R. et al., "Effects of Acid Precipitation", Environmental Science and Technology, Vol. 16, No. 3, Mar. 1982, pp. 162A-169A.

Acid precipitation has become a widespread problem due to the ubiquitous nature of the pollution sources contributing to acid precipitation and the ease with which the emissions causing it are carried by moving air masses. Serious symptoms of acidification have been documented in lakes of New York's Adirondack Mountains, far removed from any industrial activity. While precipitation is most acidic in the northeast U.S., the geographic extent of the problem encompasses the southeast and portions of the midwest and far west. A growing body of evidence suggests that acid precipitation adversely affects public welfare through loss of fish and other aquatic life, increased leaching of nutrient cations from the soil, possible reductions in crop and forest productivity, and the release of heavy metals or nutrients from rocks, soils and lake and stream bottom sediments. Research on aspects of acid rain is reviewed. The sensitivity of surface water bodies to acid precipitation was evaluated, and historical changes in surface waters were documented.

Glass, N.R., Glass, G.E. and Rennie, P.J., "Effects of Acid Precipitation", Environmental Science and Technology, Vol. 13, No. 11, Nov. 1979, pp. 1350-1352.

A dramatic change from 1950-1970 in pH has occurred in Canada and the Northeastern U.S. New data indicates pH changes as early as 1930; 60-70 percent of the acidity may be attributed to H_2SO_4 and 30-40 percent to HNO_3. Sulfuric acid comes from SO_x from stationary sources, and HNO_3 from NO_x from mobile sources. Presently all states (in the U.S.) east of the Mississippi are affected to some degree by acid rain as well as Seattle, San Francisco, and Los Angeles. In the Adirondack area (U.S.) there are more than 100 lakes now fishless due to lake acidification by acid rain. Soil acidification, calcium removal, aluminum and manganese solubilization, tree growth reduction, crop quality and quantity reduction, elimination of useful soil microorganisms, and heavy metal elements selectively exchanged for more beneficial mono and divalent cations may result from acid rain. Crops may be affected by changes in leaf surfaces and nutrient changes in the soil.

Glass, N.R., Glass, G.E. and Rennie, P.J., "Effects of Acid Precipitation in North America", Environment International, Vol. 4, No. 5-6, 1980, pp. 443-452.

Acid rain is but one aspect of the broader problem of atmospheric deposition which includes snow, fog, and dry deposition of material. First noticed and studied in the Scandinavian countries, acid precipitation has now been well documented in the United States, first in the northeast and more recently throughout much of the United States east of the Mississippi River. Numerous streams and lakes in regions with poorly buffered soils have become devoid of fish, have an impoverished aquatic flora and fauna, and are changing toward conditions of low aquatic productivity. Evidence also indicates that acid precipitation may cause damage to forest growth, crop production, wildlife, and man-made materials such as buildings, metals, paints, and statuary.

Glass, N.R., Likens, G.E. and Dochinger, L.S., "The Ecological Effects of Atmospheric Deposition", EPA Energy/Environment 3rd National Conference, June 1-2, 1978, U.S. Environmental Protection Agency, Washington, D.C., pp. 113-120.

Possible effects of increased atmospheric deposition of acidic substances caused by added reliance on coal as a fuel source are reviewed. Acidic precipitation, once considered a problem exclusive to the eastern U.S., has occurred on several occasions in western states. Acid rain can result in damage to aquatic ecosystems, crops, and forest areas. Sulfates and nitrates can also damage metals, paints, statuary, buildings, and other structures.

Gorham, E. and McFee, W.W., "Effects of Acid Deposition Upon Outputs from Terrestrial to Aquatic Ecosystems", Effects of Acid Rain Precipitation on Terrestrial Ecosystems, 1980, Plenum Press, New York, New York, pp. 465-480.

This article is a review/discussion of the interactions between terrestrial and aquatic ecosystems as influenced by acid rain. The components of acid precipitation, materials capable of being leached from soils and retained by soils when exposed to acid rain, and factors influencing the response of aquatic systems to atmospheric deposition are discussed.

Graham, M.S. and Wood, C.M., "Toxicity of Environmental Acid to the Rainbow Trout: Interactions of Water Hardness, Acid Type, and Exercise", Canadian Journal of Zoology, Vol. 59, No. 8, Aug. 1981, pp. 1518-1526.

The effects of water hardness (14 or 140 mg/l CaCO3), acid type (HCl vs. H2SO4), and activity level (rest vs. exercise) on acid toxicity to rainbow trout fingerlings were studied at 15°C in 7-day exposures. The 7-day LC50 pH levels were 4.1-4.5. Sulfuric acid was less toxic in hard water than in soft water at all pH levels during rest and exercise. Hard water decreased the toxicity of HCl at pH less than 3.5 and increased toxicity at pH above 3.8. Sulfuric acid was less toxic than HCl in hard and soft waters except for soft waters above pH 3.8. Exercise increased H2SO4 toxicity in both hard and soft waters except in soft water at pH less than 3.3. Critical swimming speeds were not significantly different between hard and soft water at the same pH in the ranges of 3.5-4.2 and neutral. However, below pH 4.6 in soft water and 4.4 in hard water, critical swimming speeds decreased 4 percent per 0.1 pH units. Any extrapolation from laboratory toxicity data to field survival criteria should consider the interactions described above.

Grodzinska, K. et al., "First International Symposium on Acid Precipitation and the Forest Ecosystem", Water, Air, and Soil Pollution, Vol. 8, No. 1, May 1977, pp. 3-146.

The correlation between acidification of tree bark and sulfur dioxide air pollution in the Bialowieza Forest in Poland served as an accurate bioindicator of forest pollution levels. The effects of rain acidified with sulfuric acid on: tree growth in eastern America, coniferous forest ecosystems in Norway, nitrogen fixation in western Washington coniferous forests, and host-parasite interactions in a simulated forest environment are described. Under the auspices of the Clean Air Act of 1970, EPA is currently developing regulatory programs for the control of acid precipitation in the U.S. The results of studies of pine, spruce, and aspen forests periodically subjected to sulfur gas emissions are presented, as are results of investigations where sulfuric acid was directly applied to various tree species. Using simple analytical techniques, the amounts of air pollutants accumulated in winter snow were determined, and results were correlated with lichen survival on trees.

Haines, B. and Waide, J., "Predicting Potential Impacts of Acid Rain on Elemental Cycling in a Southern Appalachian Deciduous Forest at Coweeta", Effects of Acid Rain Precipitation on Terrestrial Ecosystems, 1980, Plenum Press, New York, New York, pp. 335-339.

This paper describes investigations conducted at the Coweeta Hydrologic Laboratory in South Carolina involving forest (tree) responses to acid rain effects such as: (1) ion loss from leaves; (2) leaching rates of leaves; (3) root ion uptake; (4) relative/similar root reactions of different tree species; (5) soil nutrient losses; and (6) recharge curves for soil losses. The authors state that these are long-term studies.

Haines, T.A., "Acidic Precipitation and Its Consequences for Aquatic Ecosystems: A Review", Transactions of the American Fisheries Society, Vol. 110, No. 6, Nov. 1981, pp. 669-707.

Current understanding of the acid rain phenomenon, adverse effects of acid rain on aquatic ecosystems, and solutions to the problem are discussed. Increases in sulfuric and nitric acid aerosols have produced highly acidic precipitation which has affected surface waters. Contaminated waters have decreased alkalinity and pH and increased metals and organic compounds. Fish, decomposers, algae, macrophytes, and invertebrates have been affected primarily by hydrogen ion concentration changes but also by acid stress. Reduction of sulfur and nitrogen oxides, reduction of water acidity, and the breeding of resistant species are possible solutions.

Harriman, R. and Morrison, B.R., "Ecology of Streams Draining Forested and Non-Forested Catchments in an Area of Central Scotland Subject to Acid Precipitation", Hydrobiologia, Vol. 88, No. 3, May 1982, pp. 251-263.

Precipitation, water, invertebrates, and fish were studied in 12 streams flowing through forested and nonforested catchments west of Aberfoyle, Scotland. Bedrock in this region is predominantly quartzite, schists, and slates with little buffering capacity. Precipitation has an annual mean pH of 4.2-4.5. The most common tree species is Sitka spruce (Picea sitchensis Carriere). Forest streams were more acid than adjacent moorland streams (mature forest, 4.16-4.75; young forest, 4.83-5.25; and moorland, 4.90-5.80). Concentrations of aluminum were also higher in old forested streams (100-350 micrograms per liter) than in young forest and moorland streams (less than 100 micrograms per liter). Manganese concentrations were 90 micrograms per liter in forest streams and 30 micrograms per liter in nonforested streams. Among the mature forest streams only one had brown trout (Salmo trutta L.); all nonforested streams had trout. Planted salmon eggs (Salmo salar L.) died within a few weeks in forested streams but survived in nonforested streams. The only mayfly nymph found in acid streams during summer was Siphlonurus lacustris Eaton. In winter samples mayfly nymphs, Heptagenia lateralis Curtis, were found in one forest stream, but several species were present in nonforested streams. Data suggest that spruce forests collect acid pollutants and denude the soil of neutralizing elements, producing higher acidity in streams draining forested catchments.

Harte, J., Holdren, J. and Tonnesson, K., "Potential for Acid-Precipitation Damage to Lakes of the Sierra Nevada, California", OWRT-A-081-CAL(1), Apr. 1983, Energy and Resources Group, University of California, Berkeley, California.

Acid precipitation in the western United States is now being measured and monitoring networks are being set up to determine the nature and extent of its occurrence. In California, two recent studies have documented instances of acidic deposition. It would appear that in the State the nitrate component often outweighs the sulfate contribution in rainfall. One of the areas of California potentially sensitive to acidic deposition is the Sierra Nevada, located along the eastern boundary. A report on sensitive areas in North America identifies the Sierra as a region characterized by poorly buffered soils and granite based lakes. For this investigation selected subalpine lakes of the western slope of the Sierra were chosen for study, to establish baseline water quality which would allow for the identification of chemical and biological changes due to acidic deposition. These experiments were particularly concerned with recording changes in concentrations of micronutrients which might be leached from lake sediments with increasing acidification. This phenomenon is particularly important to study in the light of finds on the importance of aluminum leaching the Northeast which has led to toxic effects on biota in Adirondack lakes.

Havas, P. and Huttuneu, S., "Some Special Features of the Ecophysiological Effects of Air Pollution on Coniferous During the Winter", Effects of Acid Precipitation on Terrestrial Ecosystems, 1980, Plenum Press, New York, New York, pp. 123-131.

Finland receives 550,000 tons of sulfur/yr. The retarding of growth rates and damage to forests have been recorded in some areas of Finland, but the general effect in Finland is small. The greatest damage is produced in the winter depending on the quantity and quality pollutants accumulating in the needles of coniferous trees. Water balance is an important factor in the growth cycle of trees. The more sulfur present in the needles the greater the water deficiency (in the spring). Greater photosynthesis stress on the needles is also observed in the spring in more polluted areas.

Heagle, A.S. et al., "Response of Soybeans to Simulated Acid Rain in the Field", Journal of Environmental Quality, Vol. 12, No. 4, Oct.-Dec. 1983, pp. 538-543.

Soybeans were grown in the field during 1979 and 1980 to determine whether the acidity of simulated rainfall would affect plant injury, growth, or yield, soil chemistry, soil nematode populations, and Rhizobium nodulation of roots. Simulated rain at pH 2.8 or 2.4 caused small amounts of foliar injury evident as bifacial white or tan lesions primarily on young leaves. However, plant growth, pod yield, foliar elemental content, seed protein content, seed oil content, Rhizobium nodulation of roots, and populations of parasitic nematodes were not affected. Trends toward lower soil pH, and less Ca, Mg, and K occurred with increased acidity of simulated rain. These effects were statistically significant only for soil pH, Mg, and K at the pH 2.4 treatment.

Hendrey, G.R. et al., "Acid Precipitation: Some Hydrobiological Changes", Ambio, Vol. 5, No. 5-6, 1976, pp. 224-228.

Evidence of changes in freshwater communities of micro-decomposers, algae, aquatic macrophytes, zooplankton, and zoobenthos due to acid precipitation is reported. Acid precipitation reduces species numbers and changes the biomass of some plant and animal groups. Decomposition of leaf litter and other organic substrates is hampered, nutrient recycling is retarded, and nitrification is inhibited at pH levels observed in acid-stressed waters.

Hendrey, G.R. et al., "Geological and Hydrochemical Sensitivity of the Eastern United States to Acid Precipitation", EPA-600/3-80-024, Jan. 1980, U.S. Environmental Protection Agency, Washington, D.C.

A new analysis of bedrock geology maps of the eastern U.S. constitutes a simple model for predicting areas which might be impacted by acid precipitation and it allows much greater resolution for detecting sensitivity than has previously been available for the region. Map accuracy has been verified by examining current alkalinities and pH's of waters in several test states, including Maine, New Hampshire, New York, Virginia and North Carolina. In regions predicted to be highly sensitive, alkalinities in upstream sites were generally low, less than 200 microequivalents per liter. Many areas of the eastern U.S. are pinpointed in which some of the surface waters, especially upstream reaches, may be sensitive to acidification. Pre-1970 data were compared to post-1975 data, revealing marked declines in both alkalinity and pH of sensitive waters of two states tested, North Carolina, where pH and alkalinity have decreased in 80 percent of 38 streams (p less than 0.001) and New Hampshire, where pH in 90 percent of 49 streams and lakes has decreased (p less than 0.001) since 1949. These sites are predicted to be sensitive by the geological map on the basis of their earlier alkalinity values. Thus this mapping of sensitive areas is validated by the observed temporal trends. The map is to be improved by the addition of a soils component. Impacts of acidification on aquatic biota are reviewed and a Norwegian model of impacts on fish was calibrated and verified using North American data.

Hendrey, G.R. and Kaplan, E., "Identification of Fresh Waters Susceptible to Acidification", BNL-31000, 1982, Brookhaven National Laboratory, Upton, New York.

This research evaluates the sensitivity of freshwaters of the eastern United States to acidification, with emphasis on headwater areas; the extent to which they may have been impacted already; the most likely extent of future impacts; and to the extent possible, the impact of acidification on ground waters based on existing data bases. Subordinate objectives included: (1) acquire existing water quality data from sources outside of STORET (municipal-, county-, and state-agency files), this is intended to include both current and historical records; (2) prepare an inventory of acidified freshwaters; (3) investigate historical trends in pH, alkalinity, or other appropriate water chemistry data indicative of acidification; (4) apply existing, appropriate models or develop models, to be used in forecasting trends in freshwater chemistry under various scenarios of precipitation chemistry; (5) select several appropriate areas, including but not limited to New England and the Appalachian Mountain States, for investigation of relationships between watershed variables and the effects of acid deposition on water chemistry; and (6) interact with various agencies to obtain current and historical ground water data from existing data bases, analyze these data for possible indications of acidification, and identify ground waters of low pH.

Hendrey, G.R. and Lipfert, F.W., "Acid Precipitation and the Aquatic Environment", May 1980, Brookhaven National Laboratory, Upton, New York.

A review is presented of the problem of acid rain in the eastern United States. The historical development of acid rain, its causes and the mechanisms of its formation, and its impact on aquatic ecosystems are discussed.

Hendrey, G.R. and Vertucci, J.A., "Benthic Plant Communities in Acidic Lake Colden, New York: Sphagnum and the Algal Mat", Mar. 1980, Brookhaven National Laboratory, Upton, New York.

Lake Colden, in the central Adirondack Mountains of New York State, is botanically similar to acidified lakes in Sweden. Acidification of some Swedish lakes has been associated with an expansion of Sphagnum, primarily in shallow, sheltered littoral areas but also to depths of 18 m. During a brief botanical survey on 24-25 July 1979, a dense meadow of Sphagnum pylaesii was observed around much of the shoreline of Lake Colden. Plant community composition was determined by a visual estimate of cover along a single typical transect and through underwater photography on 28-29 August 1979. Water samples were collected and returned to the laboratory for analyses several days later. Sample pH was determined by potentiometry and alkalinity by multiple end point titrations. Biomass samples were also taken of the Sphagnum mat community and dry weight was determined. The chemical content of plant tissue was analyzed.

Henriksen, A. and Kirkhusmo, L.A., "Acidification of Groundwater in Norway", Nordic Hydrology, Vol. 13, No. 3, 1982, pp. 183-192.

The chemistry of ground- and surface-water in areas of similar geologic settings are compared by plotting the equivalent concentrations of each component in an X-Y diagram. The deviations of each chemical component from the line through origin and the plot of the two ionic sums give information about the differences in composition of the two types of water. Assuming geologic conditions of the ground- and surface-water are similar when the plots of calcium and magnesium will fall close to the ionic sum line, it is found that the alkalinity is relatively higher and the sulfate concentration is relatively lower in ground water than in surface water. These observations indicate that in areas influenced by acid precipitation the ground water is acidified less than the surface water. The ground water stations studied so far show a regional tendency to lower pH-values in areas where the regional lake surveys show low pH in surface waters.

Hileman, B., "Acid Precipitation", Environmental Science and Technology, Vol. 15, No. 10, Oct. 1981, pp. 1119-1124.

When acid precipitation falls on areas that lack calcium carbonate, the acid is not neutralized and the runoff may acidify nearby lakes and streams killing fish and other aquatic organisms. Significant portions of Ontario, Quebec, and the eastern United States receive precipitation that is about 40 times more acidic than background values. Acid deposition into the atmosphere is caused mainly by emissions of SO_2 and nitrogen oxides from the burning of fossil fuels. The sulfate ion also exchanges readily with calcium ions in soil leaving hydrogen ions behind to acidify the soil and water contained in a watershed. It has been difficult to assign a relative importance to man-made and natural sources as causative agents for acid precipitation. Acid waters tend to dissolve aluminum compounds, thus releasing aluminum ions which are very toxic to fish. Often in acidic lakes, filamentous algae and fungi proliferate. The loss of fish and many amphibians from acidified waters may affect populations of birds that feed on these organisms. Acid rain leaches many major minerals and trace metals from soils. It may also retard the decay of leaves slowing the rate of return of nutrients to the soil. Acid rain may cause drinking water supplies to become contaminated by heavy metals. Some emission control strategies and regulatory measures to control the problem are discussed.

Hoffman, W.A., Lindberg, S.E. and Turner, R.R., "Precipitation Acidity: The Role of the Forest Canopy in Acid Exchange", Journal of Environmental Quality, Vol. 9, No. 1, 1980, pp. 95-101.

Individual rain events were sampled above and below the forest canopy in Walker Branch Watershed, Tennessee, from August 1977 to June 1978 for the purpose of analysis of acidity and organic content. Strong acid content correlated with sulfate concentrations. Total acid concentrations were conserved as rain penetrated chestnut oak (Quercus prinus L.) canopies, although the strong/weak acid ratio declined substantially. Although weak acids constituted at least 30 percent of the total acidity of incident rain and increased to over 50 percent in throughfall, the increase could not be entirely attributed to organic acids. The results of the study support the fundamental importance of ion-exchange models in systematizing rain-vegetation interactions.

Hultberg, H. et al., "First International Symposium on Acid Precipitation and the Forest Ecosystem", Water, Air, and Soil Pollution, Vol. 7, No. 3, Mar. 1977, pp. 279-406.

Ion separation of acid air pollutants out of snow causes sudden, deep pH drops in lakes and running waters at an early stage of snowmelting. These pH drops have drastic effects on fish populations and are considered the main cause of changes in the microflora already at an early stage of acidification. Effects of acid precipitation on: salamanders, various aquatic ecosystems, and the soils of forests surrounding the lakes are discussed. Seasonal patterns in acidity of precipitation and their implications for forest stream ecosystems are examined. The acidity of throughfall precipitation is found to increase with the filtering of sulfur and nitrogen from the atmosphere by trees. As a consequence of acidification, losses of nutrients occur in the soils, and the growth patterns of the trees are affected. Methods of dealing with acid rainfall and the damaging effects of forest and aquatic ecosystems are described.

Hutchinson, T.C., "Effects of Acid Leaching on Cation Loss from Soils", Effects of Acid Rain Precipitation on Terrestrial Ecosystems, 1980, Plenum Press, New York, New York, pp. 481-497.

This paper presents data obtained from studies involving soils in the Smoking Hills area of Canada and soils collected near Sudbury, Canada. Chemical analysis data for the soils and stream water in the areas is given. Mobilization and leaching of heavy metals were observed to increase with increases in acid precipitation; losses of "base" elements also can occur, increased input of heavy metals to aquatic systems through soil-leaching inputs can also occur along with limited survival through developed tolerance of some plant species in affected areas.

Hutchinson, T.C. et al., "First International Symposium on Acid Precipitation and the Forest Ecosystem", Water, Air, and Soil Pollution, Vol. 7, No. 4, Apr. 1977, pp. 421-555.

Effects on the acid rainfall and heavy metal particulates resulting from the massive nickel, copper, and other metallic smelting operations in the Sudbury region of Ontario on the nearby Boreal Forest ecosystem are described. The short-term effects of a simulated acid rain on the growth-nutrient relationships of eastern white pine seedlings are reported. Soil acidification and its effects on forest vegetation are discussed. Chemical and biological relationships relevant to the effect of acid rainfall on the soil plant system reveal a distinct correlation between level of acidity and plant growth. Techniques for the wet and dry removal of particles and gases from the atmosphere to relieve the effects of acidification on forest soils and vegetation are shown to be dependent on other than atmospheric processes.

Irving, P.M., "Acidic Precipitation Effects on Crops: A Review and Analysis of Research", Journal of Environmental Quality, Vol. 12, No. 4, Oct.-Dec. 1983, pp. 442-453.

The majority of crop species studied in field and controlled environment experiments exhibited no effect on growth nor yield as a result of simulated acidic rain. The growth and yield of some crops, however, was negatively affected by acidic rain; others exhibited a positive response. This analysis of the current literature concludes that the effects of acidic precipitation on crops appear to be minimal and that when responses are observed, they may be positive or negative. More complex experimental designs and analyses may be necessary in order to examine and describe the possible subtle responses of agricultural systems to acidic precipitation.

Jacks, G. and Knutsson, G., "Susceptibility to Acidification of Ground Water in Different Parts of Sweden", KHM-TR-11, Oct. 1981, Statens Vattenfallsverk, Stockholm, Sweden.

The progressive acidification of lakes along with the recent finding that a considerable part of the lake water has passed through the soil, has raised fears that ground water is being affected as well. Acid ground water has also been detected in some places in southwestern Sweden. In order to shed light on the possible acidification of ground water a study has been performed on ground water supplies for communities in three counties, Kronoberg, Kopparberg and Vaesternorrland. Time series of ground water chemistry has been used for the purpose. In each county, 20-30 ground water supplies were selected for closer study. The county of Kronoberg has a considerably larger number of water supplies with acid ground water than the two other counties. The reason is likely to be twofold, the more acid precipitation in southern Sweden in combination with the more shallow aquifers in the county of Kronoberg. In the counties of Kopparberg and Vaesternorrland no apparent effects of acid precipitation could be traced. In the county of Kronoberg on the other hand, it is obvious that marked changes have taken place. Shallow wells in the county of Kronoberg show a decrease in alkalinity with time, as well as an increase in carbonic acid. There is no significant change in pH. Thus shallow wells, especially in the county of Kronoberg, show a low and decreasing buffering capacity.

Jacobson, J.S., "Experimental Studies on the Phototoxicity of Acidic Precipitation: The United States Experience", Effects of Acid Rain Precipitation on Terrestrial Ecosystems, 1980, Plenum Press, New York, New York, pp. 151-160.

> The pH or hydrogen ion concentration of precipitation appears to be the major factor in phototoxicity. Responses of plants (and parts of the plant system) to acid rain are discussed as well as the "threshold" effect level. Adverse effects are generally observed at pH levels of less than 4.0.

Johnsen, I., "Regional and Local Effects of Air Pollution, Mainly Sulphur Dioxide, on Lichens and Bryophytes in Denmark", Effects of Acid Rain Precipitation on Terrestrial Ecosystems, 1980, Plenum Press, New York, New York, pp. 133-140.

> Adverse effects of pollution (local and regional) have been observed for lichens and bryophytes in Denmark. No specific pollutant reaction has been observed to date. The synergistic-antagonistic effects of sulfur dioxide, ozone, pesticides, fluorides, nitrogen oxides, and heavy metals are known to be contributors to these adverse effects, however, exact mechanisms have not been defined to date.

Johnson, A.H. et al., "Recent Changes in Patterns of Tree Growth Rate in the New Jersey Pinelands: A Possible Effect of Acid Rain", Journal of Environmental Quality, Vol. 10, No. 4, Oct.-Dec. 1981, pp. 427-430.

> Increment cores from pitch pine, shortleaf pine, and loblolly pine indicate an abnormal decrease in growth rates over the past 25 years. Stream pH, a reliable index of precipitation pH, and a strong statistical relationship between stream pH and growth rates, suggests that acid precipitation may have been a growth-limiting factor for the past two decades. Other factors such as drought, fire, pests and atmospheric oxidants do not appear to be responsible for the two decades of abnormally slow growth.

Johnson, A.H. and Siccama, T.G., "Acid Deposition and Forest Decline", Environmental Science and Technology, Vol. 17, No. 7, July 1983, pp. 294A-305A.

> Available evidence does not show a clear cause and effect relationship between acid deposition and forest decline and dieback in the United States. Of particular concern is the possibility that acid deposition could have caused or may eventually cause soil changes detrimental to forest vegetation, either by stripping nutrients from the soil or by mobilizing phytotoxic elements. Such changes would constitute a problem that could be extremely expensive to correct. In this paper data is presented on the decline of red spruce in the U.S., hypotheses are summarized regarding spruce and fir dieback in central Europe, and, in light of the available information, the authors evaluate the hypotheses concerning cause.

Johnson, A.H. and Siccama, T.G., "Decline of Red Spruce in the Northern Appalachians: Assessing the Possible Role of Acid Deposition", Tappi, Vol. 67, No. 1, Jan. 1984, pp. 68-72.

High-elevation spruce-fir forests of the eastern United States receive particularly high rates of acidic deposition (up to 4 keq of H+/ha/yr), vegetation is exposed to highly acidic cloud moisture for up to 2000 h/yr, and very high levels of trace metals have accumulated. Red spruce (Picea rubens) in the northern Appalachians have died in abnormally large numbers over the past two decades without obvious cause. Considerable attention has been focused on the possible role of acid deposition in the decline. Several plausible mechanisms have been offered by a variety of researchers, but to confirm any of these mechanisms requires additional research.

Johnson, D.W., "Site Susceptibility to Leaching by H2SO4 in Acid Rainfall", Effects of Acid Rain Precipitation on Terrestrial Ecosystems, 1980, Plenum Press, New York, New York, pp. 525-535.

This paper reviews the mechanisms of sulfate adsorption in soils, soil properties in relation to sulfate adsorption and susceptibility to leaching by H2SO4. Cation leaching from soils as caused by atmospheric inputs of sulfuric acid only occurs if the sulfate is mobile in the soil or can displace other mobile anions. Soils rich in sesquioxides are likely to be resistent to leaching by sulfuric acid.

Johnson, D.W., "Acid Rain and Forest Productivity", 1981a, Oak Ridge National Laboratory, Oak Ridge, Tennessee.

Acid rain can cause increases or decreases in forest productivity depending on site nutrient status and the duration and rate of inputs. Some acid irrigation studies have shown short-term growth increases due to increased N availability, yet long-term growth reductions remain theoretically possible because of cation depletion and toxic aluminum accumulations in soils. It cannot be overemphasized that the problem of assessing acid rain effects on forest productivity is one of quantification. Generalization about acid rain effects without reference to the amount of acidic input and site nutrient conditions are hazardous.

Johnson, D.W., "The Natural Acidity of Some Unpolluted Waters in Southeastern Alaska and Potential Impacts of Acid Rain", Water, Air, and Soil Pollution, Vol. 16, No. 2, Aug. 1981b, pp. 243-252.

Near Petersburg, Alaska, natural, unpolluted waters draining surface soils and bog soils containing organic acids display pH values below 4.7. Nearby streams have pH values around 6 or slightly below. Studies were carried out to determine the role of carbonic and organic acids in forest soil leaching processes at this site as compared to others in Washington and Costa Rica. Water samples were collected and analyzed for color, pH, alkalinity and various minerals, anions and cations. Projections of the effects of acid rains and organic acids were formulated using a worst-case model. Results showed that the greatest potential declines in pH would occur in the already acidic waters, while very little change would occur in the near neutral waters. In neither case do the predicted changes in acidity present cause for alarm, since the occurrence of acidic precipitation is not common in this area at present. The findings of this forest watershed study may serve as background data in the event of future pollution by sulfate inputs.

Johnson, D.W., "Effects of Acid Precipitation on Elemental Transport from Terrestrial to Aquatic Ecosystems", 1981c, Oak Ridge National Laboratory, Oak Ridge, Tennessee.

Significant progress has been made in terrestrial-aquatic transport methodology. Several techniques and conceptual frameworks are available for assessment of acid rain effects on these transport processes. Using the anion mobility model, for instance, it is possible to assess the relative effects of acid rain versus natural, internal acid production on elemental transfer rates. However, further research is needed in order to quantify these effects on a regional scale for sensitivity assessments. Sensitivity must be defined as to what ecosystem and what effects are being considered. Criteria have been developed, but there is a major need for more information on natural acid production.

Johnson, D.W. and Richter, D.D., "Effects of Atmospheric Deposition on Forest Nutrient Cycles", Tappi, Vol. 67, No. 1, Jan. 1984, pp. 82-85.

Atmospheric depositions of nitrogen, sulfur, and hydrogen ions have probably increased over the last few decades in many parts of the world. Because most forests are deficient in nitrogen, such increases can be beneficial. Little is known about the changes (if any) in atmospheric deposition of phosphorous and base cations, but there are indications that base cation inputs have declined recently. This, coupled with atmospheric H+ inputs (acid deposition) and internal H+ generation within forest ecosystems, results in a net loss of cations from most forest ecosystems. The extent to which an individual cation is leached depends upon the relative abundance of that cation on soil exchange sites and its selectivity for adsorption. Thus, scarce or tightly bound cations may be conserved despite very intense leaching rates, while more abundant or loosely bound cations are leached. In extremely acid soils, all base cations may be conserved, while hydrogen and aluminum ions are leached.

Johnson, D.W., Turner, J. and Kelly, J.M., "Effects of Acid Rain on Forest Nutrient Status", Water Resources Research, Vol. 18, No. 3, June 1982, pp. 449-461.

This paper presents an extensive literature review that deals with the assessment of the effects of acidic atmospheric inputs on forest nutrient status within the context of natural, internal acid production by carbonic and organic acids as well as the nutrient inputs and drains by management practices such as harvesting, fire and fertilization.

Johnson, N.M., "Acid Rain: Neutralization Within the Hubbard Brook Ecosystem and Regional Implications", Science, Vol. 204, No. 4392, May 4, 1979, pp. 497-499.

The neutralization of strong acids from precipitation is largely accomplished (75 percent) in the soil zone by rapid reaction with basic aluminum salts and biological matter. On a regional basis, acidified and aluminum-rich lakes and streams in New England are confined mainly to low-order watersheds.

Johnson, N.M. et al., "Acid Rain, Dissolved Aluminum and Chemical Weathering at the Hubbard Brook Experimental Forest, New Hampshire", Geochimica et Cosmochimica Acta, Vol. 45, No. 9, Sept. 1981, pp. 1421-1437.

The effects of acid rain were studied in the Hubbard Brook Experimental Forest, where water quality and hydrologic data has been regularly sampled since 1963. Water samples were also collected at 9 sites along Falls Brook (first through fifth order stream) from February 1975 to February 1978. The watershed contains bedrock of high grade metamorphic rocks and granites overlain with glacial till composed of granites and schists; no carbonates are present. In Falls Brook the upper reaches were more acid, richer in aluminum, ammonia, and nitrate, and diminished in alkali and alkaline earth cations, total fluoride, silica, and polyphenols, compared with the lower reaches. Sulfate and chloride were constant. A study of the aluminum chemistry of the stream showed that the upper reaches contained more inorganic monomeric aluminum and less organically complexed aluminum than the lower reaches. Evidence suggested that neutralization of acid rain occurs by a two-step chemical reaction. First, strong acids are rapidly but incompletely neutralized by dissolution of a reactive aluminum phase in the soil zone. Waters rich in Al with pH 4.7-5.2 are produced. Total acidity, including both hydrogen ion and Al acidity, is then slowly neutralized by basic cations released by decomposition of silicate minerals. Most of the neutralization is accomplished within first and second order basins.

Kaplan, E., Thode, Jr., H.C. and Protas, A., "Rocks, Soils and Water Quality-- Relationships and Implications for Effects of Acid Precipitation on Surface Water in the Northeastern United States", Environmental Science and Technology, Vol. 15, No. 5, May 1981, pp. 539-544.

Soils were more important than bedrock in determining regional sensitivities to acid precipitation. This conclusion is contrary to hypotheses advanced previously. Data on rocks, soils, land use, and water quality was collected from 283 counties in New England and the Middle Atlantic States. Four of the nine possible soil classes were found: alfisols, inceptisols, spodosols, and ultisols. Bedrocks were classified as intrusive igneous, metamorphic, consolidated sedimentary, and uncon- solidated and weakly consolidated sedimentary. The area of concern contained no extrusive igneous rocks. Cluster analysis was applied to the rock and soil data. Path analysis produced two models, one relating rock types to soil classes, and the second, the effect of rocks and soils on water quality. This showed that the only rock which contributed to water quality was consolidated sedimentary. The presence of spodosols, ultisols, and inceptisols indicated surface waters of lower alkalinity and greater susceptibility to acidification from acid rain. Areas with larger percentages of alfisols have greater resistance to the effects of acid precipitation.

Killham, K., Firestone, M.K. and McColl, J.G., "Acid Rain and Soil Microbial Activity: Effects and Their Mechanisms", Journal of Environmental Quality, Vol. 12, No. 1, Jan.-Mar. 1983, pp. 133-137.

A Sierran forest soil planted with Ponderosa pine seedlings was exposed to simulated rain with ionic composition reflecting that found in northern California. The soils were collected in two samples (top 1 cm and 4 to 5 cm), which were assayed separately for respiration and enzyme activities. Changes in microbial activity were most significant in surface soils. Only the pH 2.0 input caused inhibition of both respiration and enzyme activities. The overall microbial response to the pH 3.0 and 4.0 acid regimes was one of stimulation, although the response of individual enzymes was more varied. Although changes in C-availability in the exposed soils are documented, changes in the supply of N are evaluated as the major mechanism through which simulated acid rain affects soil microbial activity.

Klein, T.M., Kreitinger, J.P. and Alexander, M., "Nitrate Formation in Acid Forest Soils from the Adirondacks", Journal of the Soil Science Society of America, Vol. 47, No. 3, May-June 1983, pp. 506-508.

Nitrate formation in three forest soils from the Adirondacks region of New York was studied in the laboratory. The organic and surface mineral layers of the soils had pH values ranging from 3.6 to 4.1. Nitrate was formed when the soils were treated with artificial rain at pH 3.5, 4.1, or 5.6. Compared to simulated rain at pH 5.6, simulated rain at pH 3.5 enhanced nitrate formation in one soil and inhibited it in two other soils. The rate of nitrate accumulation was about 10 times higher in the organic horizon than in the mineral horizon, and nitrate formation was not enhanced by ammonium additions. Nitrate formation in soil suspensions was dependent on the amount of soil in the suspension, and none was formed if little soil was present.

Klopatek, J.M., Harris, W.F. and Olson, R.J., "Regional Ecological Assessment Approach to Atmospheric Deposition: Effects on Soil Systems", 1979, Oak Ridge National Laboratory, Oak Ridge, Tennessee.

A regional ecological overview of the potential effects of acid precipitation on soils is presented. Computer maps of soil pH, CEC, base saturation, and base content in the eastern United States are displayed using county level data. These maps are then overlain with a computer map of the hydrogen ion loading and resultant maps of acid sensitive and acid insensitive soils are presented. Of 1572 counties in the eastern U.S., only 117 are classified as being acid sensitive. A number of qualifications concerning the data and the implications for management are discussed.

Kucera, V., "Effects of Sulfur Dioxide and Acid Precipitation on Metals and Anti-Rust Painted Steel", Ambio, Vol. 5, No. 5-6, 1976, pp. 248-254.

Data on the corrosion rates of unprotected carbon steel, zinc and galvanized steel, nickel and nickel-plated steel, copper, aluminum, and anti-rust painted steel due to sulfur dioxide and acid precipitation in Sweden are reported. Corrosion rates are significantly higher in polluted urban atmospheres than in rural atmospheres because of the high concentrations of airborne sulfur pollutants in urbanized areas. Economic damage is significant in the case of galvanized, nickel-plated, and painted steel, and painted wood. Damage in the U.S. in 1970 is estimated at about $7.10/yr/person, while the corresponding Swedish figure is $4.30/yr/person.

Lee, J.J. et al., "Effect of Simulated Sulfuric Acid Rain on Yield, Growth, and Foliar Injury of Several Crops", Environmental and Experimental Botany, Vol. 21, No. 2, 1981, pp. 171-185.

This study was designed to reveal patterns of response of major United States crops to sulfuric acid rain. Potted plants were grown in field chambers and exposed to simulated sulfuric acid rain (pH 3.0, 3.5, or 4.0) or to a control rain (pH 5.6). At harvest, the weights of the marketable portion, total aboveground portion and roots were determined for 28 crops. Of these, marketable yield production was inhibited for 5 crops (radish, beet, carrot, mustard greens, and broccoli), stimulated for 6 crops (tomato, green pepper, strawberry, alfalfa, orchardgrass, and timothy), and ambiguously affected for 1 crop (potato). In addition, stem and leaf production of sweet corn was stimulated. Visible injury of tomatoes might have decreased their marketability.

Lee, J.J., Neely, G.E. and Perrigan, S.C., "Sulfuric Acid Rain Effects on Crop Yield and Foliar Injury", EPA-600/3-80-016, Jan. 1980, U.S. Environmental Protection Agency, Washington, D.C.

A study was undertaken to determine the relative sensitivity of major U.S. crops to sulfuric acid rain. Plants were grown under controlled environmental conditions and exposed to simulated acid rain of three sulfuric acid concentrations (pH 3.0, 3.5, 4.0) or to a control rain (pH 5.7). Injury to foliage and effects on yield were common responses to acid rain. However, foliar injury was not a good indicator of effects on yield.

Lee, J.J. and Weber, D.E., "The Effect of Simulated Acid Rain on Seedling Emergence and Growth of Eleven Woody Species", Forest Science, Vol. 3, No. 25, 1979, pp. 393-398.

Seeds of eleven woody species were exposed to 2.3 cm/wk of simulated sulfuric acid rain at pH values of 3.0, 3.5, or 4.0, or to a simulated control rain at approximately pH 5.6. All treatments also contained a neutral mixture of cations and anions based on concentrations reported for Hubbard Brook, New Hampshire. Seeds or seedlings were subject to ambient conditions, except for precipitation. Ambient rainfall was excluded by a partial covering which allowed some dry deposition. Seeds were planted in winter, 1977; seedlings were harvested the following summer. The dry weights of the tops and roots of each seedling were recorded.

Lee, J.J. and Weber, D.E., "Effects of Sulfuric Acid Rain on Major Cation and Sulfate Concentrations of Water Percolating Through Two Model Hardwood Forests", Journal of Environmental Quality, Vol. 11, No. 1, Jan.-Mar. 1982, pp. 57-64.

Acid precipitation falls on vast areas of forested land, including most of the eastern deciduous forest of the United States. Forest productivity, ground water quality, and surface waters might all be affected. To document and quantify ecosystem response to the onset of acid precipitation, simulated sulfuric acid (H_2SO_4) rain was applied to model forest plots of sugar maple (Acer saccharum) and red alder (Alnus rubra). One set of four plots (two alder and two maple) received a control rain consisting of a stock solution equilibrated with atmospheric CO_2 to approximately pH 5.7. Final results showed that a hardwood canopy and litter layer can alter the input of chemicals to the soil from acid rain.

Likens, G.E. et al., "Biogeochemistry of a Forested Ecosystem", Report No. W80-00328, 1977, Cornell University - Section of Ecology and Systematics, Ithaca, New York.

An in-depth analysis is presented of the biogeochemistry of any terrestrial ecosystem based upon the well-known "Hubbard Brook" ecosystem studies. Long-term data is brought together for precipitation, stream-water chemistry, hydrology and weathering, and the dynamics of atmospheric gases and water as they flow through the system are considered. Illustrated are the ways in which the ecosystem is affected by the three major biogeochemical vectors of the earth: air, water, and organisms. In turn, it is shown how the system moderates and changes inputs and how it affects biogeochemical cycles by its outputs. Acid precipitation is an important example of the ways in which inadvertent human activities influence atmospheric inputs in remote areas. Ecosystem control over biogeochemical functions is highly predictable and relatively repeatable from year to year. The original data from the Hubbard Brook studies are compared with data from diverse ecosystems throughout the world.

Likens, G.E., Bormann, F.H. and Eaton, J.S., "Variations in Precipitation in Streamwater Chemistry at the Hubbard Brook Experimental Forest During 1964 to 1977", Effects of Acid Rain Precipitation on Terrestrial Ecosystems, 1980, Plenum Press, New York, New York, pp. 443-464.

This paper presents data on precipitation input to the Hubbard Brook Experimental Forest in New Hampshire during the period from 1964 to 1977. The chemical quality and quantity of precipitation and streamwater values are given in summary form, trends in precipitation chemistry, and annual (and monthly) input. Acid rain increases were observed in this area.

Lindberg, S.E. and Harriss, R.C., "The Role of Atmospheric Deposition in an Eastern U.S. Deciduous Forest", Water, Air, and Soil Pollution, Vol. 16, No. 1, July 1981, pp. 13-31.

Deposition rates and processes were investigated for Mn, Zn, Cd, Pb, and $SO_4(2-)$ in a forest canopy and its underlying soils. The following were determined: aerosol composition and solubility, empirical measurement of dry deposition, precipitation chemistry, and rates and processes of deposition to the forest. Atmospheric deposition supplied from 14% Mn to about 40% Zn, Cd, and $SO_4(2-)$, to about 99% Pb to the forest floor. Measured water solubility indicated that these metals may be readily mobilized following deposition. Dry deposition contributed about 20% of Cd and Zn, about 35% of SO(2-), about 55% of Pb, and about 90% of Mn. Wet deposition rates for single events were one to four times greater than dry deposition rates. Thus, man's activities which influence the atmosphere may lead to a shift in the exposure of vegetation to certain nutrients and trace contaminants. Heavy metals can cause visible injury to plants, and toxic internal concentrations may exist in plants with no external damage.

Lindberg, S.E., Shriner, D.S. and Hoffman, Jr., W.A., "Interaction of Wet and Dry Deposition with the Forest Canopy", 1981, Oak Ridge National Laboratory, Oak Ridge, Tennessee.

Negative growth impacts and direct injury associated with wet deposition of pollutants appear to be linked primarily with the associated hydrogen ion concentrations. Beneficial effects may also accrue to plants as a result of wet deposition of nutrient elements such as nitrogen and sulfur. The results of a field study indicate the importance of atmospheric deposition in the forest environment. Dry deposition constituted approximately 20 to 60 percent of the total annual atmospheric input of sulfate, Cd, and Zn; however, wet deposition rates for single events exceeded dry deposition rates by one to four orders of magnitude. Interaction between moisture (dew, fog, rain) and previously dry-deposited particles on leaves resulted in concentrations of elements in solution orders of magnitude above concentrations in rain alone and should not be ignored in assessing the potential effects of acid rain. Interception of acid rain by the canopy also resulted in uptake of free H+ and release of weak acids due to decomposition of some components of the leaf cuticle. Factors affecting particle retention on leaf surfaces are critical parameters influencing both biological effects and the potential for particles to interact with wet-deposited pollutants and dry-deposited gases.

Logan, R.M., Derby, J.C. and Duncan, L.C., "Acid Precipitation and Lake Susceptibility in the Central Washington Cascades", Environmental Science and Technology, Vol. 16, No. 11, Nov. 1982, pp. 771-775.

Precipitation samples were collected at five sites across the Cascade Mountains in the region of Snoqualmie Pass, 25-65 mi east and windward of Seattle, Washington, from January to July 1981. The volume weighted average pH was 4.71, with sulfate the dominant anion. The average strong acid compositions (sulfate and nitrate) at the five sites were 57 to 62 equivalent % sulfuric acid. Strong acid levels were higher in summer than in winter. Sulfate deposition for 1981 in this region was estimated to be 16.0 kg per ha per year, corrected for sea salt. This compares with 16.3 kg per ha per year sulfate deposition calculated from 1978 data. Therefore, the Mt. St. Helens emissions apparently bypassed these collection sites. Twenty-nine remote lakes in the Alpine Lakes Wilderness Area 3-20 miles north of the precipitation collectors were sampled in the summer of 1981. None were acid (pH 5.62-7.46, median 6.86). Alkalinities were low (4.0-190, median 57 microeq per liter). Specific conductances averaged 5.8 micro S per cm. Two reservoirs had pH of 7.62 and 7.83, alkalinities of 246 and 346 micro eq per liter, and specific conductances of 3.95 and 38.5 micro S per cm. All lakes, though not at present acid, were classified as susceptible to acidification.

Malmer, N., "Acid Precipitation: Chemical Changes in the Soil", Ambio, Vol. 5, No. 5-6, 1976, pp. 231-235.

Data on the chemical changes in soil due to increasing acid precipitation in Scandinavia are reported. Studies indicate that acid precipitation results in decreased soil pH, decreased base saturation, and increased leaching. Adverse effects may be compensated through an increase in weathering. The time required and the intensities of the effects for weathering, and the variation in susceptibility among soils need further investigation.

Matziris, D.I. and Nakos, G., "Effect of Simulated Acid Rain on Juvenile Characteristics of Aleppo Pine (Pinus-Halepensis Mill)", Forest Ecology and Management, Vol. 1, No. 3, 1977, pp. 267-272.

A study was undertaken to determine the effect of simulated acid rain on growth and other characteristics of Aleppo pine and to investigate intraspecific variability. The effect of acid rain on soil properties was also examined. During one growing season, seedlings from 1 yr old half sib families of Aleppo pine were irrigated with simulated acid rains of pH 3.1 and 3.5. A control group was irrigated with pH 5.1 water. Seedlings that were treated with acid rain with a pH of 3.1 reached a mean height of 22.6 cm, which was 8.2 percent shorter than the mean height of the control seedlings. Irrigation with acid rain increased the mortality of the seedlings, negatively influenced the formation of terminal buds, and dissolved and leached considerable amounts of calcium carbonate from the soil.

Mayer, R. and Ulrich, B., "Input to Soil, Especially the Influence of Vegetation in Intercepting and Modifying Inputs--A Review", Effects of Acid Rain Precipitation on Terrestrial Ecosystems, 1980, Plenum Press, New York, New York, pp. 173-182.

This article discusses the input of elements into soil from wet and dry deposition and from vegetation.

McColl, J.G., "Effects of Acid Rain on Plants and Soils in California", ARB-R-81/148, Sept. 1981, California State Air Resources Board, Sacramento, California.

Effects of acid rain on some California plants and soils were studied. Plants growing in soil were treated with simulated rain on varying acidity. Direct foliar damage was not apparent, other than under extreme conditions which are not normally experienced in the field. Sugar beet was the most sensitive of the agronomic species tested. Germination of Douglas-fir seed was inhibited under severe acid conditions. Similarly, growth of two-year-old conifer seedlings showed little deleterious effects, except under the most severe treatments. Acid rain affects plant productivity (positively and negatively), and the effect for a given input acid was largely predicated by the soil in which the plants were growing. A simple, reliable laboratory method was developed for determining potential sensitivity of soils to leaching by acid rain.

McKinley, V.L. and Vestal, J.R., "Effects of Acid on Plant Litter Decomposition in an Arctic Environment", Applied and Environmental Biology, Vol. 43, No. 5, May 1982, pp. 1188-1195.

The effects of acidic pH on the microbial decomposition of plant litter were studied during the summer of 1980 in microcosms from Toolik Lake, Alaska. Toolik Lake is a large, deep water lake on the north slope of the Brooks Range of mountains within the Trans-Alaska Pipeline corridor. It is very oligotrophic and its waters have very little buffering capacity. Microbial decomposition of Carex in response to acidification was evaluated directly by determinations of weight loss and changes in C/N ratios of the litter over time. Microbial biomass and activity associated with the litter were evaluated by the measurement of adenosine triphosphatase concentrations and by determination of the rates of acetate incorporation into microbial lipids, respectively. The community structure of the microbiota on the litter surface was examined by scanning electron microscopy and the effects of acid on the mineralization of Carex lignocellulose were also examined. Microbial activities associated with Carex litter were significantly reduced within 2 days at pH values of 3.0 and 4.0, but not 5.0, 5.5, or 6.0. ATP levels were significantly reduced at pH 3.0, but not at the other pH's tested. After 18 days microbial activity significantly correlated with weight loss, nitrogen content, and C/N ratios of the litter but did not correlate with ATP levels. Fungi present at ambient pH did not become dominant at pH below 5.5, diatoms were absent below pH 4.0, and bacterial numbers and extracellular slime were reduced at pH 4.0 and below. It was concluded that if the pH of the water from this slightly buffered lake were sufficiently reduced, rates of litter decomposition would be significantly reduced.

McLaughlin, S.B. et al., "Interactive Effects of Acid Rain and Gaseous Air Pollutants on Natural Terrestrial Vegetation", 1983, Oak Ridge National Laboratory, Oak Ridge, Tennessee.

Results of the multi-elemental analysis of tree rings have indicated that accumulation patterns of trace elements differ widely between elements and between tree species. Many elements show patterns of increasing concentration in recent tissues with some elements such as zinc and chromium showing steady increases (in the Oak Ridge, Tennessee area) over the past 30 years. Highest levels of aluminum, zinc, copper, and chromium occur in most recent tissues of shortleaf pine and are at levels of potential concern for adverse effects to metabolically active tissues.

McLaughlin, S.B., West, D.C. and Blasing, T.J., "Measuring Effects of Air Pollution Stress on Forest Productivity: Perspectives, Problems, and Approaches", Tappi, Vol. 67, No. 1, Jan. 1984, pp. 74-80.

Changing levels and patterns of emissions of atmospheric pollutants in recent decades have resulted in increased exposure of extensive forests in Europe and North America to both gaseous pollutants and acid rain. Reports of decreased growth and increased mortality of forest trees in areas receiving high rates of deposition of atmospheric pollutants have emphasized the need to better understand and quantify both the mechanisms and kinetics of potential changes in forest productivity. The complex chemical nature of combined pollutant exposures, and the fact that these changes may involve both direct effects to vegetation, as well as indirect and possibly beneficial effects mediated by a wide variety of soil processes, make quantification of such effects particularly challenging. A variety of productivity studies in Europe and the U.S. have been evaluated. Dendroecological approaches, which combine detailed analysis of long-term growth trends as influenced by tree age, competitive status, climatic variables, and pollutant levels, offer great promise for quantifying both the temporal and spatial relationships of productivity changes, and their relationship to patterns of deposition of atmospheric pollutants.

Mitchell, M.J., Landers, D.H. and Brodowski, D.F., "Sulfur Constituents of Sediments and Their Relationship to Lake Acidification", Water, Air, and Soil Pollution, Vol. 16, No. 3, Oct. 1981, pp. 351-359.

The organic and inorganic sulfur fraction in three New York lake sediments were determined. These lakes, subjected to acid precipitation, differed in buffering capacity and pH. The pH and acid neutralizing capacity (in meq per liter) respectively were: Oneida, 7.2-9.4 and 1700; South, 4.9-6.5 and 2-10; and Deer, 5.9-6.8 and 90-124. Total sulfur in sediments was lowest in Oneida (0.1-0.3% dry mass). South (0.25-0.65%) and Deer (0.3-0.5%) sediments had higher S levels near the outlets. Sulfides were highest in Oneida (1% of total S) which had the lowest redox potential (-80 to -110 mV). South and Deer contained about 0.2% sulfides. Inorganic S varied from 8% of the total S in Deer to 22% in Oneida. Ester sulfates were 80% of the total S in Oneida and Deer, and 45% in South. Oxygen consumption was highest in Oneida and lowest in South sediments. The low ester sulfate and higher carbon/nitrogen ratio in South sediments suggest that lake acidification may inhibit decomposition.

Mortvedt, J.J., "Impacts of Acid Deposition on Micronutrient Cycling in Agro-Ecosystems", Environmental and Experimental Botany, Vol. 23, No. 3, 1983, pp. 243-249.

Plant availability of all micronutrients except Mo increases with increasing soil acidity. Toxicities due to Mn and Al in extremely acid soils may result in decreased crop yields. Changes in soil pH due to acid depositions are minimal in most agricultural soils because of relatively high buffering capacities of these soils. Modern farming practices such as liming and return of crop residues also may override depositional effects. Therefore, micronutrient cycling in most agro-ecosystems should not be significantly affected by acid depositions. The long-term effects of acid depositions are not known. Plant micronutrients include B, Cu, Fe, Mn, Mo and Zn. Although Cl and Co are also micronutrients, their supply in most agro-ecosystems is usually sufficient. Micronutrient cycling in agro-ecosystems is a complex series of processes involving micronutrient additions to and removal from soils and changes in the chemical forms, solubilities and consequent availabilities of micronutrients to plants induced by changes in the ecosystem. Crop removal and erosion of surface soils result in losses of micronutrients from agricultural soils. Return of crop residues to soils mitigates such losses by returning some of the plant-contained micronutrients and by reducing the potential for erosion. High rates of acid deposition on poorly buffered soils over a period of time may result in measurable changes in soil acidity. Unless a soil is near the critical pH level for Mn or Al toxicity (usually about pH 5.0), soil acidification due to acid precipitation should not have an adverse effect on plant growth. Decreases in soil pH would increase plant uptake of all micronutrients except Mo, but a large fraction of the increased uptake usually would be recycled to soil with the return of crop residues.

National Technical Information Service, "Acid Precipitation: Effects on Terrestrial Ecosystems--1976-November, 1982 (Citations from the Energy Data Base)", Nov. 1982a, U.S. Department of Commerce, Springfield, Virginia.

This bibliography contains citations concerning acid rain and its effects on terrestrial ecosystems.   Impacts on soil structure and vegetation, including plants, hardwood and softwood trees are discussed. Future trends in the geological distribution and deposition of acid rain are considered.   This document contains 180 citations fully indexed and includes a title list.

National Technical Information Service, "Acid Precipitation: Effects on the Aquatic Ecosystem--1977-November, 1982 (Citations from the Energy Data Base)", Nov. 1982b, U.S. Department of Commerce, Springfield, Virginia.

This bibliography contains citations concerning acid rain and its effect on aquatic ecosystems. Physiological and reproductive effects on fish and other aquatic populations are considered. Exposure pathways, factors affecting the acidification of lakes and ponds, and the known and projected consequences are included.   This document contains 152 citations fully indexed and includes a title list.

Nelson, P.O. and Delwiche, G.K., "Sensitivity of Oregon's Cascade Lakes to Acid Precipitation", WRRI-85, Sept. 1983, Water Resources Research Institute, Oregon State University, Corvallis, Oregon.

Susceptibility to acidification and present extent of anthropogenic acidification were studied for 63 Oregon Cascade lakes in 1982. Chemical parameters included pH, conductivity, alkalinity, calcium, magnesium, sodium, potassium, dissolved silica, chloride and sulfate.   Major ion balances were calculated. Analyses for total organic carbon, fluoride and aluminum were performed in some cases. The lakes have extremely dilute ionic compositions. Average conductivity was 17.0 umhos/cm. Average laboratory-equilibrated pH was 6.96. Alkalinities averaged 137.6 ueq/l; 25 lakes had alkalinities of less than 50 ueq/l. Calcium was the major cation (36 percent) and bicarbonate was the major anion (82 percent) present. Aluminum and sulfate concentrations were extremely low. The carbonate system was found to be the dominant buffer system.   Extremely low aluminum and total organic carbon concentrations indicate the minor contribution of aluminum hydrolysis species and weak organic acids to the overall acid neutralizing capacity of these waters.   The calcium-to-bicarbonate ratio and the low dissolved aluminum and sulfate concentrations show that anthropogenic acidification is undetectable to date. Oregon's Cascade lakes do not presently have an acid precipitation problem but are extremely susceptible to acidification. The lakes can serve as a chemical benchmark to compare with acidified lakes of similar volcanic geology.

Nilssen, J.P., "Acidification of a Small Watershed in Southern Norway and Some Characteristics of Acidic Aquatic Environments", Internationale Revue der Gesamten Hydrobiologie, Vol. 65, No. 2, 1980, pp. 177-207.

Freshwater lakes in a region of southern Norway which receives a substantial amount of atmospheric pollutants were studied for acidity and biological characteristics. Precipitation in the area has a pH of 4.2-4.5.

The acid originates further south in Europe, but prevailing winds bring this pollution to Norway. In some lakes which have lost buffering capacity, spring thaw and autumn rain are accompanied by increases in conductivity and in sulfate deposition, and by a conspicuous drop in pH. Decreasing pH, increasing conductivity, and increasing total hardness have been observed in a river draining the area. Fish, daphnia, and chaoborus have disappeared from the most acidic lakes, and these lakes are characterized by increasing abundance of Juncus bulbosus and macrophytes covered by filamentous algae. Pelagic zooplankton are mainly of the littoral species, and there are no leeches or gastropods. Brown trout is the fish most resistant to acid, while Eurasian perch were absent from most of the acid lakes. Moderately acidic lakes contained abundant food for carnivorous fish.

Nilsson, S.I., "Ion Adsorption Isotherms in Predicting Leaching Losses from Soils Due to Increased Inputs of 'Hydrogen Ions'--A Case Study", Effects of Acid Rain Precipitation on Terrestrial Ecosystems, 1980, Plenum Press, New York, New York, pp. 537-551.

This study was conducted at Lisselbo, Sweden and involved field studies with laboratory analysis applied to predictions concerning changes in acid-base variables in soils and changes in nitrogen mobilization rates. Calculations, equations, and adsorption isotherms are presented. Adsorption isotherm studies provide a more sensitive and accurate means of evaluating the effects of acid precipitation on soils.

Norton, S.A. et al., "Responses of Northern New England Lakes to Atmospheric Inputs of Acids and Heavy Metals", OWRT A-048-ME(1), July 1981, Land and Water Resources Center, University of Maine, Orono, Maine.

Ninety-four low-humic lakes in northern New England (82 in Maine) located in largely noncalcareous terrain were studied to detect possible changes in pH and associated effects. Modern colorimetric and electrode pH measurements yield equivalent results plus or minus 0.1 pH unit, permitting comparison of modern electrode pH measurements with older colorimetric measurements. Eighty-five percent of the lakes are more acidic now. Studies of present water chemistry indicate that Fe, Mn, Pb, Zn, and Al concentrations tend to be higher with decreasing pH for the set of lakes; and Na, K, Ca, Mg, and alkalinity tend to be higher with increasing pH for the set of lakes. Surface sediment chemistry indicates that (at the 95 percent significance level) the concentrations of Na20, K20, MgO, and organics are highest at higher lake pH values; Cu increases with decreasing pH. Calcium oxide, FeO, MnO, A1203, TiO2, Pb, and Zn concentrations do not relate to pH. The relative abundances of taxa of diatom remains in profundal surface sediments from a subset of 25 lakes are related to surface water pH's in the range of 4.4 to 7.0. In a cluster analysis, the diatom assemblages from lakes with pH less than 5.7 separate from those with pH greater than 5.8. Greatest taxonomic diversity occurs at pH 5.8. Some unknown factors associated with lake altitude (52-1009 m) accounts for scatter in the pH-diatom relationship. In studies of cladoceran associations in profundal surface sediments from 18 of the lakes with a pH range of 4.4 to 6.9, it was not possible to uncouple the effects of pH and altitude (78-1123 m). Cladoceran stratigraphy in sediment cores from three high-altitude lakes indicates community changes concurrent with changes in diatom assemblages postulated to be caused by lake acidification.

Nriagu, J.O., "Isotopic Variation as an Index of Sulphur Pollution in Lakes Around Sudbury, Ontario", Nature, Vol. 273, No. 5659, May 18, 1978, pp. 223-224.

The sulfate (SO_4^{-2}) content of 120 lakes was determined gravimetrically as $BaSO_4$. The $BaSO_4$ was thermally decomposed to liberate sulfur dioxide (SO_2) and the ^{32}S. In the preliminary survey, the SO_4^{-2} concentration in the lakes was highly correlated inversely with the distance from the smelter at Copper Cliff. Sulfate levels were high up to a distance of 60 km. The 41 lakes in which the SO_4^{-2} concentrations fall close to the regression line (SO_4^{-2} vs. distance from smelter) were sampled again the next year. On the basis of the observed pH, the lakes can be divided into those which are well buffered and unresponsive to the input of acid precipitation and those which are poorly buffered and more easily influenced by smelter emissions. The ^{34}S ratio data for the well buffered lakes show considerable scatter, suggesting derivation of the S from several sources, complex interplay between the point source influence and the transformations of the S in the watershed or lakes, or a combination of these factors. The homogeneous nature of the ^{34}S ratio around the stack implies that the data for lakes which derive their S mostly from atmospheric sources will also be fairly uniform, as was in fact observed. The weak correlation between ^{34}S ratio and pH reflects the decoupling of the S and proton cycles within these lakes. Unlike the conservative SO_4^{-2}, protons are highly active and do not simply accumulate in lakes, presumably because of the synergistic influence of atmospheric carbon dioxide.

Olson, R.A., "Impacts of Acid Deposition on N and S Cycling", Environmental and Experimental Botany, Vol. 23, No. 3, 1983, pp. 211-223.

Modern concern on the environmental impact of "acid rain" centers especially on the oxides of N and S. More appropriately expressed as "acid deposition", the topic has been related largely to impacts on human health and man-made structures with focus only in recent years on possible damage to soils and agricultural production. Likely deleterious effects on soils include reduction in pH, increased activity of various elements with toxicity potential, accelerated mineral weathering and leaching of reserve soil nutrients and undesirable changes in the soil microflora. Areas with greatest potential for soil damage from deposition are in close proximity and downwind from large coal-burning plants, especially where shallow and/or coarse textured soils exist on inaccessible mountainous positions with forest cover. For most of the agricultural regions of this country and the world, however, the N and S fallout products are more beneficial in supplying needed plant nutrients than they are detrimental. Their acidification capacity is insignificant in comparison with that of natural N and S transformations in soils or that induced by conventional fertilizer programs of farmers. The slight acidification imposed over the long term can improve calcareous and alkali soils as growth media and is readily corrected by liming of the acid soils made more acid than desirable for crop production.

Olson, R.J., Johnson, D.W. and Shriner, D.S., "Regional Assessment of Potential Sensitivity of Soils in the Eastern United States to Acid Precipitation", ORNL/TM-8374, Dec. 1982, Oak Ridge National Laboratory, Oak Ridge, Tennessee.

Areas in the eastern United States are evaluated for their sensitivity to acid deposition by combining county-level information on soil chemistry, bedrock geology, terrain characteristics, and land-use information. The report presents three sets of sensitivity maps that represent continuing refinements in the sensitivity criteria and available data bases. Criteria were developed in cooperation with the Canada-United States Transboundary Working Group to obtain comparable sensitivity maps for both countries. The final analysis covered the eastern 37 states and excluded the 1,013 counties that are predominantly agricultural or urban. Soils are characterized for their potential to undergo acidification from acid deposition. The criteria are moderate pH and low cation exchange capacity. The one soil type meeting these criteria occurs extensively only in 16 counties in Nebraska. The soils characterization map also shows low pH soils (potential for aluminum leaching) and peat soils (potential for acidifying precipitation). Areas are also classified for their potential to reduce the acidity of acid deposition prior to the transfer of acid inputs to aquatic systems. Low soil pH, low soil sulfate adsorption capacity, bedrock with no buffering capacity, and steep terrain are factors associated with low potential to reduce acidity. Eight percent of the 2,660 counties in the east were found to have low potential to reduce acidity with an additional 20 percent having moderate potential. Areas occur in northern Minnesota, Wisconsin, and Michigan; the New England states; parts of New York and Pennsylvania; the Appalachian mountains; and Florida.

Overrein, L.N., Seip, H.M. and Tollan, A., "Acid Precipitation: Effects on Forest and Fish", NP-2902584, Dec. 1980, Norges Landbrugshoegskole, Norway.

This interdisciplinary research program was launched in 1972 in response to concern in Scandinavian countries that acid precipitation was causing changes to the natural environment. A major hypothesis was that anthropogenic release of sulfur oxides and other pollutants may alter geobiochemical and biochemical cycles with consequences for the biota. The main research efforts were directed towards possible threats to forest and freshwater fish. Results of the entire program are summarized. Information is presented under the following section headings: emissions and transport; atmospheric deposits in Norway; water acidification--status and trends; chemical modifications of precipitation in contact with soil and vegetation; snow and snowmelt; land-use changes and acidification; effects of acid precipitation on soil productivity and plant growth; and effects of acid water on aquatic life.

Petersen, L., "Podzolization: Mechanisms and Possible Effects of Acid Precipitation", Effects of Acid Rain Precipitation on Terrestrial Ecosystems, 1980a, Plenum Press, New York, New York, pp. 223-238.

A brief review of the mechanisms of podzolization (involving iron and aluminum) is given. A laboratory experiment involved the leaching of soil extracts through ion exchange columns to determine the capacity of water-soluble organic (soil) matter to react with certain metal ions. This experiment indicates that podzolization may be amplified under conditions of acid precipitation.

Petersen, L., "Sensitivity of Different Soils to Acid Precipitation", Effects of Acid Rain Precipitation on Terrestrial Ecosystems, 1980b, Plenum Press, New York, New York, pp. 573-577.

Soil orders and classifications according to response to acid precipitation are given with a brief description of soil type and content.

Pfeiffer, M.H. and Festa, P.J., "Acidity Status of Lakes in the Adirondack Region of New York in Relation to Fish Resources", Aug. 1980, Bureau of Fisheries, Department of Environmental Conservation, State of New York, Albany, New York.

As a result of extensive glacial activity, the Adirondack region of northeastern New York State contains approximately 2,900 individual lakes and ponds, encompassing approximately 282,000 surface acres. Many of these surface waters have low alkalinities due to a carbonate-poor geology; therefore, they are particularly sensitive to the high acid ion deposition associated with the region's airshed. Since 1975, pH and alkalinity measurements have been made on 849 ponded waters throughout the region to determine the scope of water quality impacts associated with acid ion deposition and to provide a baseline inventory for indexing future measurements. The present condition of surface waters is described on the basis of summertime, one-meter depth, and pH measurements obtained with a pH meter under air-carbon dioxide-equilibrium conditions. Twenty-five percent of the waters in the survey, comprising 10,460 surface acres, registered pH readings below 5.0. Comparisons of historic and post-1974 acidities are made where data points from comparable methodologies exist. Relationships between meter pH, colorimetric pH, alkalinity, conductivity, calcium, lake surface area, lake surface elevation, and geographical location are discussed. Changes in fish species composition and sportfishing yields observed in waters exhibiting increased acidity are reviewed.

Pitblado, J.R., Keller, W. and Conroy, N.I., "A Classification and Description of Some Northeastern Ontario Lakes Influenced by Acid Precipitation", Journal of Great Lakes Research, Vol. 6, No. 3, 1980, pp. 247-257.

Statistical analysis of water chemistry data (23 variables for 187 lakes in the Sudbury, Ontario, area) showed that most of the chemical variability was attributable to 4 components: nutrient status, buffering status, atmospheric deposition status, and sodium chloride status. Seven distinct groups of lakes were obvious. Group 1 lakes, scattered throughout the study area, were high in nutrients and chlorophyll, reflecting cultural eutrophication. Groups 2 and 3 were dilute lakes of the Precambrian Shield, had an inverse relationship to the atmospheric deposition status and buffering status dimensions, and were low in smelter-produced ions. They differed in nutrient and chlorophyll-a status, with Group 2 having the higher values. Groups 4, 5, and 6 reflected the impact of airborne pollutants from smelting operations in Sudbury. Group 5 lakes, closest to the smelters, had high acidity, low buffering capacity, and high concentrations of smelter-related metals. Group 6 lakes showed a lesser impact, and Group 4, limited impact, being more buffered and productive. Group 7, a group of anomalous lakes sharing higher pH and high ionic strength, were influenced by limestone bedrock or surficial material as well as urban runoff.

Pough, F.H., "Mechanisms by Which Acid Precipitation Produces Embryonic Death in Aquatic Vertebrates", OWRT-A-077-NY(2), Mar. 1981, Cornell University, Ithaca, New York.

Fourteen species of amphibians show a general similarity in their tolerance of acid media during embryonic development. More than 85 percent mortality is produced by pH's of 3.7 to 3.9 and more than 50 percent mortality occurs at pH's of 4.0 or less. Similar values have been reported for fishes. The sensitivity of amphibian embryos to acidity is greater in late stages of their development than it is during the initial cleavage of the embryos. The teratogenic effects of acidity appear to be the result of damage to the superficial tissues of the embryo. A similar response occurs in fish embryos. Because of the similarity of sensitivity and response to acidity of fishes and amphibians, the latter animals are suitable experimental models for investigations of the details of acid resistance. Controlled breedings of African clawed frogs (Xenopus laevis) indicated that the offspring of some pairs of parents were more resistant to acidity than those of other pairs. Wild populations of spotted salamanders (Ambystoma maculatum) breeding in some ponds in the Ithaca, New York, region have probably been exposed to increasingly acid conditions for the past three decades (10 or more generations). In the most acid ponds more than 70 percent of the embryos die before hatching. Despite the intensity and duration of this selection, it was not possible to demonstrate any difference in sensitivity to acidity between eggs collected from acidic and neutral breeding sites.

Puckett, L.J., "Acid Rain, Air Pollution, and Tree Growth in Southeastern New York", Journal of Environmental Quality, Vol. 11, No. 3, July-Sept. 1982, pp. 376-381.

Whether dendroecological analyses could be used to detect changes in the relationship of tree growth to climate that might have resulted from chronic exposure to components of the acid rain-air pollution complex was determined. Tree-ring indices of white pine, pitch pine, and chestnut oak were regressed against orthogonally transformed values of temperature and precipitation in order to derive a response-function relationship. Results of the regression analyses for three time periods suggest that the relationship of tree growth to climate has been altered. Statistical tests of the temperature and precipitation data suggest that this change was nonclimatic. Temporally, the shift in growth response appears to correspond with the suspected increase in acid rain and air pollution in the Shawangunk Mountain area of southeastern New York in the early 1950's.

Raynal, D.J., "Actual and Potential Effects of Acid Precipitation on a Forest Ecosystem in the Adirondack Mountains", NYSERDA-80-28, Dec. 1980, New York State Energy Research and Development Authority, Albany, New York.

This report characterizes several important forest ecosystem components and relates their sensitivity to acid precipitation through field studies, laboratory experiments and simulations. The study characterizes ion concentrations due to bulk precipitation and forest canopy throughfall, examines soil nutrient leaching patterns, estimates the effects of nitrate input from precipitation, assesses the benefits of applying lime to soils, and documents population level and species diversity changes of soil microorganisms. It also characterizes the regeneration dynamics and growth response of forest trees.

Raynal, D.J., Roman, J.R. and Eichenlaub, W.M., "Response of Tree Seedlings to Acid Precipitation--II--Effect of Simulated Acidified Canopy Throughfall on Sugar Maple Seedling Growth", Environmental and Experimental Botany, Vol. 22, No. 3, 1982, pp. 385-392.

Sugar maple seedling radical growth in a laboratory growth apparatus was significantly reduced after exposure to simulated acidified canopy throughfall at pH 3.0 and below. Seedlings exposed to low pH were susceptible to bacterial infection; survival of seedlings transplanted to soil declined with increasing acidity of simulated canopy throughfall. Extension growth and leaf weight gain of established potted seedlings subjected to acidic throughfall was dependent on the soil nutrient supplying capacity. Under nutrient-limited conditions, throughfall of pH 3.0 promoted seedling growth, although causing foliar damage. At higher fertility levels, reduction in growth was found only after the pH 2.0 treatment. While these studies indicate that sugar maple seedlings are potentially susceptible to direct and indirect effects of acid precipitation, evaluation of the findings in relation to natural conditions is complicated by the capacity of vegetation and the forest floor to buffer the seedling environment from extreme pH changes, the direct effects of acidity on pathogenic microorganisms and the predisposition of seedlings to infection, the episodic nature and varying acidity of precipitation, and the differential sensitivity of laboratory and field grown seedlings to acidic deposition.

Reuss, J.D., "Chemical/Biological Relationships Relevant to Ecological Effects of Acid Rainfall", June 1975, Corvallis Environmental Research Laboratory, U.S. Environmental Protection Agency, Corvallis, Oregon.

Measurement and interpretation of rainfall acidity is examined in terms of effects on the soil-plant system. The theory of carbon dioxide-bicarbonate equilibria and its effect on rainfall acidity is given. The relationship of a cation-anion balance model of acidity in rainfall to plant nutrient uptake processes is discussed. The flux of hydrogen ions due to plant uptake processes and sulfur and nitrogen cycling is considered. Hydrogen ion is produced by oxidation of reduced sulfur and nitrogen compounds mineralized during decomposition of organic matter. The soil acidifying potential due to the oxidation of the ammonium ion in rainfall is apparently of a similar magnitude as that due to the direct acidity inputs.

Richter, D.D., Johnson, D.W. and Todd, D.E., "Atmospheric Sulfur Deposition, Neutralization, and Ion Leaching in Two Deciduous Forest Ecosystems", Journal of Environmental Quality, Vol. 12, No. 2, Apr.-June 1983, pp. 263-270.

In the 1981 water year, bulk precipitation was primarily a solution of dilute H_2SO_4, and $SO_4(2-)$ was the dominant anion in throughfall and soil leachates in two eastern Tennessee deciduous forests. Ecosystem inputs of $SO_4(2-)$, which included dry deposition of forest canopies, may have been up to 40 percent greater than input estimates based on atmospheric deposition sampling in open areas. Volume-weighted mean annual pH of bulk precipitation was 4.3; of throughfall 4.8; and of leachates from O2, Al, and B21 soil horizons about 6.0. At both sites, strong acids in precipitation were largely neutralized prior to rainwater's infiltration into mineral soil. Base cations that exchanged with $H(+)$ (hydrogen ions) in acid precipitation were almost entirely supplied by forest canopies and litter layers, and did not come directly from exchangeable mineral soil pools. However, even in these infertile soils with low cation exchange capacities, exchangeable bases were more than two orders of magnitude greater than annual $H(+)$ input in bulk precipitation, and represented a substantial reserve for base cations in canopies and litter layers that exchanged with $H(+)$ in acid rain. Furthermore, inputs of $H(+)$ from acid precipitation were equal to about 0.4 percent of the base cations that are biologically cycling and immediately available in these ecosystems. Both soils are base-poor Udults and classified as sensitive to acid rain, but the deposition, cycling, and soil data presented in this report indicate that leaching remains a process affecting cation reserves and soil development only over the very long term.

Roberts, T.M. et al., "Effects of Sulphur Deposition on Litter Decomposition and Nutrient Leaching in Coniferous Forest Soils", Effects of Acid Rain Precipitation on Terrestrial Ecosystems, 1980, Plenum Press, New York, New York, pp. 381-393.

This paper describes the results observed of naturally occurring acid rain conditions and simulated conditions in the Delamere Forest in central England. Experimental methods are described. These studies showed that acid rain events of pH 3-3.5 produced small changes in decomposition rates and nitrogen availability, with little effect on soil leaching.

Rorison, I.H., "The Effects of Soil Acidity on Nutrient Availability and Plant Response", Effects of Acid Rain Precipitation on Terrestrial Ecosystems, 1980, Plenum Press, New York, New York, pp. 283-304.

Acidic soils have relatively low levels of essential major nutrients and high levels of cations and trace amounts of essential elements. Raising of pH may cause stunting of plants due to increased uptake of nutrients and decreases in availability of trace elements. This article discusses the constraints of acidic soils, individual factors (chemicals) to be considered, the effects of plants on soil pH, the tolerance mechanisms of plants and soils, and nitrogen-aluminum interactions.

Scherbatskoy, T. and Klein, R.M., "Response of Spruce and Birch Foliage to Leaching by Acidic Mists", Journal of Environmental Quality, Vol. 12, No. 2, Apr.-June 1983, pp. 189-195.

Seedlings of yellow birch (Betula alleghaniensis Britt.) and white spruce (Picea glauca (Moench) Voss) were exposed to mists consisting of distilled water at pH 5.7, or sulfuric acid in distilled water at pH 4.3 or 2.8. Misting treatments, 4 h each, were applied once, twice or three times, separated by 72 h. Leaf tissue chlorophyll concentrations were determined, and foliar leachate was analyzed for $K(+)$, Ca $(2+)$, $NO3(-)$,

H2PO4(-), total carbohydrate, total protein, and total amino acids. Decreased leaching of carbohydrate, protein, K(+), and H2PO4(-) occurred with increasing number of mistings, indicating that these substances were not readily resupplied to leaching sites. Protein leaching was reduced by misting at pH 2.8. Leaching of amino acids, K(+), and Ca(2+) increased with decreasing mist pH, and leachate pH was higher than the pH of applied mists, suggesting that cation exchange plays a role in foliar leaching by acidic solution. Amino acid leaching from birch increased with increasing number of mistings at pH 2.8. There was no strong effect on chlorophyll concentrations, and leaching of NO3(-) and H2PO4(-) did not vary with mist pH.

Schindler, D.W. et al., "Effects of Acidification on Mobilization of Heavy Metals and Radionuclides from the Sediments of a Freshwater Lake", Canadian Journal of Fisheries and Aquatic Sciences, Vol. 37, No. 3, Mar. 1980, pp. 373-377.

Large (10 m) diameter enclosures were sealed to the sediments in 2-2.5 m of water in Lake 223. Two tubes were held at control pH (6.7-6.8), one was lowered to pH 5.7 and one to pH 5.1, using H_2SO_4. Aluminum, manganese, zinc, and iron were released from lake sediments at pH 5 and 6. Concentrations of zinc in the overlying water column exceeded 300 micrograms/l. Radioisotopes of several heavy metals added to the water of the enclosure showed the following: all metals were removed from the water at log-linear rates, with half-times of 5-25 d. Acidification caused several metals to become more soluble, including Fe-59, Co-60, Mn-54, and Zn-65. Solubility of V-48 and Hg-203 decreased with increasing acidity. Acidification also slowed the loss to sediments of Mn-54 and Zn-65. Losses of Ba-133, Se-75, Cs-134, and V-48 were more rapid under acid conditions. The fractions of any isotope retained by a 0.45-micrometer filter, activated charcoal, and mixed-bed ion exchange resin remained constant throughout the experiment at any given pH.

Schindler, D.W. and Turner, M.A., "Biological, Chemical and Physical Responses of Lakes to Experimental Acidification", Water, Air, and Soil Pollution, Vol. 18, No. 1-3, 1982, pp. 259-271.

Work is summarized on biological, physical and chemical changes which occur early in the experimental acidification of a small lake. Examples of interaction are offered. The acidification was carried out over a 5 year period, and data collected are compared to a two year pre-acidification stage. Significant changes included increased transparency, rates of hypolimnion heating and rates of thermocline deepening; increased concentrations of Mn, Na, Zn, Al, and chlorophyll; decreased concentrations of suspended C, total dissolved N, Fe, and chloride; increases in Chlorophyta but decreases in Chrysophyta; the disappearance of the opossum shrimp Mysis relicta and the fathead minnow Pimephales promelas; the appearance of epidemics of the filamentous alga Mougeotea; decreased fitness and decline in numbers of Orconectes virilis; and increased embryonic mortality of the lake trout Salvelinus namaycush. Sulfur budgets for two lakes experimentally acidified with sulfuric acid revealed that an average of 1/4 to 1/3 of added sulfate is sedimented, presumably as FeS, reducing the efficiency of acidification. The sedimentation occurs under both oxic and anoxic conditions. The utility of whole-ecosystem mass balance studies of S in experimental and observational mass balance studies is discussed.

Schnitzer, M., "Effect of Low pH on the Chemical Structure and Reactions of Humic Substances", Effects of Acid Rain Precipitation on Terrestrial Ecosystems, 1980, Plenum Press, New York, New York, pp. 203-222.

This is a study of humic acid and fulvic acid compounds and their reactions to conditions of increased acidity. Long-term exposures to increased acid conditions cause reactions of HA and FA with each other and the soil to produce conditions unfavorable to biological activity and also to produce reduced soil fertility.

Schofield, C.L., "Acid Precipitation: Effects on Fish", Ambio, Vol. 5, No. 5-6, 1976, pp. 228-231.

Available information on the effects of acid precipitation on freshwater fish is summarized. Extensive depletion of valuable fish stocks has occurred in Norway, Sweden, and parts of eastern North America due to acid precipitation of fresh water. Extinction is due to chronic reproductive failure from excessive acid during sensitive stages of the life-cycle. Size, age, acclimation history, genetic background, and ionic strength of the water interact in complex ways to determine relative acid tolerance in fish. Species differ in acid tolerance.

Schofield, C.L., "Effects of Acid Rain on Lakes", Acid Rain: Proceedings of ASCE National Convention, 1979, American Society of Civil Engineers, New York, New York, pp. 55-69.

The total alkalinity, cation and anion concentrations, and heavy metal concentrations in surface water and their response to acid rain input are discussed. Freshwater lakes in areas having soft water, water with low total alkalinity, and acid and/or "thin" soils are particularly sensitive to acid rain loading. Excessive loadings of acid rain in the form of precipitation or snowmelt produce severe and rapid onset of adverse effects in freshwater lakes.

Sears, S.O. and Langmuir, D., "Sorption and Mineral Equilibria Controls on Moisture Chemistry in a C-Horizon Soil", Journal of Hydrology, Vol. 56, No. 3/4, Apr. 1982, pp. 287-308.

The chemistry of soil moisture was studied in central Pennsylvania sandy loam soils over a 12 month period. The dolomite bedrock depth was 6-14 m, and depth to the water table was 60-90 m. A total of 146 samples were collected at 1-9 m depths with suction lysimeters. The soil moisture chemistry was described by Ca greater than Na greater than Mg = K, and bicarbonate greater than chloride greater than sulfate = nitrate. The pH was 5.20-6.74. Dissolved silica was 15-117 (average 54) mg/l as SiO_2. Specific conductance varied from 20 to 400 micro S per cm. Total dissolved solids in the soil water increased with decreasing precipitation and increasing temperature. This was a result of evapotranspiration, which concentrated the salts in the soil water, reduced soil relative permeabilities, and increased the contact time between minerals and water. The increased acidity of precipitation in warmer months and drier periods also contributed to the increase in total dissolved solids. Water chemistry was not controlled by the solubility equilibria involving calcite, dolomite, quartz, amorphous silica, and the illite-kaolinite reaction. Regression analysis of dissolved ion activities showed that the cations were strongly buffered by Donnan equilibrium, with selective adsorption in the order K greater than H greater than Mg greater than Ca. Soil moisture was underestimated with respect to quartz and probably kaolinite. Sorption equilibria actively controlled major cation concentrations at all depths of the soil. Solution-mineral equilibria did not limit the cation concentration. Continued buildup of dissolved species and a solubility equilibrium are prevented by the short residence time of the water in the C horizon.

Shinn, J.H., "Critical Survey of Measurements of Foliar Deposition of Airborne Sulfates and Nitrates", Air Pollution Control Association Meeting, Houston, Texas, Mar. 1978, Lawrence Livermore Laboratory, Livermore, California.

This paper reviews the available data on processes affecting foliar deposition and retention of particles from the point of view of determining the effects of sulfates and nitrates on plants. The particle characteristics of size, shape, frequency distribution, and chemical speciation from typical air monitoring data are discussed in terms of dry deposition velocity and foliar collection and retention; first principles as well as empirical results are examined as methods of calculating dry deposition. Wet deposition of sulfates is examined in terms of foliar collection, foliar retention, scavenging, and the dependency upon precipitation characteristics. It is shown that the particle collection efficiency and the interception coefficient can vary over an order of magnitude due to either the effects of plant type or the misrepresentation of the roughness, Z_0, of the vegetation canopy. Particle diameter has little effect on collection efficiency in the submicron-size range but great effect at larger diameters. In wet deposition, the scavenging efficiency is very low for particles less than 4 mu m in diameter, and the scavenging coefficient is so highly dependent upon rainfall rate that it is practically case-specific. More experimental data and improvements are needed before the sulfate and nitrate effects can be accurately assessed.

Shriner, D.S., "Effects of Simulated Acidic Rain on Host-Parasite Interactions in Plant Diseases", Phytopathology, Vol. 68, No. 2, Feb. 1978, pp. 213-219.

A study was initiated to determine whether simulated rain acidified with sulfuric acid influences disease development in plants, and, if so, to suggest possible mechanisms that might account for the interaction. Experimental materials and methods used are described. The effects of simulated rain acidified with H_2SO_4 were studied in five host-parasite systems. Plants were exposed in greenhouses or fields to simulated rain of pH 3.2 or 6.0 in amounts and intervals common to weather patterns of North Carolina. Simulated acid rain resulted in: an 86 percent inhibition of the number of Telia produced by Cronartium Fusiforme on Willow Oak; a 66 percent inhibition in the reproduction of root-knot nematode on field-grown kidney beans; a 29 percent decrease in the percentage of leaf area of field-grown kidney beans affected by Uromyces Phaseoli; and either stimulated of inhibited development of Halo blight on kidney beans, depending on the state of the disease cycle in which the treatments were applied. Results indicated that acidity of rain was an environmental parameter that should be of concern to plant pathologists and ecologists.

Shriner, D.S. and Cowling, E.B., "Effects of Rainfall Acidification on Plant Pathogens", Contract No. 2-7405-ENG-26, 1978, NATO Institute on Acid Rain Effects, Toronto, Canada.

Wind-blown rain, rain splash, and films of free moisture play important roles in the epidemiology of many plant diseases. The chemical nature of the aqueous microenvironment at the infection court is a potentially significant factor in the successful dissemination, establishment, and survival of plant pathogenic microorganisms. Acidic rainfall has a potential for influencing not only the pathogen, but also the host organism, and the host-pathogen complex. Although host-pathogen interactions add a degree of complexity to the study of abiotic environmental stress of plants, it is hoped, through the use of a combination of general concepts, theoretical postulations, and experimental data, to describe the potential role that rainfall acidity may play in the often subtle balance between populations of plants and populations of plant pathogens. The direct effects of acidic precipitation on vegetation are becoming increasingly better understood. The indirect consequences of both acute and chronic exposure of vegetation to acidic precipitation are very complex, however. Their effect is variable in time, and involves a variety of potential interactions which are only partially understood.

Sigma Research, Inc., Proceedings: Ecological Effects of Acid Precipitation, Feb. 1982, Richland, Washington.

As acid precipitation has increased in prominence as an environmental issue, it has also increased in prevalence as a research project. Agencies in the public and private sectors--in the United States and overseas--have initiated programs of study; however, there is no centralized mechanism to integrate these efforts or promulgate the data obtained through them. Recognizing that such a situation can foster duplication of effort--with the resultant waste of limited resources--the Electric Power Research Institute and the U.S. Environmental Protection Agency co-sponsored a workshop on the ecological effects of acid precipitation. Participants from the United States, Canada, Norway, Sweden, and the United Kingdom exchanged information about their acid precipitation programs; areas of emphasis, approaches and techniques being used, and planned directions. They also discussed possible methods of coordinating their activities and defined further research needs. Among the key recommendations made by workshop participants were that definition and study of linkages between system inputs and responses should be emphasized, effects should be quantified, and small workshops (10 to 20 participants) to review specific topics should be held periodically.

Singh, B.R., Abrahamsen, G. and Stuanes, A., "Effect of Simulated Acid Rain on Sulfate Movement in Acid Forest Soils", Journal of the Soil Science Society of America, Vol. 44, No. 1, Jan.-Feb. 1980, pp. 75-80.

The effect of simulated acid rain on sulfate mobility in iron-podzol and semipodzol (Typic Udipsamments) forest soils of southern Norway was studied. The study was carried out with lysimeters with undisturbed soil. The lysimeters were watered with "rain" having pH 5.6 and 4.3. It was found that sulfate mobility was higher in the semipodzol than in the iron-podzol, and it was dependent on their sulfate adsorption capacities which in their turn were dependent on Al contents of these soils. Sulfate losses from applied 35S increased with increasing volume and decreasing pH of the "rain". The element losses were also higher in the semipodzol reflecting further higher mobility of sulfate in this soil. The total sums of cations, on equivalent basis, in the leachate from the semipodzol were nearly equal at pH 5.6 and 4.3 (0.480 and 0.485 meq/liter, respectively). In the iron-podzol, however, total sums of cations at pH 5.6 and 4.3 differed slightly (0.314 and 0.387 meq/liter, respectively).

Sparks, D.L. and Curtis, C.R., "An Assessment of Acid Rain on Leaching of Elements from Delaware Soils into Groundwater", OWRT A-053-DEL(1), 1983, College of Agricultural Sciences, University of Delaware, Newark, Delaware.

The effects of acid rain on the kinetics of elemental release from three major Delaware soil types were investigated. All three soils contained low clay and organic matter contents which would indicate rather low buffering capacities. The kinetics of elemental release were evaluated by leaching the soils with simulated acid rain at pH levels of 2.5, 3.4, 4.8, and 5.6. Leachates were collected until an apparent equilibrium in elemental release was obtained. In all soils and at all acid rain pH levels, quantities of basic elemental release were in the order Ca, K, Mg, Na. The amount of heavy metals and Si release were very low. As the pH of the acid rain decreased, the total quantity of each released element increased. In all cases, a rapid initial release of each element was followed by slower release rates. The kinetics of Al release were characterized by a similar time trend. The greatest Al release rates occurred at pH 2.5 and from the soil type with the lowest clay and organic matter contents and thus the poorest buffering capacity. The Al that was released at pH 2.5 from the soils would probably be either monomeric or polymeric Al; however, some release could have occurred from the clay structures. Anionic release, including Ca(-), $SO_4(=)$, and $NO_3(-)$, was evaluated and found to be negligible.

Sposito, G., Page, A.L. and Frink, M.E., "Effects of Acid Precipitation on Soil Leachate Quality--Computer Calculations", EPA-600/3-80-015, Jan. 1980, U.S. Environmental Protection Agency, Washington, D.C.

The multipurpose computer program GEOCHEM was employed to calculate the equilibrium speciation in 23 examples of acid precipitation from New Hampshire, New York, and Maine, and in the same number of mixtures of acid precipitation with minerals characteristic of soils in the same three states. Between 100 and 200 soluble inorganic and organic complexes were taken into account in each speciation calculation. The calculations performed on the acid precipitation samples showed that the metals (including heavy metals) and the sulfate, chloride, and nitrate ligands would be almost entirely in their free ionic forms, while the phosphate, carbonate, ammonia, and organic ligands would be in their protonated forms. This result was independent of the geographic location of the acid precipitation and the month of the year in which the sample was collected. The speciation calculations on the precipitation-soil mineral mixtures showed that aluminum and iron levels in a soil solution affected by acid precipitation would be signficantly higher than in one whose chemistry is dominated by carbonic acid. The higher levels found were caused by the lower pH value of acid precipitation as well as by complexes formed with inorganic and organic ligands. It was also shown that soil cation exchangers would absorb preferentially heavy metals, such as Cd and Pb, which are found in acid precipitation.

Sprules, W.G., "Midsummer Crustacean Zooplankton Communities in Acid-Stressed Lakes", Fisheries Research Board of Canada Journal, Vol. 32, No. 3, Mar. 1975, pp. 389-399.

The distribution of limnetic crustacean zooplankton species and species associations in 47 industrially acidified lakes of the La Cloche Mountains, Canada, is examined. The pH, which ranged from 3.8-7.0, and, to a lesser extent, land area and depth are the major determinants of the structure of the communities. The pH has a great effect on the zooplankton communities, primarily in lakes with pH below 5.0 where many species are completely eliminated.

Stednick, J.D. and Johnson, D.W., "Natural Acidity of Waters in Podzolized Soils and Potential Impacts from Acid Precipitation", 1982, Oak Ridge National Laboratory, Oak Ridge, Tennessee.

Nutrient movements through sites in southeast Alaska and Washington were documented to determine net changes in chemical composition of precipitation water as it passed through a forest soil and became stream-flow. These sites were not subject to acid precipitation (rainfall pH 5.8 to 7.2), yet soil water was acidified to 4.2 by natural organic acid-forming processes in the podzol soils. Organic acids precipitated in the subsoils, allowing a pH increase. Streamwater pH ranged from 6.5 to 7.2 indicating a natural buffering capacity that may exceed any additional acid input from acid rain. Precipitation composition was dominated by calcium, magnesium, sodium, and chloride due to the proximity of the ocean at the southeast Alaska site. Anionic constituents of the precipitation were dominated by bicarbonate at the Washington site. Soil podzolization processes concurrently increased solution color and iron concentrations in the litter and surface horizons leachates. The anion flux through the soil profile was dominated by chloride and sulfate at the southeast Alaska site, whereas at the Washington site anion flux appeared to be dominated by organic acids. Electroneutrality calculations indicated a cation deficit for the southeast Alaska site.

Strayer, R. and Alexander, M., "Effects of Simulated Acid Rain on Glucose Mineralization and Some Physicochemical Properties of Forest Soils", Journal of Environmental Quality, Vol. 18, No. 4, Dec. 1981, pp. 460-465.

Samples of forest soils were exposed to a continuous application of 100 cm of simulated acid rain (pH 3.2-4.1) at 5 cm/hr, or to intermittent 1-hr applications of 5 cm of simulated acid rain three times per week for 7 weeks. The major effects of the simulated acid rain were localized at the top of the soil and included lower pH values and glucose mineralization rates; the acid rain penetrated further in the more acid soils. Glucose mineralization in the test soils (pH values of 4.4-7.1) was inhibited by the continuous exposure to simulated acid rain at pH 3.2 but not at pH 4.1. The extent of inhibition depended on the soil and the initial glucose concentration. These data suggest that acid rain may have a significant impact on microbial activity.

Strayer, R.F., Chyi-Jiin, L. and Alexander, M., "Effect of Simulated Acid Rain on Nitrification and Nitrogen Mineralization in Forest Soils", Journal of Environmental Quality, Vol. 10, No. 4, Oct.-Dec. 1981, pp. 547-551.

Columns containing samples of forest soils were leached with either a continuous application of 100 cm of simulated acid rain (pH 3.2-4.1) at 5 cm/hr or an intermittent 1.5-hr application of 1.2 cm of simulated acid rain twice weekly for 19 weeks. The upper 1.0- to 1.5-cm portions of soil from treated columns were used to determine the changes in inorganic N levels in the soil. Nitrification of added ammonium was inhibited following continuous exposure of soil to simulated acid rain of pH 4.1-3.2. The extent of the inhibition was directly related to the acidity of the simulated rain solutions. The production of inorganic N in the absence of added ammonium was either stimulated or unaffected following continuous treatment of soils with pH 3.2 simulated acid rain. The addition of nitrapyrin caused a decrease in nitrification in water-treated soil but had little effect on nitrification in soil treated with pH 3.2 simulated acid rain.

Tamm, C.O., "Acid Precipitation: Biological Effects in Soil and on Forest Vegetation", Ambio, Vol. 5, No. 5-6, 1976, pp. 235-239.

Information on the effects of acid precipitation on forest ecosystems, particularly forests on coarse-textured soils with a low base status like the ones of Scandinavia, is reviewed. While decreased forest growth due to acid precipitation has been convincingly demonstrated, some soil organisms and processes are sensitive to acid applications at levels comparable to amounts received in 5 to 10 years in southern Scandinavia. A model for the effects of increased strong acid deposition in a forest ecosystem is presented.

Taylor, F.B. and Symons, G.E., "Effects of Acid Rain on Water Supplies in the Northeast", Journal of the American Water Works Association, Vol. 76, No. 3, Mar. 1984, pp. 34-41.

Results of the first study concerning the impact of acid precipitation on drinking water are reported in terms of health effects in humans as measured by U.S. Environmental Protection Agency maximum contaminant levels. The study focused on sampling surface water and ground water supplies in the New England states but also included other sites in the Northeast and the Appalachians. No adverse effects on human health were demonstrated, although the highly corrosive nature of New England waters may be at least partly attributable to acidic deposition in poorly buffered watersheds and aquifers.

Tetra Tech, Inc., "Integrated Lake-Watershed Acidification Study (ILWAS): Contributions to the International Conference on the Ecological Impact of Acid Precipitation", May 1981, Lafayette, California.

The Integrated Lake-Watershed Acidification Study (ILWAS) was initiated to study and detail lake acidification processes for three lake watershed basins in the Adirondack Park region of New York. The three basins (Woods, Sagamore, and Panther), receive similar amounts of acid deposition yet observable pH values for the lakes are very dissimilar indicating unequal acid neutralizing capacities among the watersheds.

This volume contains a compilation of seven papers. Relevant topics include: a characterization of the geology, hydrology, limnology and vegetation of the three study sites, an analysis of acid precipitation quality and quantity, the effects of vegetative canopy, the effects of snowmelt, the effects of winter lake stratification, comparison of heavy metal transport, examination of acidic sources other than direct precipitation, assessment of lake acidification during spring thaw, and integration of all acidification components with a mathematical model.

Torrenueva, A.L., "Effects of Acid-Rain on Terrestrial Ecosystems", Research Review of Ontario Hydro, No. 2, May 1981, pp. 49-56.

The current state of knowledge on the effects of acid precipitation on vegetation and soils is assessed. Research studies, mostly from North America and Scandinavia, dealing with the effects of simulated acid rain on vegetation, soil chemistry, soil biology and soil biochemical processes are reviewed with particular emphasis on northern temperate forest ecosystems. The studies reviewed to date have not revealed any effects of acid precipitation on vegetation and soils under natural field conditions. The coniferous forests of Ontario do not seem to be in immediate danger from acid precipitation because of ample nutrient reserve in the soil and the relatively well-buffered acid forest soil. However, more biogeochemical studies are needed in other forest ecosystems such as hardwoods.

Tschupp, E.J., "Effect on Groundwater Resources of Precipitation Modified by Air Pollution", 55th Annual Meeting of the American Geophysical Union, Apr. 8-12, 1974, Washington, D.C.

The inadvertant modification of precipitation by air pollutants has a significant effect on the quality of the water that enters underground aquifers each year as annual recharge. Most pollutants, both particulates and gases, are soluble in precipitation and are carried into the soil with the precipitation that does not run off. The most pronounced effects are those associated with the increased acidity of precipitation caused by atmospheric pollutants. The increased acidity causes accelerated weathering and chemical reactions as the precipitation passes through the soil and rock in the process of recharging an aquifer. The net effect on the ground water resource is the reduction of water quality because of increased mineralization. The fate of some of the air pollutants in the process of recharging the aquifers is described and estimates of the quantity of pollutants annually participating in the ground water recharge process are made on a national basis.

Tukey, Jr., H.B., "Some Effects of Rain and Mist on Plants, with Implications for Acid Precipitation", Effects of Acid Rain Precipitation on Terrestrial Ecosystems, 1980, Plenum Press, New York, New York, pp. 141-150.

This is a summary paper of the factors that may produce injury to plant species. They include the physical condition and age of the plant and environmental factors such as temperature, light, soil content, and quantity and quality of precipitation. The plant surface and root systems as well as seasonal variations play major roles in adaptation to environmental stresses such as acid rain.

Turk, J.T. and Adams, D.B., "Sensitivity to Acidification of Lakes in the Flat Tops Wilderness Area, Colorado", <u>Water Resources Research</u>, Vol. 19, No. 2, Apr. 1983, pp. 346-350.

   The use of regionally plentiful coal and oil shale resources in the western United States may result in an increase of atmospheric emissions in and upwind of northwestern Colorado, as may smelting and an increase in population. Such emissions have been previously associated with water quality degradation from acid rain. Lakes of the Flat Tops Wilderness Area in northwestern Colorado are sensitive to acidification if precipitation in the area becomes as acidic as that of the northeastern United States. About 50 percent of the titration alkalinity in the most poorly buffered lakes is from dilution of the titrant acid rather than neutralization. Alkalinity is readily predicted from watershed characteristics including lake altitude, fraction of exposed bedrock, and drainage area. Alkalinity values as small as 70 micro eq/l occur in the higher-altitude lakes. On the basis of alkalinity titration curves for the lakes studied, an alkalinity of 200 micro eq/l or greater is necessary to adequately protect a lake from precipitation as acidic as that of the northeastern United States. Most lakes at altitudes of 3380 m or greater are predicted to have alkalinity values of 200 micro eq/l or less. About 370 lakes having a total surface area of about 157 ha are calculated to be sensitive to acidification.

Tveite, B. and Abrahamsen, G., "Effects of Artificial Acid Rain on the Growth and Nutrient Status of Trees", <u>Effects of Acid Rain Precipitation on Terrestrial Ecosystems</u>, 1980, Plenum Press, New York, New York, pp. 305-318.

   This paper gives results of growth measurements in four field irrigation experiments with artificial acid rain within southern Norway. Foliar analysis were carried out in three of the experiments. Height and diameter growth were stimulated by increased "rain" acidity in a Scots pine sapling stand. The reason for this is probably increased nitrogen uptake from the soil. A beneficial effect of sulphur application either alone, or in combination with increased nitrogen uptake is also possible. In the other experiments no treatment effects on height or diameter growth were found.

Tyler, G., "Leaching Rates of Heavy Metal Ions in Forest Soil", <u>Water, Air, and Soil Pollution</u>, Vol. 9, No. 2, Feb. 1978, pp. 137-149.

   Pure organic MOR horizons from two spruce forest soils developed on Archaean Till in southeastern Sweden were studied. The objectives were to describe differences in leaching among eight heavy metal elements at various pH levels of artificial precipitation and to predict residence times. One soil was heavily polluted by zinc and copper from a brass foundry. Very long residence times, particularly of lead, in forest soil make accumulation possible even at a comparatively low deposition rate.

Ulrich, B., Mayer, R. and Khanna, P.K., "Chemical Changes Due to Acid Precipitation in a Loess-Derived Soil in Central Europe", <u>Soil Science</u>, Vol. 130, No. 4, Oct. 1980, pp. 193-199.

A Fagus silvatica forest in the Solling highlands, Federal Republic of Germany, showed increasing Al concentrations in the soil during a study of the effect of acid precipitation on the ecosystem between 1966-1979. A noticeable change occurred in 1973. Before this, carbon and nitrogen stores in the forest floor were increasing, but afterwards, they decreased. This indicates a shift from one decomposition process to another as a response to acid precipitation. The pH has remained between 3 and 4, decreasing slightly with time. Aluminum and Fe ion concentrations increased between 1966 and 1973 to 2 mg/l and 3 mg/l of equilibrium soil solution, with the buffer reaction reaching deeper into the soil. Acid precipitation has induced soil internal H(+) ion production at a rate of 2.9-5.5 kmole H(+) per hectare per year, partly from changes in N nutrition and partly from N deficiency. Proton consumption by silicate weathering is estimated at 0.2-2 kmole H(+) per hectare per year, meaning that the buffering capacity of this soil is reaching its capacity. Since this forest has reached a critical stage in less than 20 years of acid rain, further acid precipitation will produce serious results here as well as in all of central Europe, especially the ridges and plateaus of the medium highlands.

U.S. Fish and Wildlife Service, "Effects of Acid Precipitation on Aquatic Resources: Results of Modeling Workshops", FWS/OBS-80/40.12, July 1982, Eastern Energy and Land Use Team, Kearneysville, West Virginia.

This publication reports on a series of Adaptive Environmental Assessment (AEA) workshops on acid precipitation, sponsored by the U.S. Fish and Wildlife Service's National Power Development Group. The report identifies priority research needed for a better understanding of the acidification process and its impacts on aquatic resources of watershed ecosystems. A preliminary computer simulation model, developed through the workshop process and designed to describe and integrate the mechanisms by which acid precipitation alters stream and lake chemistry and impacts fish populations and other aquatic components, is presented.

Vangenechten, J.H., "Interrelations Between pH and Other Physiochemical Factors in Surface Waters of the Campine of Antwerp, (Belgium): With Special Reference to Acid Moorland Pools", Archiv fur Hydrobiologie, Vol. 90, No. 3, Nov. 1980, pp. 265-283.

The interrelations between pH, conductivity, and various ions in 53 surface waters near Antwerp, Belgium, were studied. The waters were divided into three classes according to pH: below 4.5, 4.5-7.0, and above 7.0. Measured and calculated conductivity agreed equally in the three types. However, in extremely acid bogs, a small excess of anions over cations was noticed, and in alkaline lakes, a large excess of anions over cations. Calcium concentrations increased significantly with pH in alkaline waters, and less significantly in more acid water. The contribution of sulfate to the total anionic content decreased sharply with pH. Although high levels of Ca and Mg ions appeared in acid waters, alkalinity (carbonate-bicarbonate) was very low. Indications are that the high acidity originates from both natural (sulfide breakdown) and industrial (sulfate-containing acid rain) sources. The excess of Na over Cl found in extremely acid bogs is unexplained. Physiochemical compositions in these Belgian surface waters were comparable with those in acidified lakes in Scandinavia and North America.

Evidence is growing of serious problems in forests of the central European region and eastern North America. In West Germany, fully 560,000 hectares of forests have recently been damaged. Trees have died in 60,000 ha of the nation's forests, and those in another 100,000 ha are reported to be seriously damaged. In eastern Europe, sulfur dioxide emission levels are extremely high. More than 500,000 ha of forests in Czechoslovakia have been severely damaged. Only recently has acid deposition emerged as a potentially sweeping threat to forests, especially in central Europe. Some of the devastation is brought about as a consequence of contact between acid waters and the plant tissues. Rainfall in the pH range of 3.0 to 4.0 has been shown to cause tissue injury which can ultimately reduce forest growth. However, rainfall at such a low pH level is not common. Preliminary evidence suggests that indirect effects resulting from acid-induced soil changes are far more important. Acid deposition upsets the natural balance in the soils. Negatively charged sulfate ions, after accumulating in soils, can be washed out by rains, carrying with them nutrients such as calcium and magnesium. Acid conditions also slow the bacterial decomposition needed for the continued production of nutrients. In the long term, pollution-induced nutrient depletion accelerates the natural forest aging process, leading to the eventual exhaustion of the soil's ability to sustain tree growth. Acid-induced forest impacts are probably magnified in some instances by the increases in rainwater acidity as it passes through the forest canopy and runs down the tree bark, adsorbing in the journey dry-deposited acids present on the surfaces of plants. It is not certain at present whether acid-afflicted forests can be saved or whether acid induced impacts on forests or forest soils can be reversed.

Wiklander, L., "The Sensitivity of Soils to Acid Precipitation", Effects of Acid Rain Precipitation on Terrestrial Ecosystems, 1980, Plenum Press, New York, New York, pp. 553-567.

This paper presents a summary and review of sources of soil acidification, acidification by pedogenic processes, soil acidification and ion exchange, and nutrient loss from cultivated soils plus the sensitivity of soils to acid precipitation.

Wood, T. and Bormann, F.H., "Increases in Foliar Leaching Caused by Acidification of an Artificial Mist", Ambio, Vol. 4, No. 4, 1975, pp. 169-172.

Pinto bean and sugar maple seedlings were exposed to artificial mists adjusted to various acidities. Compared with nutrient losses at pH 5.0, statistically significant increases in leaching of potassium, manganese, and calcium are found when the mist pH is lowered to values of 4.0, 3.3, 3.0, and 2.3. Foliar tissue damage, visible at pH levels 3.0 and 2.3, is probably responsible for the large losses found at the high acidities.

Wright, R.F. and Gjessing, E.T., "Acid Precipitation: Changes in the Chemical Composition of Lakes", Ambio, Vol. 5, No. 5-6, 1976, pp. 219-224.

Changes in the chemical composition of lakes in southern Scandinavia and eastern North America due to acid precipitation are discussed. Chemical weathering and ion exchange processes in terrestrial watersheds are inadequate to neutralize long-term deposition of acid precipitation on poorly buffered areas. Sulfate is the major anion.

Vaughan, H.H., Underwood, J.K. and Ogden, III, J.G., "Acidification of Nova Scotia Lakes I: Response of Diatom Assemblages in the Halifax Area", Water, Air, and Soil Pollution, Vol. 18, No. 1-3, 1982, pp. 353-361.

A comparison was made of diatom remains in surficial lake sediments collected in 1971 and 1980. Data from the four lakes studied reveals that all lakes have undergone similar and pronounced shifts in diatom assemblages. The percentages of alkaliphilous species have fallen since 1971, while acidophilous species have increased. These systematic shifts have occurred in spite of the fact that the four lakes are measurably different in almost every respect. These shifts may be ascribed to the influx of airborne pollutants as supported by chemical results of lacustrine and precipitation samples collected in the Halifax area. Local rates of acidification are stressing the basic biological components of lacustrine systems to the point where such changes are taking place. If deposition of acids in Nova Scotia increases in the future, it may result in relatively excessive effects on lacustrine biological communities.

Wainwright, M., Supharungsun, S. and Killham, K., "Effect of Acid Rain on the Solubility of Heavy Metal Oxides and Fluorspar (CaF$_2$) Added to Soil", Science of the Total Environment, Vol. 24, No. 1, May 1982, pp. 85-90.

Acid rain of extremely low pH (pH 2.0), collected downwind of a coking works, solubilized heavy metal oxides added to a sandy loam soil in the order of increasing solubilization CdO greater than ZnO greater than CuO greater than PbO. Artificial acid rains (pH 2.6 and 4.2) had a much less marked effect on metal oxide solubility. On the other hand, acid rain released less fluorine ions into soil solution from added CaF2 than did either artificial rain or distilled water.

Webb, A.H., "Weak Acid Concentrations and River Chemistry in the Tovdal River, Southern Norway", Water Research, Vol. 16, No. 5, May 1982, pp. 641-648.

The chemistry of water in the Tovdal River, Norway, was investigated by analysis of weekly samples from August 1978 to December 1979. The pH varied slightly over the study period, 4.64 to 5.14, and was lowest in winter. Concentrations of both the air pollution derived species (sulfates, nitrates, and ammonium) and ground water derived minerals (Ca, Mg, and K) increased steadily in river water during summer and winter to a broad peak during spring and decreased to a minimum in summer. The rapid metabolism of nitrate and ammonium produced some irregularities in their patterns. The broad peak in the spring of the acid rain constituents (sulfates and nitrates) coincided with the peak in alkalis because of leaching from the ground during the snowmelt period. Sulfate levels were higher in the fall of 1979 than in the fall of 1978, alkalis were lower, and therefore pH was lower. Total weak acid concentrations (silica, Al, and ammonium) also showed spring maximums and summer minimums. These were correlated with total organic carbon concentrations, but the inorganic weak acid species largely controlled the seasonal variations in weak acid content.

Wetstone, G.S. and Foster, S.A., "Acid Precipitation: What is it Doing to Our Forests", Environment, Vol. 25, No. 4, May 1983, pp. 10-12 and 38-40.

Concentrations of aluminum, manganese, and other heavy metals are higher in acid lakes due to enhanced mobilization of these elements in acidified areas.

APPENDIX F

CONTROL STRATEGIES

Ball, J.G. and Menzies, W.R., "Acid Rain Mitigation Study--Volume I: FGD Cost Estimates", EPA-600/2-82-070A, Sept. 1982a, U.S. Environmental Protection Agency, Research Triangle Park, North Carolina.

The report gives results of work to provide a consistent set of capital investment and operating costs for flue gas desulfurization (FGD) systems retrofitted to existing industrial boilers. The investigation of wet limestone scrubbers and lime spray drying FGD systems included: (1) the apparent discontinuities in both FGD systems capital investment and operating costs; (2) FGD retrofit factors applied to existing boilers based on published reports; and (3) differences between PEDCo Environmental, Inc. and TVA cost estimates for utility boiler FGD systems. These costing issues were examined on the bases of design scope, costing factors (for equipment installation, indirect investment, etc.), year of costs, inherent strengths and weaknesses, and published data of actual system costs. Recommendations are made for the cost bases to use in further acid rain studies.

Ball, J.G. and Menzies, W.R., "Acid Rain Mitigation Study--Volume II: FGD Cost Estimates (Appendices)", EPA-600/2-82-070B, Sept. 1982b, U.S. Environmental Protection Agency, Research Triangle Park, North Carolina.

The report gives results of work to provide a consistent set of capital investment and operating costs for flue gas desulfurization (FGD) systems retrofitted to existing industrial boilers. The investigation of wet limestone scrubbers and lime spray drying FGD systems included: (1) the apparent discontinuities in both FGD system capital investment and operating costs; (2) FGD retrofit factors applied to existing boilers based on published reports; and (3) differences between PEDCo Environmental, Inc. and TVA cost estimates for utility boiler FGD systems. These costing issues were examined on the bases of design scope, costing factors (for equipment installation, indirect investment, etc.), year of costs, inherent strengths and weaknesses, and published data of actual system costs. Recommendations are made for the cost bases to use in further acid rain studies.

Ball, J.G., Muela, C.A. and Meling, J.L., "Acid Rain Mitigation Study--Volume III: Industrial Boilers and Processes", EPA-600/2-82-070C, Sept. 1982, U.S. Environmental Protection Agency, Research Triangle Park, North Carolina.

The report gives results of a 4-month study of existing industrial sources of SO_2 emissions in the Acid Rain Mitigation Study (ARMS) region, including all the states east of the Mississippi River, as well as MN, IA, MO, AR, LA, ND, SD, NE, KS, OK, and TX. Study aims were to: (1) identify and characterize existing industrial sources of SO_2 emissions; (2) identify control techniques that can be used to reduce SO_2 emissions from these sources; and (3) estimate the SO_2 emission reduction potential and the associated costs in constant 1980 dollars based on application of these controls. Simplifying assumptions were made for the balance of the SO_2 sources studied.   In addition, since sites were not visited, the remaining useful lives of the sources were not determined, and average FGD unit retrofit factors were estimated. Recommendations concerning the use of study results are discussed.

Bengtsson, B., Dickson, W. and Nyberg, P., "Liming Acid Lakes in Sweden", Ambio, Vol. 9, No. 1, 1980, pp. 34-36.

In 1976, the Swedish government initiated liming lakes and rivers to improve the quality of water polluted by acid rains. During 1977-1979, 700 lakes and rivers were treated directly with a total of 120,000 tons of lime. A single application may be effective for 5 to 10 years. Ecological studies showed that liming acid waters greatly increased the populations of zooplankton, phytoplankton, and fish. However, pH often decreases during periods of high water flow, such as snow melt, due to stratification. The present budget for the liming project is $2.5 million, about one-tenth of the actual need. Proposed funds for 1980 and 1981 are $5 million and $7.5 million, respectively.

Blake, L.M., "Liming Acid Ponds in New York", New York Fish and Game Journal, Vol. 28, No. 2, 1981, pp. 208-214.

This paper reviews various treatments, considering changes in pH, alkalinity and fish populations as a result of liming ponds to neutralize acid conditions. Cost-benefit ratios are considered. Application of hydrated lime in New York State is limited to open-water periods for economic reasons. Hydrated lime is superior to the less potent agricultural lime, calcium carbonate, although the latter permits a longer time between treatments. No matter which type of lime is used, the lime is dumped from bags from a slowly moving boat. More lime is spread in shallow water areas than over deep water areas so that the lime is placed where the fish live and it is exposed more effectively to wave action and currents. The use of lime in remote areas is more difficult due to the high cost of transport. Possible increases in mercury content of fish after lime treatment were studied. Mercury levels were determined in brook trout from selected lakes, limed and unlimed. No increase in mercury levels was apparent in either limed or unlimed control waters. Costs for liming remote ponds range as high as $297 per acre. Experience with the dry dispersal method has shown that lime can be moved at a rate of 5 tons per hour. The average cost of this method should be about $100 per acre. It was concluded that for both accessible and remote ponds, liming is an effective and economically feasible tool which can be used to counteract the adverse impact of acid precipitation and maintain selected fisheries.

Bomberger, D.C. and Phillips, R.C., "Technological Options for Mitigation of Acid Rain", Acid Rain: Proceedings of ASCE National Convention, 1979, American Society of Civil Engineers, pp. 132-166.

This article discusses the reduction of SO_x emissions from power plants fueled by coal using such methods as mechanical coal cleaning, chemical coal cleaning, catalytic hydrosulfurication of fuel oils, coal gasification, coal liquefication, pyrolysis, indirect synthesis of liquid fuels, stack gas scrubbing for control of SO_2 emissions, lime scrubbing and regenerative scrubbing. The reduction of NO_x emissions involves combustion modification, NO_x removal from stack gases, dry and wet NO_x removal processes, and fluidized bed combustion.

Catalono, L. and Makansi, J., "Acid Rain: New SO_2 Controls Inevitable", Power, Vol. 127, No. 9, Sept. 1983, pp. 25-33.

Rainfall pH in the northeastern U.S. now averages between 4.2 and 4.6. Most scientists agree that damage will occur to sensitive ecosystems when pH falls below 4.6, and that this increase in rainfall acidity is due to anthropogenic, or man-made, sources--SO_2 and NO_x emissions from fossil-fuel-burning powerplants and smelters, as well as NO_x from mobile sources. The debate in the U.S. on how to deal with this problem is summed up. Scientific studies admit incomplete data but urge quick action. Congress considers many legislative proposals. Options available to meet acid-rain-emission limits are discussed. Controls must go beyond the use of scrubbers. Application of the advanced fluidized-bed combustion, limestone-injection multistage burners, pressurized fluidized-bed combustion, and MHD power systems is weighted.

Chadwick, M.J., "Acid Depositions and the Environment", Ambio, Vol. 12, No. 2, 1983, pp. 80-82.

Acidifying substances present in the atmosphere of major environmental significance are oxides of sulfur and nitrogen. Atmospheric sulfur occurs in compounds in gas, liquid and solid phases. Residence times for atmospheric sulfur dioxide are estimated at 1 to 2 days. Wet deposition of sulfur compounds, mainly in aerosol form as sulfate, occurs by in-cloud and below-cloud scavenging. Control strategies for reducing acidifying substances emitted to the atmosphere will need to concentrate on one or more stages of fuel use: before burning, low N and S fuel; during burning, staged, catalytic or coal-limestone combustion; waste gas venting, flue gas desulfurization and ammonia injection.

Chawla, R.C. and Varma, M.M., "Pollutant Transfer Between Air, Water and Soil: Criteria for Comprehensive Pollution Control Strategy", Journal of Environmental Systems, Vol. 11, No. 4, 1981-82, pp. 363-374.

The intra-domain interactions that influence the pollutant transfer between air, water and soil may be either due to the bulk phase transfer (e.g., acid rain) or due to interfacial transfer (e.g., pesticide evaporation from soil to air). The mathematical equations based on the general principles of mass transfer such as Fick's Law of Diffusion and Rault's Law, and equilibrium relationships can quantitatively predict the pollutant transfer from one phase to another. Thus far, pollution control efforts have been directed at each individual domain without much regard to their effect on the other domains. A basic understanding of transfer processes is presented in the paper. The problem of trihalomethanes in water is discussed to support the case for a comprehensive pollution control strategy which makes both technical and economic sense.

DePinto, J.V. and Edzwald, J.K., "An Evaluation of the Recovery of Adirondack Acid Lakes by Chemical Manipulation", OWRT-B-095-NY(1), June 1982, Office of Water Research and Technology, U.S. Department of Interior, Washington, D.C.

In order to develop alternative remedial methods for recovery of Adirondack acid lakes believed due to acid rain, a study was conducted on chemical manipulation methods for their neutralization. This study specifically addressed an evaluation of materials (calcium hydroxide and carbonate, agricultural limestone, fly ash, water treatment plant softening sludge, cement plant by-pass dust) for their neutralizing effectiveness and for establishing a neutral pH buffer system, and an evaluation of the effect of various lake recovery materials on algal growth. Laboratory continuous-flow microcosms were used as models to assess acid lake recovery. These models were filled with actual acid lake water over a layer of lake sediments, subjected to a given chemical treatment, and continuously fed water of selected quality.

Electric Power Research Institute, "Joint Symposium on Stationary Combustion NO_x Control: Proceedings", EPRI-WS-79-220(V.1), May 1981, Palo Alto, California.

The Joint Symposium on Stationary Combustion NO_x Control consisted of over 50 presentations describing recent advances in NO_x control technology, including applications for pulverized coal-fired utility boilers. Volume I of the Proceedings contains manuscripts documenting the presentations in four sessions. These papers address the regulatory aspects of NO_x emissions and NO_x control through burner design and combustion modification. NO_x Emission Issues (Session I) addresses regulatory topics such as the short-term ambient NO_2 standard, acid rain concerns, etc. The Manufacturers Update of Commercially Available Combustion Technology (Session II) describes advanced burner designs for NO_x control that are currently offered by the utility boiler manufacturers. NO_x Emission Characterization of Full-Scale Utility Powerplants (Session III) presents results of recent field tests evaluating combustion modifications and other low NO_x approaches. Low NO_x Combustion Development (Session IV) details the recent progress in the development of advanced low NO_x firing systems not yet commercially available.

Ellis, J.D. and Golomb, A., "Lake Acidity and Its Neutralization", Research Review of Ontario Hydro, No. 2, May 1981, pp. 67-74.

The literature dealing with the effects of acid deposition on aquatic ecosystems and with documentation of the efficacy of neutralization as a lake restoration technique is reviewed. A considerable body of inferential evidence exists that would suggest that acid deposition is having a detrimental effect on the chemistry and biology of acid-sensitive lakes. Lake neutralization experiments have been only partly successful in restoring low-pH lakes to a nonacid-stressed oligotrophic state.

Fortin, M. and McBean, E.A., "Management Model for Acid Rain Abatement", Atmospheric Environment, Vol. 17, No. 11, 1983, pp. 2331-2336.

A linear programming-based model is developed for examining management alternatives for acid rain abatement. The model reflects both expected values and elements of the stochastic considerations. A case study application of the model demonstrates the impact of cost sharing for the different management alternatives.

Fraser, J. et al., "Feasibility Study to Utilize Liming as a Technique to Mitigate Surface Water Acidification", EPRI-EA-2362, Apr. 1982, General Research Corporation, McLean, Virginia.

This study evaluated the feasibility of liming as a technique for mitigating surface water acidification. The study addressed mainly the application of calcium-based alkaline materials to lakes and waterways. The study addressed the types and combinations of alkaline materials, the input location, delivery modes, and approximate cost of some of the techniques. An assessment of potential chemical, physical and biological changes resulting from short- or long-term lime applications was made. Finally, recommendations were suggested for additional research needed to evaluate liming as a technique for mitigating the effects from surface water acidification.

Gilbert, A.H., "Economics of Acid Rain: An Invisible Hand of Control", International Journal of Environmental Studies, Vol. 18, No. 2, Feb. 1982, pp. 85-90.

Economics currently plays an insignificant role in the control of acid rain. Economists have fallen into the old familiar pattern of evaluating the impacts of acid rain instead of developing economic techniques for controlling the source of its emission. Power generating plants and other major producers of acid rain discharge their oxide wastes into the air in order to reduce production costs. This cost saving enables the producer to increase profits and expand production at the expense of those who suffer from acid rain.   The solution to this misallocation requires the internalization of all production costs. The price of the product will then reflect its true cost.

Golomb, D., "Acid Deposition-Precursor Emission Relationship in the Northeastern U.S.A.:   The Effectiveness of Regional Emission Reduction", Atmospheric Environment, Vol. 17, No. 7, 1983, pp. 1387-1390.

About 20 percent of each of the emitted SO_2 and NO_x are wet-deposited as sulfate and nitrate ions over the northeastern U.S.A. This leads to a mean regional precipitation acidity of 67 "mu" mol per liter (pH = 4.2), in good agreement with observations. Assuming a constant emission/deposition ratio, the average rain acidity can be predicted for various emission reduction scenarios.

Gorham, E., "What To Do About Acid Rain", Technology Review, Vol. 85, No. 7, Oct. 1982, pp. 59-66.

The problems created by acid rain are considered. The polluting gases in earth's atmosphere are traced to their sources. Sulfur and nitrogen emissions in coal-using industrial states, the pollutions in the atmosphere, and their return to lakes and streams, forests and fields with rain and snow, and also in dry microscopic particles or gases, are studied. Other substances deposited by acid rains are also considered and the reduction of acid deposition proposed.

Hastings, A. and Schaefer, M., "Controlling Nitrogen Oxides", EPA-600/8-80-004, Feb. 1980, U.S. Environmental Protection Agency, Washington, D.C.

Recent research indicates that nitrogen oxides (NO_x) could be one of the most troublesome air pollutants of the 1980's. More than 20 million metric tons of NO_x are annually polluting U.S. air as a result of the widespread combustion of fossil fuels in power plants, industrial boilers, and automobiles and trucks. Present levels of NO_x emissions already pose a significant threat to our health and environment. Nitrogen oxides are directly harmful to human health, and are precursors of photochemical oxidants such as ozone, the major component of urban smog. They can also be converted into nitric acid, one of the two principal components of acid precipitation. The EPA is actively working with other Federal agencies and the academic, industrial, and private research communities to develop viable combustion technologies which will strictly limit NO_x emissions. The authors have prepared this Research Summary to inform the public of the status of their efforts to make improved control technologies available as soon as possible.

McLean, R.A., "The Relative Contributions of Sulfuric and Nitric Acids in Acid Rain to the Acidification of the Ecosystem--Implications for Control Strategies", Journal of the Air Pollution Control Association, Vol. 31, No. 11, Nov. 1981, pp. 1184-1187.

It is proposed that the contributions of nitric acid and sulfuric acid in atmospheric deposition to the acidification of the terrestrial and aquatic ecosystems are not equivalent and that, except possibly during the spring thaw, nitric acid contributes considerably less to the acidification of the ecosystem. A major part of the assessment of the impact of acid rain on aquatic and terrestrial ecosystems involves the examination, by means of ion budgets, of the fate of each of the ionic constituents in atmospheric deposition in soils, surface waters, and other places. It was concluded that much of the nitric acid in acid rain is decomposed in the soils and waterways, and in the process acid is consumed. Therefore, unlike sulfuric acid in acid rain, nitric acid is not a significant contributor to long term acidification of the environment. Even so, during the spring thaw, in areas which have been heavily impacted by acid rain for a number of years, nitric acid which has concentrated in the snow pack over the winter may cause ecological damage, especially to fish populations. Though there is little doubt that tighter control strategies are needed to diminish the effects of acid rain on remote ecosystems, the existing control strategies, which have put more emphasis on the control of emissions of sulfur oxides than nitrogen oxides, have a reasonable scientific basis given present limited knowledge of their effects on the ecosystem.

PEDCo-Environmental, Inc., "Acid Rain: Control Strategies for Utility Boilers--Volume II--Appendices", DOE-METC-82-42-V.2, May 1981, Cincinnati, Ohio.

These appendices provide a detailed breakdown of cost estimates for SO_2 and NO_x control strategies applied to the top 50 coal-fired SO_2-NO_x emitters as defined in the summary report: Acid Rain Control Strategies for Utility Boilers, Volume 1. Appendix A contains costs for the eastern and western low-sulfur coal strategies; Appendix B, costs for the 90 percent and 60 percent (partial scrubbing) SO_2 control attained through the use of wet flue gas desulfurization; Appendix C, costs for dry flue gas desulfurization (applicable to only three plants); and Appendix D, NO_x control costs for low-NO_x burners and overfire air. All costs are in 1980 dollars, and treatment of annual operating and maintenance and fixed charges is as uniform as possible to maximize comparisons on the same or similar basis.

Rohde, W., "Reviving Acidified Lakes", Ambio, Vol. 10, No. 4, 1981, pp. 195-196.

The progressive acidification of lakes can be somewhat counteracted by the addition of neutralizing materials, but even very acid bodies of water are objects of natural colonization within a short time. Sweden has treated about 1000 lakes by adding calcium carbonate or other neutralizing materials, but the treatment is expensive and must be repeated. When the pH reaches 4 or 5, almost none of the original organisms are able to survive. Japan's volcanic lakes with pH's of 2-5 have existed long enough for acid-tolerant and acidophillic organisms to become established. These lakes include bacteria, fungi, algae, animals, macrophytes, and one species of fish. Thus, water devoid of life is the object of colonization, and research should be undertaken to determine the best autotrophic algae to introduce as pioneers. From naturally acid waters, plankton and benthic algae could be collected and studied, and eventually used to inoculate large bodies of water. At present, the only alternative is to apply neutralizing agents.

Streets, D.G. et al., "Analysis of Proposed Legislation to Control Acid Rain", ANL/EES-TM-209, Jan. 1983, Argonne National Laboratory, Argonne, Illinois.

This report reviews the activities of the 97th Congress of the United States related to the proposed establishment of an acid rain control program for the nation. Fourteen bills were introduced that address acid rain or the long-range transport of air pollutants. This report analyzes the emissions reductions and costs required by the five major bills: Mitchell (S. 1706), Moynihan (S. 1709), D'Amours (H.R. 4816), Moffett (H.R. 4829), and the Senate Committee (S. 3041). The emissions reductions range between 6.5 and 12.6 million tons of sulfur dioxide per year, at a cost of $2.6 to $5.4 billion per year. Impacts on specific midwestern states are discussed. In an appendix, the report reproduces copies of the five major bills referred to above.

Sverdrup, H., "Lake Liming", Chemistry Series, Vol. 22, No. 1, 1983, pp. 12-18.

Based on knowledge available from experiments and literature, a theory is presented concerning the dissolution of calcite which is sinking in a lake. The basic equation for the dissolution is based on a diffusion model for the dissolution. A model is presented also for evaluating different calcite powders offered commercially for liming operations.

Szabo, M., Shah, Y. and Abraham, J., "Acid Rain: Control Strategies for Coal-Fired Utility Boilers--Volume I", DOE/METC-82-42-V.1, May 1982, U.S. Department of Energy, Washington, D.C.

This report presents a detailed evaluation of the cost and effectiveness of conventional controls for emissions of sulfur oxides (SO_x) and nitrogen oxides (NO_x) from coal-fired utility boilers. The cost and control efficiency data are based on analyses of the 50 U.S. utility plants emitting these pollutants in the greatest quantities in 1979 (the 50 highest emitters for each pollutant, with some overlap of plants emitting large amounts of both). These plant-specific data can be used to verify other cost-effectiveness models, such as those from ICF and Teknekron, that were developed on a generic basis. The methodology and results of this evaluation have been used for review of assumptions and cost estimates prepared by the U.S. EPA for participation in the National Acid Rain Assessment Plan (NARAP) and the U.S. Canadian Transboundary Group. The study is based on the premise that coal-fired utility power plants in the midwestern United States are the major contributors to the acid rain problem in the Northeast. This premise and other important factors in the acid rain problem are addressed in a companion report assessing the overall acid rain issue. The assessment concludes that reducing SO_x and NO_x emissions from midwestern coal-fired power plants may not significantly reduce the acidity of rain, even at the cost of billions of dollars for controls. In fact, local sources of SO_x and NO_x, chiefly oil-fired boilers and automobiles in the Northeast, may contribute more significantly to the acid rain occurring there than previously realized. In view of the current uncertainty about the relative importance of various possible contributors to the acid rain problem, the utility evaluation reported here, which is detailed and plant-specific, should prove valuable in further investigations.

U.S. Environmental Protection Agency, "Controlling Sulfur Oxides", EPA-600/8-80-029, Aug. 1980, Washington, D.C.

This Research Summary describes EPA's program to develop new and improve existing technologies for sulfur oxides control. As society increasingly turns to coal as the primary utility and industrial fuel, while trying to deal with the problems of acid precipitation, visibility degradation, and unhealthy air, consideration will have to be given to the fact that most of the sulfur oxides which will be emitted over the next two decades will come from plants existing today. If we vigorously pursue the successful demonstration of control technologies and then take advantage of them, especially those which can reduce emissions from existing plants, the adverse health and environmental effects of the troublesome sulfur oxides can be significantly reduced.

Wiesenfeld, J.R. and Kreiss, W.T., "Source-Receptor Relationship in Acid Precipitation: Implications for Generation of Electric Power from Coal", PD-LJ-82-268R, 1982, Physical Dynamics, Inc., La Jolla, California.

This document reports the conclusions of a technical panel convened to examine the source-receptor relationship in acid precipitation. While a decrease in acid deposition is certainly a worthy goal for pursuit, the ecological consequences with or without such decreases are uncertain. Even more difficult to predict, however, is how best to limit deposition of acidifying species by the control of emissions. Although control costs using various strategies may be estimated to about a factor of 2, assessment of the benefits to be derived (in terms of deposition abatement) from the imposition of such controls remains a matter of significant uncertainty. It therefore becomes imperative that a far better understanding of the source-receptor relationship be gained in order to optimize the benefits of a major emissions reduced program so that the likely benefits of alternative control strategies may be estimated with reasonable accuracy. An attempt was made to avoid detailed descriptions of all processes mediating emission and deposition. Specific questions identified by the workshop panel as being of direct importance to the refinement of the source-receptor relationship are related. Promising research areas that might reasonably be expected to yield significant information concerning these scientific uncertainties are described.

Wright, R.F. and Henriksen, A., "Restoration of Norwegian Lakes by Reduction in Sulphur Deposition", Nature, Vol. 305, No. 5933, Sept. 1983, pp. 422-424.

Acid precipitation, causing the acidification of freshwaters and hence damage to fisheries, affects more than 33,000 sq km of southern Norway. As a first step in dealing with the acidification problem, Norway, Sweden and Finland have proposed that emissions of sulfur in Europe be reduced by 30 percent by 1993. Using data from a 1974-75 survey of major-ion chemistry and fish populations in 683 lakes in southern Norway, a 30 percent reduction in sulfur deposition will restore chemical conditions such that 22 percent of the lakes now experiencing fisheries problems should be able to support fish. The method of prediction derives from empirical relationships between sulfur deposition, water chemistry and fisheries status for several thousand lakes and rivers in Europe and North America. The method takes into account lake-to-lake differences in sensitivity, changes in base cation concentrations, and the buffering of dissolved aluminum at pH readings less than 5.4 and of bicarbonate at pH levels greater than 5.4.

Zawadzki Ltd., "Preliminary Study: Use of Low-Sulfur Coal and Coal Cleaning in Control of Acid Rain", DOE/MC/14784-T1, May 1981, U.S. Department of Energy, Washington, D.C.

This preliminary study was undertaken as part of a larger DOE assessment of the acid rain problem and the feasibility of various control techniques. From among the many strategies that are proposed for control of acid rain, this study deals with two: (1) use of naturally occurring low-sulfur coal; and (2) cleaning of raw coal in preparation plants prior to firing. The distribution and ownership of the U.S. coal reserves are discussed with emphasis on the reserves having low sulfur content. Some of the basic constraints on the availability and use of low-sulfur coals by utilities are enumerated. A preliminary estimate of the sulfur reduction potential of U.S. coals achievable by coal preparation is presented. Also included are a brief analysis of coal cleaning costs and the effects of coal cleaning on other aspects of acid rain control.

APPENDIX G

DRY DEPOSITION

Dolske, D.A. and Sievering, H., "Trace Element Loading of Southern Lake Michigan by Dry Deposition of Atmospheric Aerosol", Water, Air, and Soil Pollution, Vol. 12, No. 4, Nov. 1979, pp. 485-502.

Aerosol samples and meteorological data were collected at a mid-southern Lake Michigan site (87 deg 00 min W, 42 deg 00 min N) from May through September, 1977. Hi-volume samplers with cellulose fiber filters and a digital meteorological data recording system were operated on board the U.S. EPA's R/V Roger R. Simons during four intensive sampling periods. Aerosol samples were analyzed by atomic absorption spectroscopy for 17 trace elements. A diabatic drag coefficient method was used to determine aerosol deposition velocity overlake. A mean value of 0.5 cm/s was found for the aerosol diameters between 0.1 and 2.0 micrometers. By relating the observed trace element concentrations and depository velocity to a long-term climatological record, annual dry deposition loadings to the southern basin for nine elements were estimated. For four elements, Fe, Mn, Pb, and Zn, dry deposition loadings to the southern basin alone of at least 500,000, 30,000, 250,000, and 100,000 kg/yr were found. For Fe and Mn, these loadings represent about 15 percent of the total of all inputs to Lake Michigan. For Zn and Pb, about one-third to one-half of the annual loading from all sources is from dry deposition of atmospheric aerosol.

Dolske, D.A. and Sievering, H., "Nutrient Loading of Southern Lake Michigan by Dry Deposition of Atmospheric Aerosol", Journal of Great Lakes Research, Vol. 6, No. 3, 1980, pp. 184-194.

Dry deposition accounted for 15 percent or more of all atmospheric nutrient inputs of total phosphorus and nitrate-nitrite nitrogen and for 3 percent of total aerosol mass over southern Lake Michigan. High-volume samplers with cellulose fiber filters were used to collect 42 aerosol samples aboard the U.S. EPA's R/V Roger R. Simons at 87 degrees 00 min West and 42 degrees 00 min North during May to September, 1977. Micrometeorological data was documented simultaneously. Aerosol deposition velocity was estimated at 0.65 cm per sec using a diabatic drag coefficient method. Dry decomposition loading for this area of the lake was 150,000-180,000 kg per year for P and 3.5 million to 5.1 million kg per year for N. The aerosol P and N was strongly associated with fine particles, less than 1.0 micrometers in diameter. Nutrients deposited in this form are immediately available to the photosynthetically active regions: overlake, west shore (Milwaukee), Chicago/Gary, Michigan City, and east shore.

Droppo, J.G. and Doran, J.C., "Dry Deposition of Air Pollutants", 1979, Battelle Pacific Northwest Laboratories, Richland, Washington.

Dry removal of air pollutants on natural surfaces represents a significant sink for various gases and particles. A survey of data applicable to the computation of dry removal rates is given. Included are both radioactive and nonradioactive materials: iodine, SO_2, O_3, NO_x, and NO particles. This summary incorporates the results of recent field studies by the authors.

Garland, J.A., "Dry Deposition of Gaseous Pollutants", 1979, AERE, Harwell, England.

Deposition by direct interaction with the earth's surface does not require the interaction of falling precipitation and is therefore called dry deposition. Possible economic effects on crops, natural vegetation and building materials have stimulated studies of deposition of a few pollutants to a limited range of surfaces, but understanding the life cycle of atmospheric pollutants requires a knowledge of the behavior over the sea and areas of no economic significance. This paper is intended as a description of the current understanding of dry deposition. It discusses the deposition process, methods of measurement, deposition velocities, and uptake by the sea.

Hicks, B.B., "Dry Deposition of Acid Particles to Natural Surfaces", 1979, Argonne National Laboratory, Argonne, Illinois.

Monitoring programs conducted over the northeastern continental USA during the past few years have indicated that sulfate particles present in air near the surface are often acidic. These particles, which are typically small and hygroscopic, might be expected to attach themselves to foliage, thus imparting a strong but very localized dose of acid. At this time, the efficiency with which small particles attach themselves to leaf surfaces and the conditions under which they might be re-emitted by abrasion, for example, are largely unknown, and so a considerable uncertainty must be associated with any evaluation of the net effect. Application of deposition velocities in the range presently advocated for sulfate particles suggests acid fluxes by dry deposition that average about two orders of magnitude less than those probably resulting from rainfall. This should not be interpreted as an indication that dry deposition effects can be neglected, since it is clear that acid particles might reside on surfaces for considerable times, perhaps until washed off by rain or sufficiently diluted by dewfall.

Hicks, B.B., Wesely, M.L. and Durham, J.L., "Critique of Methods to Measure Dry Deposition--Concise Summary of Workshop", EPA-600/D-82-155, Jan. 1982, U.S. Environmental Protection Agency, Research Triangle Park, North Carolina.

At the Workshop on Dry Deposition Methodology, held December 4 and 5, 1979, at the Argonne National Laboratory in Argonne, Illinois, dry deposition measurement techniques were assessed for routine monitoring use. A majority opinion was reached that commonly used techniques such as surrogate surfaces and collection vessels are not sufficiently accurate for use in networks, because the highly varied properties of the natural surfaces of interest cannot be simulated adequately. Further research was recommended on dry deposition parameters in order to estimate dry deposition rates, if possible, from measurements of atmospheric concentrations at a single height, together with observations of surface properties and micrometeorological parameters. The ability to perform such investigations in the field is critically dependent upon advances in chemical and physical capabilities to provide methods with standard relative errors of less than 1 percent for a single instrument on successive measurements, or with time responses of less than 1 second. These requirements are not being achieved for many pollutant species. At present, the most promising methods for monitoring are eddy accumulation, modified Bowen ratio, and variance. Regardless of the method employed, monitoring sites should be chosen that are representative of the surrounding areas in terms of surface properties, meteorological conditions, and pollutant characteristics.

Ibrahim, M., Barrie, L.A. and Fanaki, F., "Experimental and Theoretical Investigation of the Dry Deposition of Particles to Snow, Pine Trees and Artificial Collectors", Atmospheric Environment, Vol. 17, No. 4, 1983, pp. 781-788.

    The study described was conducted to develop a model of particle deposition to a smooth surface, such as water or snow, and then to compare the model with measurements. Another objective was to investigate particulate deposition to smooth surfaces.

Lindberg, S.E., Turner, R.R. and Lovett, G.M., "Processes of Atmospheric Deposition of Metals and Acids to Forests", 1982, Oak Ridge National Laboratory, Oak Ridge, Tennessee.

    This study quantified the atmospheric deposition of the water-leachable fractions of trace metals to upper canopy foliage, determined the major mechanisms of deposition, and assessed the interactions between deposited metals and acid rain for several forested areas. Atmospheric sources contributed significantly to the annual flux of these elements to the forest floor in Walker Branch Watershed, Tennessee; deposition supplied from 14% (Mn) to approximately 40% (Cd, Zn) to approximately 99% (Pb) to this flux. The measured water solubility of these metals in suspended and deposited particles indicated that they may be readily dissolved following deposition; however, only Pb appeared to be absorbed directly at the leaf surface. Dry deposition constituted a major fraction of the total annual atmospheric input of Cd and Zn (approximately 20%), Pb, (approximately 55%), and Mn (approximately 90%). In southern forests, interactions between dry deposition and acid rain can result in concentrations of metals and H+ on leaves considerably higher than in rain alone. In the subalpine forests of New England, cloud deposition can also contribute to increased ion concentration on leaves.

Robinson, E., Lamb, B. and Chockalingam, M.P., "Determination of Dry Deposition Rates for Ozone", EPA-600/3-82-042, Apr. 1982, U.S. Environmental Protection Agency, Research Triangle Park, North Carolina.

    The report presents ozone (O3) velocity deposition (V(d)) measurements over three different vegetation types at three different geographic locations in the United States. The purpose of this study was to relate V(d) measurements of ozone to more commonly measured meteorological parameters over a wide range of ambient conditions. In this way, a general calculation procedure could be developed for ozone V(d) to be used in pollutant transport models.

Sehmel, G.A., "Particle and Gas Dry Deposition: A Review", Atmospheric Environment, Vol. 14, No. 9, 1980, pp. 983-1011.

    Published numerical values of particle and gas dry deposition velocities are summarized, but results have not been generalized. The deposition velocities for particles range over three orders of magnitude and the deposition velocities for gases range over four orders of magnitude.   For numerical prediction purposes, a model developed by Sehmel and Hodgson is recommended.

Sheih, C.M., Wesely, M.L. and Hicks, B.B., "Estimated Dry Deposition Velocities of Sulfur Over the Eastern United States and Surrounding Regions", Atmospheric Environment, Vol. 13, No. 10, Oct. 1979, pp. 1361-1368.

Surface deposition velocities of sulfur dioxide and sulfate particles over the eastern half of the United States, southern Ontario, and nearby oceanic regions are computed from equations developed in recent field experiments, for use in studies of regional-scale atmospheric sulfur pollution.   Surface roughness scale lengths and resistances to pollutant uptake by the surface are estimated from consideration of land-use characteristics and the likely biological status of the vegetation. Midsummer conditions are assumed, but other seasons can be easily considered.   Average deposition velocities for grid cells corresponding to half-degree increments of longitude and latitude are presented for a range of atmospheric stabilities.

Sickles, J.E., Bach, W.D. and Spiller, L.L., "Comparison of Several Techniques for Determining Dry Deposition Flux", EPA-600/D-83-057, June 1983, U.S. Environmental Protection Agency, Research Triangle Park, North Carolina.

Over the period from 1/22/81 through 5/4/82, measurements were conducted to permit comparison of several techniques for determining dry deposition flux of nitrates and sulfates.   Direct flux estimates were made by using actual leaf surfaces and foliar wash, and by exposing and washing three surrogate surfaces: bucket, petri dish, and cellulose filter.   Indirect flux estimates were made using SSI high-volume sampling along with meteorological measurements.   The time scale for the direct methods was nominally one month.   The indirect method used 24-hour particulate sampling and three-hour meteorological observations to calculate average flux values for the exposure periods of the direct methods.   This permitted comparison of direct and indirect methods on a similar time scale.

Sievering, H. et al., "An Experimental Study of Lake Loading by Aerosol Transport and Dry Deposition in the Southern Lake Michigan Basin", EPA-905/4-79-016, July 1979, U.S. Environmental Protection Agency, Chicago, Illinois.

A Lake Michigan experimental program to assess the contribution to Great Lakes loading by atmospheric transport and dry deposition of aerosol was conducted.   A midlake and nearshore trace element and nutrients data base with associated meteorology capable of establishing a climatology for mass transfer to Lake Michigan was collected during 1977 and 1978.   Significant data for Al, Ca, Cu, Fe, Mg, Mn, Pb, Ti, Zn, total P, NO_3 and SO_4 were obtained.   A strong linear dependence upon atmospheric thermal stability in the variability of all 12 aerosol constituents was found, but no linear dependence upon wind speed was found.   Bulk deposition velocities as a function of overlake climatology were used to calculate dry deposition atmospheric loadings to Lake Michigan.

APPENDIX H

PRECIPITATION METALS AND ORGANICS

Andersen, A., "Atmospheric Heavy Metal Deposition in the Copenhagen Area", Environmental Pollution, Vol. 17, No. 2, 1978, pp. 133-151.

   Atmospheric dry and wet deposition (bulk precipitation) of the heavy metals Cu, Pb, Zn, Ni, V and Fe over the Copenhagen area was measured by sampling in plastic funnels from 17 stations during a 12-month period. Epigeic bryophytes from 100 stations in the area were analyzed for the heavy metals Cu, Pb, Cd, Zn, Ni, V, Cr and Fe. Samples of the epiphytic lichen Lecanora conizaeoides Nyl. ex Cromb. from 25 stations were analyzed for Cu, Pb, Zn, Cr and Fe. There was a linear correlation between bulk precipitation and heavy metal concentration in lichens and bryophytes. An exponential correlation was found between bulk precipitation and heavy metal concentrations in soil. Regional variation of the heavy metal levels in the Copenhagen area was described and three subareas with high metal burdens were distinguished. The heavy metal gradients from a secondary smelter in one of these subareas were steepest in soil compared with lichens and bryophytes.

Andren, A.W., Lindberg, S.E. and Bates, L.C., "Atmospheric Input and Geochemical Cycling of Selected Trace Elements in Walker Branch Watershed", Report No. ESD-728, June 1975, Oak Ridge National Laboratory, Oak Ridge, Tennessee.

   Integrated studies of atmospheric and aquatic chemistry were used to quantify the cycling of selected trace elements in Walker Branch Watershed, Oak Ridge, Tennessee. The relative elemental contribution of wet deposition and aerosol impaction have been evaluated. Data for the concentrations of Cd, Cr, Cu, Hg, Mn, Ni, Pb, and Zn in rain and their input (in g/ha) for the period June, 1973-July, 1974, are presented. Cadmium, Cu, Hg, and Pb in rain are enriched with respect to soil by 1 to 3 orders of magnitude and must thus have a nonsoil source. Elemental ratios from fly ash are comparable to rain ratios for Cu, Zn, and Cd. Arsenic, Co, Cu, Hg, Se, V, and Zn possess enrichment factors very close to those of fly ash and can thus tentatively be attributed to three coal-fired steam plants in the vicinity of Oak Ridge. Lead is attributed to automotive emissions. Data on stream water trace element transport and speciation in dissolved and particulate forms for a 6-month period and a summary mass balance calculation are also reported. Concentrations of trace elements in the dissolved fraction were relatively constant; thus, export of dissolved forms is a simple function of discharge. The particulate fraction represents a significant transport mechanism for Cr, Hg, and Mn. Budget calculations indicate that the watershed efficiently retains Pb (97-98 percent of the input), Cd (94-95 percent), and Cu (82-84 percent), while it less readily accumulates Cr (59 percent), Mn (57 percent), Zn (73-75 percent), and Hg (69-75 percent).

Aten, C.F., Bourke, J.B. and Walton, J.C., "Heavy Metal Content of Rainwater in Geneva, New York During Late 1982", Bulletin of Environmental Contamination and Toxicology, Vol. 31, No. 5, Nov. 1983, pp. 574-581.

   Many previous studies of metal ions in the atmosphere have collected samples of particles by filtering air, but these studies do not show directly what falls out of the atmosphere at any particular place. Further, no correlation with the precipitation acidity is possible in such studies. This paper reports on a study to provide a more or less complete determination of the precipitation-borne pollution fallout in a rural setting that has not been included in the established networks. Rainwater samples were collected at a relatively isolated site near Geneva, New York, from September through December, 1982, and at another similar site starting in late November. The samples were analyzed for 19 metallic (or metalloid) elements, nitrate, sulfate, pH and conductance.

Beavington, F., "Some Aspects of Contamination of Herbage with Copper, Zinc, and Iron", Environmental Pollution, Vol. 8, No. 1, Jan. 1975, pp. 65-71.

Samples of white clover and paspalum were taken from an industrial area (copper smelter and steel works) and a rural area of Australia and analyzed for copper, zinc, and iron contents. Mean levels of copper, zinc, and iron for all herbage collected in the industrial area were 38.0, 70.4, and 723 ppm as compared with levels of 9.8, 28.6, and 129 ppm, respectively, for the rural samples. Zinc and copper levels correlated significantly with distance from the smelter, and iron levels with distance from the steelworks. Higher concentrations of all three contaminants were found in white clover than in paspalum.

Berry, W.L. and Wallace, A., "Trace Elements in the Environment: Their Role and Potential Toxicity as Related to Fossil Fuels. A Preliminary Study", Report No. UCLA-12-946, Jan. 1974, National Technical Information Service, U.S. Department of Commerce, Springfield, Virginia.

The aim has been first, to provide a background of chemical and biological information (so far as this is available) about the nature, occurrence and effects of trace elements; second, to review information about the ecological behavior of trace elements; and third, in the light of the foregoing, to identify the probable hazards of trace element emissions to the environment, particularly as a result of the use of fossil fuels for power generation. The emphasis throughout has been to seek gaps in our knowledge of these matters, and thus to uncover areas in which research would be especially timely. In general, the ecology of trace elements is very delicately balanced, and small additions of some trace elements from polluting sources can significantly alter an existing ecosystem. Some animals in particular are very sensitive to changes in concentration of trace elements, and the limits between adequacy and toxicity are very often narrow. Trace elements, at both low and high levels, are also of great importance to the health of man. The pathways and rates of movement of trace elements between each component of any soil-water-plant-animal system have not, so far, been well defined. These pathways must be more clearly understood and quantified to predict the fate and effects of any trace element pollutant in an ecosystem.

Budd, W.W. et al., "Aluminum in Precipitation, Streams, and Shallow Groundwater in the New Jersey Pine Barrens", Water Resources Research, Vol. 17, No. 4, Aug. 1981, pp. 1179-1183.

Aluminum appears in waters of the New Jersey Pine Barrens at concentrations of 0-200 micrograms per liter (volume weighted average, 105 micrograms per liter) in precipitation; 100-800 micrograms per liter (volume weighted average, 345 micrograms per liter) in streams; and 0-3000 micrograms per liter (volume weighted average, 468 micrograms per liter, and excluding one very acid well, 329 micrograms per liter) in ground water. These levels are 10 times greater than in most terrestrial waters, except for the acidified waters of New England and New York. The pH of precipitation in this study averaged 4.0; of ground water, 4.6; and of streams, 4.1. The total (acid reactive) aluminum deposited in the McDonald's Branch Basin was 140 mg per sq meter per year over the study period, May 1978 to May 1980. Stream and ground water outputs were 149 and 110 mg per sq meters per year, respectively. Increased precipitation, streamflow, and decomposing organic matter caused higher Al inputs and outputs in summer. In streams, Al may be transported as an organometallic complex with dissolved organic matter, humic and fulvic acids. Ground water Al concentrations depend on gibbsite solubility in mineral soils and are pH dependent.

Bufalini, J.J. et al., "Hydrogen Peroxide Formation from the Photooxidation of Formaldehyde and Its Presence in Rainwater", Journal of Environmental Science and Health, Vol. 14, No. 2, 1979, pp. 135-141.

The photooxidation of formaldehyde with sunlamps (E(max) = 3100 A) produces hydrogen peroxide (H_2O_2) at varying concentrations depending upon the amount of water vapor present. It is postulated that the variable production of H_2O_2 is a result of condensation on the reactor surface. Rainwater samples were also analyzed for H_2O_2. Summer rain in the Research Triangle area of North Carolina contained as much as 200 ppb (w/w) of H_2O_2. Contrastingly, rainwater collected in December contained much less H_2O_2, usually 2-8 ppb. The lower concentrations found in the colder conditions is attributed to a decrease in photochemical activity.

Carpenter, K.E., "The Lethal Action of Soluble Metallic Salts on Fishes", British Journal of Experimental Biology, Vol. 4, No. 4, 1927, pp. 378-390.

The lethal action of soluble metallic salts on fishes was examined on the basis of mathematical analysis and physiological experiments. Investigation of the effects of lead salts on fishes determined a survival curve with no theoretical threshold of toxic concentration. The lethal efficacy of the solution varied in inverse proportion to the actual size of the fish and was directly dependent on the absolute quantity of lead salt. The role of the metallic ion in the lethal process was examined. The effective action of lead salts on fishes was purely external in character; the precipitation of an organic lead compound clogged the gills and inhibited their respiratory function, leading to death by suffocation. The effects of soluble salts of zinc, iron, copper, cadmium, and mercury were also studied.

Costeque, L.M. and Hutchinson, T.C., "The Ecological Consequences of Soil Pollution by Metallic Dust from the Sudbury Smelters", Proceedings of Institute of Environmental Sciences Annual Technical Meeting, 1972, Institute of Environmental Sciences, Mt. Prospect, Illinois, pp. 540-545.

The ecological consequences of heavy metal damage are probably being masked by the damage due to sulfur dioxide emitted by the metal smelters of Sudbury, Ontario, in Canada. Soil and vegetation east and south of the source, plant leaves, and dust and rainfall were sampled for analysis of copper, nickel, cobalt, zinc, manganese, lead, and iron content. Elevated levels of nickel were detected up to 31 mi from the smelters and toxic water levels extended to 10 mi. The soil contamination has a pattern indicative of an airborne smelter source. A significant reduction of pH occurred in the soil within 1½ mi of the smelter. A pH of 2.2 was recorded in one instance, suggesting the presence of free sulfuric acid. Nickel levels were 2835 ppm at 0.5 mi from the smelter, 1522 ppm at 4-5 mi, 306 ppm at 12 mi, and 83 ppm at 31 mi. Copper followed the same pattern from 1528 ppm to 31 ppm; cobalt decreased from 127 ppm to 19 ppm. Average soil levels should be 40 ppm for Ni, 20 ppm for Cu, and 8 ppm for Co.

Crecelius, E.A., Johnson, C.J. and Hofer, G.C., "Contamination of Soils Near a Copper Smelter by Arsenic, Antimony, and Lead", Water, Air, and Soil Pollution, Vol. 3, No. 3, Sept. 1974, pp. 337-342.

Stack dust from a large copper smelter near Tacoma, Washington has contaminated soil in the area with arsenic, antimony, and lead. Within 5 km of the smelter, 380 ppm (dry weight) arsenic, 200 ppm antimony, and 540 ppm lead were measured in the surface soil (0-3 cm). Plants grown in these soils may be affected. The consumption of plants coated with this heavy metal-rich dust may be of concern. Standard orchard leaves in the area were collected. The following mean concentrations and standard deviations were determined for orchard leaves: 10.5 + or - 0.5 ppm As and 3.5 + or - 0.5 ppm Sb.

Dorn, R.C. et al., "Study of Lead, Copper, Zinc, and Cadmium Contamination of Food Chains of Man", EPA-R3-73-034, Dec. 1972, University of Missouri, Columbia, Missouri.

The amount of soil, vegetation, meat, and milk contamination by cadmium, copper, lead, and zinc were estimated on a farm exposed to stack emissions from a lead smelter, lead ore concentrate spillage from trucks, and dust from stockpiled ore at the smelter; these values were compared with those from a farm outside the lead production area. Within 1 year, the lead concentrations in soil samples collected 60 ft, 140 ft, and 220 ft from the highway on the test farm increased 219 percent, 257 percent, and 284 percent, respectively. High lead and cadmium concentrations were found in dust fall and air filter samples from the test farm, and test farm dust fall and soil samples also showed high concentrations of copper. The highest airborne, suspended Cd, Cu, Pb, and Zn concentrations were observed in winter on the test farm; this corresponded with high dust fall, soil, and vegetation levels and the time of greatest increase in lead assimilation by a test cow. There was a general dilution in the lead concentrations at each step of the food chain from soil to cattle tissues on the control farm, while biomagnification of the lead in the grass roots was observed on the test farm. On an equivalent basis, the test cow's blood Pb concentration at the end of the study period was 1/246 the test farm soil concentration and the control cow's blood Pb was 1/187 the control farm soil concentration. The liver, kidney, muscle, and milk of the test cow contained very small amounts of cadmium, and the lead levels were 2.35 micrograms/g, 3.75 micrograms/g, 0.19 micrograms/g, and 13 micrograms/g, respectively. The Pb concentrations in the milk of the test cow were 1.9 times those in the milk of the control cow, and the Pb concentrations in the milk from a cow exposed only to lead ore concentrate spillage from trucks and background sources were intermediate between those of the test and control animals.

Dorn, R.C. et al., "Environmental Contamination by Lead, Cadmium, Zinc, and Copper in a New Lead-Producing Area", Environmental Research, Vol. 9, No. 2, Apr. 1975, pp. 159-172.

An inventory of plant and soil materials in the environs of the Oak Ridge Gaseous Diffusion Plant provided quantitative evidence for the transfer of chromium and zinc to vegetation from a cooling tower operation. Chromium concentrations in foliage ranged from several orders of magnitude above background levels to near background at 1609 m from the tower. Concentrations of extractable zinc ranged from 5-72 ppm at distances of 180 and 15 m respectively, indicating no increase above background at 180 m. Potted tobacco plants placed downwind from the plume accumulated chromium, reaching equilibrium and attaining maximum concentrations (237 + or - 18 ppm) after 4-6 wk of exposure. While the tobacco plants (highly sensitive to chromium) developed symptoms of toxicity to drift, similar effects were not observed for native plant species at similar concentrations and downwind distances. Vegetation analyses along a horizontal gradient indicated that the equilibrium concentration in foliage and litter remained relatively constant, provided that additions of treatment chemicals to makeup water and drift emission rates remained unchanged. Distribution coefficients for hexavalent chromium added in solution to soils showed that little chromium was absorbed. Values for zinc indicated that the soil acted as a reservoir for the small quantities derived from drift.

Eisenreich, J.J., Looney, B.B. and Thornton, J.D., "Assessment of Airborne Organic Contaminants in the Great Lakes Ecosystem", Nov. 1980, Department of Civil and Mineral Engineering, University of Minnesota, Minneapolis, Minnesota.

No systematic survey of the identity, concentrations and frequency of occurrence of true organics has been performed in the Great Lakes Basin ecosystem; however, a review of literature and past research was conducted for both dry and wet deposition to assess current needs regarding the identification and control of organic airborne contaminants. Atmospheric deposition was found to be greater for Lakes Superior, Michigan, and Huron than for Lakes Erie and Ontario, as a result of general mass air circulation patterns, a large lake/basin area, and higher atmospheric concentrations. While atmospheric deposition of airborne PCBs is the most serious known toxic organic problem affecting Great Lakes water quality, other trace organics for which the atmospheric pathway is important include DDT-group pesticides, alpha-BHC, beta-endosulfan, methoxychlor, PAHs, and phthalate esters. The report recommends that a study be made of land-based sampling vs. over-lake sampling, the sources of airborne trace organics, and the transport for organics from urban areas to the lakes. In addition, the atmospheric pathway for the input of trace organics should be compared to other input pathways including tributary inflow, municipal and industrial discharges, and erosion.

Eisenreich, J.J., Looney, B.B. and Thornton, J.D., "Airborne Organic Contaminants in the Great Lakes Ecosystem", Environmental Science and Technology, Vol. 15, No. 1, Jan. 1981, pp. 30-38.

Pollution of the Great Lakes by airborne organic contaminants is significant and leads to accumulation in the food chains because, with the exception of Lake Erie, concentrations of suspended solids available for pollutant removal are less than 1 mg per liter. In addition, major pollution sources are upwind, surface areas and surface area to basin ratios are large, water and chemical residence times are long, and redistribution occurs through mixing. Trace organics in the air may enter the water by adsorption onto particulate matter, partitioning across the air-water interface, wet deposition through rain and snow, and dry deposition by aerosols. A table lists total annual deposition of airborne trace organics to each of the five Great Lakes for polychlorinated biphenyls (PCB), several pesticides, and several polynuclear aromatic hydrocarbons. The chemical with the most information available is PCB. Total sources in Lake Superior are estimated at 8,000-9,700 kg per year, with 6,600-8,300 of this coming from the atmosphere. Sinks are 1,100-1,700 kg per year, leaving an annual addition of 6,300-8,000 per year. Water concentration is increasing at the rate of 0.1-0.5 ng per liter per year. Although PCB production was prohibited in 1979 and their disposal was regulated, PCBs are widespread in the environment and are capable of codistillation, volatilization from landfills and incomplete incineration.

Eriksson, E., "Aluminum in Groundwater: Possible Solution Equilibria", Nordic Hydrology, Vol. 12, No. 1, 1981, pp. 43-50.

The chemistry of inorganic aluminum compounds in soils and rocks below pH 7.5 was reviewed because of high concentrations of aluminum found in acid ground water on the west coast of Sweden. Acid precipitation falls in this area, so the solubility of aluminum compounds contained in soils and rocks is of interest. Studies are cited on the chemistry of hydroxides, Al-silicate minerals, amorphous aluminum silicates, and basic aluminum sulfate. The chemical analysis of acid ground water is discussed, based on two sets of experimental data. The acidification process probably proceeds as follows. When precipitation turns acid, this attacks the cation exchange reservoir in the soil. When this is depleted, aluminum hydroxides are attacked, and when sulfate concentrations are high enough, aluminum oxides are transformed into basic sulfates. When the process has continued long enough, aluminum reaches the ground water table.

Evans, R.D. and Rigler, F.H., "Calculation of the Total Anthropogenic Lead in the Sediments of a Rural Ontario Lake", Environmental Science and Technology, Vol. 14, No. 2, Feb. 1980, pp. 216-218.

Sediment cores from a rural Ontario lake (Bob Lake, latitude $44^\circ 55'$N, longitude $78^\circ 47'$W) were analyzed for total anthropogenic lead content. A high correlation between deposition of anthropogenic lead and sediment depth facilitated calculation of total fallout of anthropogenic lead in Bob Lake. The total fallout of 820 mg m^{-2} is large for a rural area in comparison to that calculated for the Lake Michigan-Chicago area (390 mg m^{-2}).

Franzin, W.G., McFarlane, G.A. and Lutz, A., "Atmospheric Fallout in the Vicinity of a Base Metal Smelter at Flin Flon, Manitoba, Canada", Environmental Science and Technology, Vol. 13, No. 12, Dec. 1979, pp. 1513-1525.

Atmospheric fallout in the vicinity of a base metal smelter at Flin Flon, Manitoba, Canada, was monitored with bulk precipitation collections over one year (y) and by winter snow samples collected from the surfaces of frozen lakes. Simple correlation analysis of the data obtained from both types of collection indicated that the smelter was a major source of Zn, Cd, Pb, As, and Cu in bulk precipitation in the Flin Flon area. Metal deposition with respect to distance from the smelter fit curves of the type $y=ax^b$. Integration of the deposition curves to selected levels (1 $mg.m^{-2}.y^{-1}$ for Zn, Pb, and Cu; 0.1 $mg.m^{-2}.y^{-1}$ for As; and 0.01 $mg/m^{-2}.y^{-1}$ for Cd) gave the following radii for affected zones: Zn, 131-264; Cd, 113-284; Pb, 44-87; As, 37-68; and Cu, 33-60 km. Total deposition within these zones was as follows: Zn, 850-1616; Cd, 2.0-8.0; Pb, 14-55; As, 0.4-4.0; and Cu, 14-48 t. Direct comparison of the two collection methods, possible for only one winter period, showed that bulk precipitation collections yielded estimates of metals deposition that were, on the average, only 55 percent of estimates determined from winter snow samples from the same location and time period.

Freedman, B. and Hutchinson, T.C., "Smelter Pollution Near Sudbury, Ontario, Canada, and Effects on Forest Litter Decomposition", Effects of Acid Rain Precipitation on Terrestrial Ecosystems, 1980, Plenum Press, New York, New York, pp. 395-424.

This paper presents data concerning observations of inputs of iron, nickel, copper, and sulfate-sulfation rates in forest sites near the Sudbury smelter during 1976-1977. Increases in metals and sulphur levels in soils and vegetation at forest sites close to the smelter were observed with accumulations of litter standing crop, lower rates of leaf decomposition, lower populations in soil micofungi and micoarthropods, and lower rates of soil metabolic activity.

Glotfelty, D.E., "The Atmosphere as a Sink for Applied Pesticides", Journal of the Air Pollution Control Association, Vol. 28, No. 9, Sept. 1978, pp. 917-922.

The role of the atmosphere as a sink for applied pesticides is reviewed. There is evidence that even nonvolatile pesticides enter the atmosphere by application drift, soil erosion, or post-application volatilization. Dominant reaction mechanisms of pesticides in the atmosphere include photochemical degradation and reaction with atmospheric oxidants. These processes are described. The atmosphere becomes a sink only when the rate of chemical degradation limits the residence time of the pesticide. Atmospheric degradation rates and reaction pathways are not fully known. Pesticides may be removed from the atmosphere by surface deposition, washout, or chemical reaction. Uncertainties regarding the significance of the atmosphere as a sink for pesticides and the persistence and transport of compounds will need further study.

Hanson, D.W., Norton, S.A. and Williams, J.S., "Modern and Paleolimnological Evidence for Accelerated Leaching and Metal Accumulation in Soils in New England, Caused by Atmospheric Deposition", Water, Air, and Soil Pollution, Vol. 18, No. 1-3, 1982, pp. 227-230.

In the absence of historic data about the chemistry of soils, it is possible to evaluate the long-term effects of acidic precipitation by examining comparable organic soil litter along a pH gradient or by examining the chemistry of lake sediments deposited over a long period of time. Such examinations include long-term water quality data such as total dissolved solids, concentrations of specific metals and changes in conductivity; cation exchange capacity and base saturation values for soils located on precipitation pH gradients; lysimeter studies; and chemical analysis of organic soils on precipitation pH and metal gradients. Such evidence indicates that lead has been accumulating at an accelerating rate in New England soils, and therefore in lake sediments, for at least 100 years, and that zinc has been accumulating at an accelerating rate in New England soils which are circumneutral. In acidic or acidified soils, there is a net loss of zinc due to solution by acidic precipitation. The zinc concentration in lake sediments in acidifying drainage basins decreases as the ambient pH decreases. Accelerated leaching of potassium, calcium, magnesium, and manganese has occurred over the last 100 years, at least in areas with originally acidic soils, and more recently in areas where soils have become acidified or are circumneutral.

Harder, H.W. et al., "Rainfall Input of Toxaphene to a South Carolina Estuary", Estuaries, Vol. 3, No. 2, June 1980, pp. 142-147.

Toxaphene in rainfall over a South Carolina salt marsh increased from background levels to variable high levels (12-497, mean 157 ng per kg) in July and August, the season of heaviest agricultural use. Mean concentration of polychlorinated biphenyls was 7.5 ng per kg; DDT, 1.4 ng per kg; and DDE, 0.5 ng per kg. Airborne toxaphene concentrations ranged from 0.33-7.2 ng per kg air. Estimated rainfall input of toxaphene during June-September, 1977, for the 26 sq km North Inlet estuary was 1.2 kg. Separate experiments showed that dry deposition of toxaphene accounted for less than 15 percent of the total.

Heit, M. et al., "Anthropogenic Trace Elements and Polycyclic Aromatic Hydrocarbon Levels in Sediment Cores from Two Lakes in the Adirondack Acid Lake Region", Water, Air, and Soil Pollution, Vol. 15, No. 4, May 1981, pp. 441-464.

The degree to which anthropogenic trace elements and polycyclic aromatic hydrocarbons have been deposited in remote lakes known to be affected by acid rainfall was investigated. The lakes investigated are located in the Adirondack State Park on the western slope of the Adirondack Mountains. Sagamore Lake is a brown water lake due to high levels of humic materials, while Woods Lake is generally clear. Sediment cores were taken from each of the lakes during March, 1978. Several conclusions were reached. Except for perylene, the prime source of all parental PAHs measured and the majority of trace elements appears to be combustion. All combustion products had primarily an anthropogenic origin. Levels of all parental PAHs except perylene and levels of several of the metals significantly increased in the surface sediments of both lakes compared to their background concentrations. PAH concentrations in Woods Lake were quite high, approaching levels reported for more heavily populated and industrialized areas. Metals were present at 3-4 times higher concentrations in Woods Lake than in Sagamore Lake. Metals and PAH levels decreased in concentration with depth to background levels in both lakes. This baseline depth corresponded to about 30 years ago. Lead has increased to the greatest extent of any of the metals considered relative to their baseline quality. Long distance atmospheric transport and region-wide deposition of anthropogenically derived elements and PAHs into these remote lakes appears to be more significant than input from local sources.

Hemenway, D.R., "Heavy Metals in Acid Precipitation", Dec. 1982, Vermont Water Resources Research Center, University of Vermont, Burlington, Vermont.

Precipitation samples were collected sequentially on a subevent basis. Specific contaminants in the segmented precipitation samples were measured for seven separate events. The concentration of Al, Cd, Pb, and V in the subsamples of each event analyzed were determined along with sample pH. In the last three events, sulfate, nitrate, and chloride concentrations were also measured. The subsamples for each event were filtered through a 0.1 um membrane filter in an attempt to separate particulate matter from soluble species. Filtrate concentrations were as high as 243 ug/l, 53 ug/l, 91 ug/l and 8.3 ug/l for Al, V, Pb, and Cd, respectively, for the initial subsample. The only metal investigated associated with the suspended particulate was Al with concentrations of up to 350 ug/l in the initial subsamples. Sulfate, NO_3, and Cl showed similar trends to pH during the course of an event. These concentrations were always high at the onset of precipitation, lowering as the event continued, but once again increasing, in differing degrees, in the final sub-samples. Results indicate that chemical constituents in precipitation vary greatly with time. Removal of certain contaminants may be greatly affected by scavenging efficiencies of cloud and rain drops, and aerodynamic and chemical behavior of the aerosol rather than simple washout/dilution effects.

Hemphill, D.D. and Pierce, J.O., "Accumulation of Lead and Other Heavy Metals by Vegetation in the Vicinity of Lead Smelters and Mines and Mills in Southeastern Missouri", 2nd Annual Trace Contaminants Conference Proceedings, 1974, Pacific Grove, California, pp. 325-332.

The accumulation of lead and other heavy metals in or on vegetation near lead smelters, mines, and mills in southeastern Missouri was investigated from 1971 to 1974. An area 12 by 25 mi in the New Lead Belt, containing four mines and mills and one smelter, as well as an area of 14 by 14 mi outside the New Lead Belt, with a smelter at the center, were sampled. Post oak (Quercus stellata) and shortleaf pine (Pinus echinata) foliage collected within 0.5 mi of a smelter accumulated maximum lead levels of 8,125 ppm and 11,750 ppm, respectively. Elevated lead levels in white oak and blueberry leaves were detected at distances greater than 7 mi from the smelters and at distances of 4 mi from the mines and mills. Similar patterns of elevated levels of cadmium, copper, and zinc were found. Anomalous levels of zinc at certain sampling sites did not, however, appear to be correlated with the proximity of the site to the sources of emission in the area.

Henriksen, A. and Wright, R.F., "Concentrations of Heavy Metals in Small Norwegian Lakes", Water Research, Vol. 12, 1978, pp. 101-112.

As part of regional surveys of lakes in Norway the concentrations of Zn, Pb, Cu, and Cd were measured in surface- and bottom-water samples collected from representative, small, pristine lakes (136 in southern Norway samples in October 1974, 58 resampled in March 1975, and 77 in northern Norway sampled in March 1975). The lakes, a statistically representative sample of small lakes in Norway, were chosen such that their watersheds are undisturbed. Heavy-metal concentrations in these lakes thus reflect only natural inputs and anthropogenic inputs via the atmosphere. The generally low concentrations (Zn 0.5-12.0 ugl^{-1}; Pb 0-2.0 ugl^{-1}; Cu 0-210 ugl^{-1}; Cd 0.1-0.5 ugl^{-1}) measured in lakes in central and northern Norway provide estimates of natural "background" levels. These estimates may be too high because they include the global-scale deposition of heavy metals from the atmosphere which has increased as a result of industrial activities. Concentrations of Zn and Pb in lakes in southernmost and southeastern Norway lie above these "background" levels, apparently because of atmospheric deposition associated with the acidic precipitation that falls over southern Scandinavia. Increased heavy-metal concentrations in acid lakes may also be due to increased mobilization of metals due to acidification of soil-and surface-waters.

Hirao, Y. and Patterson, C.C., "Lead Aerosol Pollution in the High Sierra Overrides Natural Mechanisms which Exclude Lead from a Food Chain", Science, Vol. 184, No. 4140, May 31, 1974, pp. 989-992.

Most of the lead contained in sedge leaves and vole mountain meadow mice within one of the most pristine, remote valleys in the U.S. is not natural, but comes from smelter fumes and gasoline exhausts. The mice were studied in Thompson Canyon in the crest of the High Sierra mountains. Isotopic compositions of Thompson Canyon leads and urban sources were used to determine the origin of the Thompson leads. About 97 percent of the lead entering the Canyon was of industrial origin, and 98 percent of the lead entering the valley remained there. Lead deposition and snow lead input, lead output, lead in humus, metals in food chain (calcium, strontium, barium, and lead), and lead in sedge leaves and vole tissues were measured. In a food chain, natural mechanisms do not allow lead to accompany the bulk of the nutritive metals as they proceed to higher tropic levels. This exclusion can be expressed quantitatively by a comparison of lead/calcium ratios at successive tropic levels. This ratio decreased by an overall factor of 200 in proceeding from rock to soil moisture to sedge to vole. This factor would have been 1200 if lead aerosols had not collected on sedge leaves and circumvented the tendency by sedge to exclude lead from the nutritive metals it absorbed from soil moisture.

Hoffman, W.A., Lindberg, S.E. and Turner, R.R., "Some Observations of Organic Constituents in Rain Above and Below a Forest Canopy", Environmental Science and Technology, Vol. 14, No. 8, Aug. 1980, pp. 999-1000.

Some organic components of rain have been determined in samples collected on an event basis above and below the canopy of a deciduous forest. Plasticizers and chlorohydrocarbons that were identified were attributed to sources external to the vegetation. Organic acids and high molecular weight "waxy" components identified were attributed to the foliage itself. Approximately equal amounts of material were ascribed to each of these sources. Low molecular weight organic acids were not found generally. Event sampling is advocated as the best approach to use for identifying sources of organic components in rain reaching the forest floor.

Hopkins, J.M., Wilson, R.H. and Smith, B.N., "Lead Levels in Plants", <u>Naturwissenschaften</u>, Vol. 59, No. 9, 1972, pp. 421-422.

The use of tetraethyl lead in gasoline has been the major factor responsible for the tremendous rise in atmospheric lead concentrations since the 1920s. As a result, lead contamination has increased along roadsides, in urban areas, and in roadside plants. Plant samples were taken along a highway two miles south of Austin, Texas in August, 1970. After drying and ashing, plant samples were analyzed with an atomic absorption spectrophotometer for lead. The levels of lead decreased with increasing distance from the highway. Plants taken 32 m northwest of the road contained less than one-half as much lead as plants adjacent to the road. More lead was found in plants on the northwest side of the road than on the southeast due to prevailing winds. Different species had different capacities to accumulate lead.

Hutchinson, T.C., "The Occurrence of Lead, Cadmium, Nickel, Vanadium, and Chloride in Soils and Vegetation of Toronto in Relation to Traffic Density", <u>Canadian Botanical Association and the American Institute of Biological Sciences, Joint Meeting</u>, June 20-24, 1971, Alberta, Canada.

From June to August, 1970, soil and plant samples were taken at eight sites around Metropolitan Toronto. The sites were selected to give a wide range of traffic density and degree of urbanization. Surface soil, soil from a 2-3 in depth, and soil from a 5-6 in depth were collected, and replicate samples were taken at distances from 0 to 30 m from the road edge. Total concentrations of lead, cadmium, nickel, vanadium, and chloride in the soils and plant samples were determined. The toxicity of particular metallic salts was determined using root elongation as a bioassay technique. High levels of lead occurred in soils and vegetation in downtown Toronto. The pattern of distribution was clearly indicative of traffic source. Cadmium showed a similar distribution, although at much lower levels. Levels of Pb and Cd in soils were inhibitory to plant growth; laboratory studies showed that Cd and Pb were taken up to a great extent by roots, and that uptake correlated to inhibition. Lead and cadmium levels in trees increased several fold over the last 100 years, and the greatest increase occurred since 1940. Vanadium and nickel accumulated in surface soils with the lowest concentrations at the rural site. Nickel correlated with distance from the highway. The source was anthropogenic. High chloride concentrations near roads were due to road salting operations. Chlorides caused damage to grasses and trees near roads.

Hutchinson, T.C. and Whitby, L.M., "A Study of Airborne Contamination of Vegetation and Soils by Heavy Metals from the Sudbury Copper-Nickel Smelters, Canada", Report EL-3, Nov. 1973, Department of Botany and the Institute of Environmental Sciences and Engineering, Toronto University, Toronto, Canada.

Soil, vegetation, and rainfall samples were collected for heavy metal analysis by atomic absorption spectrometry in an investigation of airborne contamination from copper-nickel smelters in the Sudbury basin region of Canada. Both Cu and Ni showed declining concentrations in soil and vegetation with increasing distance from the smelters. Rainfall was more acidic the closer its proximity to the smelters, with large quantities of Cu, Ni, and Fe being present in a pattern of distribution which again pointed to the smelters as a continuing source of metal pollution. Growth of the radicles of tomato and lettuce seedlings in water extracts of soils collected at various distances from a smelter was severely inhibited at distances up to 8 km from the smelter.

Jackson, T.A. et al., "Experimental Study of Trace Metal Chemistry in Soft-Water Lakes at Different pH Levels", Canadian Journal of Fisheries and Aquatic Sciences, Vol. 37, No. 3, Mar. 1980, pp. 387-402.

The biogeochemistry of Hg, Zn, Co, Fe, Mn, Cr, V, Th, Ba, Cs, As, and Se in two soft-water lakes of the Canadian Shield was investigated by means of carrier-free gamma-emitting isotopes introduced into limnocorrals in which the pH of the water was varied from 6.8 to 5.1. The residence times of the radionuclides in the water were determined, and the partitioning of the nuclides among different metal-binding agents in the water and sediments was studied with the aid of membrane filtration, dialysis, solvent extractions, and fractionation on Sephadex columns. Metal behavior varied systematically with metal properties. Metals of high crystal field stabilization energy, high electronegativity, or small ionic radius were most readily scavenged by greater than 0.45 micrometer suspended particles and dispersed colloids in the water, disappeared most rapidly from the water column, and were preferentially accumulated by sedimentary binding agents, including organic substances. Which property of a metal had the dominant effect on metal behavior depended on environmental factors, such as the ambient pH and the nature of the binding agents. Thus, Hg was removed fairly rapidly from the water at pH 6.7-6.8 owing to its high electronegativity, but was removed more slowly than any other metal at pH 5.1 owing to its large ionic radius. Acidification of lake water to pH 5.1 interfered with accumulation of Hg and other metals by organic ooze, probably owing in part to interference with the deposition or formation of the complexing agents with the 265 nm absorption band. Acidification also lowered the concentration of NaOH-extractable colloidal phosphate in the ooze but had no effect on NaOH-extractable orthophosphate content.

Kahl, J.S., "Metal Input and Mobilization in Two Acid-Stressed Lake Watersheds in Maine", Dec. 1982, Land and Water Resources Center, University of Maine, Orono, Maine.

Two small lakes in Hancock County (ME), Little Long Pond and Second Pond, were compared with respect to atmospheric deposition, aqueous chemistry, and metal mobilization from sediments. The ponds have similar granitic bedrock, elevation, watershed areas, soil composition, and atmospheric deposition of acids and metals, but they exhibit marked differences in surface water pH. Little Long Pond is acidic (pH 4.5-5.6), whereas Second Pond is circumneutral (pH 6.0-6.8) and has higher dissolved cation concentrations. These differences are caused, at least in part, by thinner soils, greater relief, and more exposed bedrock in Little Long watershed, compared to Second Pond watershed. Dated sediment chemical profiles indicate that accelerated sedimentation of Pb and Zn began in the mid-1800's in both ponds, presumably due to increased atmospheric deposition.   Little Long hypolimnetic sediments show concurrent depletion in Ca, Mn, and more recently, Zn. Experiments with limnocosms show that up to 50 percent of the Ca and Mn, and 25 percent of the Zn, in recent sediments may be leached at a pH near 4.0 in less than a year, suggesting that in-situ leaching may be an important influence on sediment chemistry in acidic waters. Nearly 10 percent of the Ca and Zn was mobilized at pH 5.0. Except for the epilimnetic Little Long cores (greater Mn release), and the hypolimnetic Little Long cores (less Ca release), the sediments from different sites and lakes responded similarly to experimental acidification in the laboratory. However, Al release by Little Long sediments was relatively more important in buffering acid inputs, whereas Ca was more important in Second Pond sediments.

Klein, D.H. and Russell, P., "Heavy Metals: Fallout Around a Power Plant", Environmental Science and Technology, Vol. 7, No. 4, Apr. 1973, pp. 357-358.

The degree of fallout of heavy metals around a coal-fired power plant was measured with respect to soil and plant material enrichment. Samples of soil, grasses, leaves, and biota were analyzed for contents of iron, titanium, zinc, cobalt, chromium, copper, nickel, cadmium, silver, and mercury, and the data were correlated with wind patterns. The soils were enriched with Cd, Co, Cr, Cu, Fe, Hg, Ni, Ti, and Zn, and the plant materials were enriched in Cd, Fe, Ni, and Zn.   Soil enrichments correlated with wind patterns and with the metal content of the coal except for Hg, which was only slightly enriched.

Komai, Y. et al., "Accumulation of Heavy Metals and Their Behavior in Soil and Plants Due to Air Pollution. Part I. Accumulation of Zinc in Soil in the South Part of Osaka Prefecture", Japan Social Science Meeting, 1972, Osaka Prefecture, Osaka, Japan, p. 132.

Accumulation of zinc, especially soluble zinc, in the surface soil in the southern part of Osaka Prefecture was thought to be due to air pollution. The determination of heavy metals in floating dust, falling dust, rain water, and surface running water revealed considerable amounts of zinc. The content of zinc in the surface soil in vegetable fields where the soil is periodically fertilized varied with the number of years of fertilization.   It was concluded that the increase of heavy metals, especially zinc, was due to air pollution. The total content of zinc in soil in which water pollution was not a factor was 400 ppm in arable land and 1,000 ppm in nonarable land. The amount of zinc soluble in 0.1 N HCl was 70 percent at most.

LaRiviere, M.G. et al., "Bibliography on Cycling of Trace Metals in Freshwater Ecosystems", Report No. PNL-2706, July 1978, National Technical Information Service, U.S. Department of Commerce, Springfield, Virginia.

This bibliography contains a listing of pertinent literature directly addressing the cycling of trace metals in freshwater ecosystems. Data on cycling, including the influences of environmental mediators, are included.

Lindberg, S.E. et al., "Atmospheric Deposition of Heavy Metals and Their Interaction with Acid Precipitation in a North American Deciduous Forest", 1981, Oak Ridge National Laboratory, Oak Ridge, Tennessee.

Atmospheric deposition dominates the landscape cycle of Pb and has a measurable influence on the cycles of Cd and Zn, but a minimal influence on the Mn cycle in a deciduous forest in the eastern United States. Rain event deposition rates are orders of magnitude greater than the intervening dry deposition rates; however, dry deposition supplies 20 to 90 percent of the total annual input to the forest. Interactions between dry deposition and acid precipitation result in concentrations of metals and H+ in the canopy 100 to 1000 times higher than rain concentrations alone.

Little, P. and Martin, M.H., "A Survey of Zinc, Lead, and Cadmium in Soil and Natural Vegetation Around a Smelting Complex", Environmental Pollution, No. 3, July 1972, pp. 241-254.

Analysis of samples of leaves and soil collected in the Avonmouth area of Severnside, Great Britain, showed the distribution of airborne zinc, lead, and cadmium to be strongly affected by prevailing wind conditions. Levels of zinc, lead, and cadmium in elm leaves collected in October, 1971, ranged from 8000, 5000, and 50 ppm dry matter close to a smelting complex, to values of about 200, 100, and less than 0.25 ppm, respectively, at distances of 10-15 km from the factory. The Avonmouth industrial complex includes the largest lead and zinc smelting plant in the world. Determinations of metal content were made using an atomic absorption spectrophotometer, and results of the analyses are presented in the form of concentration contour maps.

Little, P. and Martin, M.H., "Biological Monitoring of Heavy Metal Pollution", Environmental Pollution, Vol. 6, No. 1, Jan. 1974, pp. 1-19.

A simple, inexpensive technique is described for detailed monitoring of airborne metal pollution using Sphagnum moss suspended in fine nylon hair-nets from natural vegetation. The technique was used to monitor pollution, month by month, at 47 sampling points near a zinc and lead smelting complex at Avonmouth, near Bristol, England. Maps of the area were constructed by computer, depicting the levels and patterns of distribution of Zn, Pb, and cadmium collected by the moss bags. The results are related to wind data, localized climatic conditions, and topography. The reliability and practical significance of this simple biological monitoring technique is discussed, with particular reference to the collection and retention of metals by natural vegetation. A comparison is made between moss bags and other monitoring devices. Sphagnum bags should be at least as valid as deposit gauges, if not more so.

Moeller, P.K. and Stanley, R.L., "Origin of Lead in Surface Vegetation", Report 87, July 1970, California State Department of Public Health, Berkeley, California.

A model is presented for determining the relationship of airborne lead and soil lead concentrations to lead concentrations in pasture grasses. The major portion of atmospheric lead is derived from combustion of leaded gasoline and from fumes and dusts of industries using lead. Soil may be contaminated with lead by aerial fallout, industrial wastes, pesticides, fertilizers, and agricultural minerals. Soil composition and its acidity are important factors in whether or not lead will be available for plant uptake. On the basis of several published studies, it is concluded that not more than 15 micrograms Pb/g dry weight in forage crops is due to the uptake of lead from soil even at soil lead contents up to about 700 micrograms/g and possibly 3000 micrograms/g. Amounts of lead in herbage substantially greater than 15 micrograms/g are therefore due to aerial fallout. The ratio of lead in herbage to the rate of lead fallout was reported previously as a constant of 50 sq m days/kg herbage. This model was applied to a typical situation in which grass lead content of pastures toxic to horses was greater than about 80 micrograms/g, top soil lead contents averaged about 300 micrograms/g, and the suspended particulate lead concentration averaged about 0.5 micrograms/cu m.

Montagnini, F., Neufeld, H.S. and Uhl, C., "Heavy Metal Concentrations in Some Non-Vascular Plants in an Amazonian Rainforest", Water, Air, and Soil Pollution, Vol. 21, No. 1-4, Jan. 1984, pp. 317-321.

The presence of acid rain in the remote Amazon rainforest of southwestern Venezuela suggests the possibility of long-range transport of industrial pollutants to that region. Heavy metal concentrations were analyzed in samples of bryophytes and epiphyllous organisms growing on leaves and on bark in this forest. Concentrations of Cd, Pb, Ni, and Cr were higher in leaves with epiphylls than in leaves without epiphylls. All heavy metal concentrations in bryophytes from the Amazon Basin site were lower than in bryophytes from temperate zones. The results indicate that long-range transport of air pollutants from industrial centers to the remote Amazon Basin is occurring at only extremely low rates.

Muhlbaier, J. and Tisue, G.T., "Cadmium in the Southern Basin of Lake Michigan", Water, Air, and Soil Pollution, Vol. 15, No. 1, 1981, pp. 45-59.

A preliminary mass balance for cadmium was determined for Lake Michigan, southern basin, which covers an area of 18,000 sq km. The overall input rate of Cd exceeds the estimated loss rate by a factor of 2.5. Sources of Cd input in tons per year are as follows: rain, 4.3 (40.1%); dry deposition, 2.2 (20.8%); erosion, 1.0 (9.4%); and tributaries, 3.1 (29.2%), for a total of 10.6 tons. Sedimentation is the major (93%) sink in the system, 3.8 tons per year, with outflow of water at a Cd concentration of 20 ng per liter eliminating 0.28 tons per year or 6.9%. Cadmium levels were projected by a mathematical model. If the input rate increases more rapidly than 3% per year, the EPA's standard for protection of aquatic life, 880 ng per liter for waters of Lake Michigan's hardness, will be exceeded in 100 years.

Munshower, F.F., "Cadmium Compartmentation and Cycling in a Grassland Ecosystem in the Deer Lodge Valley, Montana", Thesis (Ph.D.), June 1972, Montana University, Missoula, Montana.

Distribution, compartmentation, and transfer of cadmium was studied in a grassland ecosystem where airborne cadmium from a zinc smelter has accumulated for approximately 50 years until permanent closure in 1969. Thus, it was possible to study the cadmium cycle under discontinued contamination conditions. The cadmium content in plant and animal tissue samples was analyzed by atomic absorption spectrophotometry. The soil reservoir of introduced Cd is restricted to the top 2-5 cm, is geographically distributed in relation to the smelter and prevailing winds, and normally is not transferred to lower soil horizons. This indicates that the only probable routes of Cd exit from this ecosystem will be by surface erosion or by cropping, and not by leaching from the soil. Analyses of plants grown in controlled Cd concentrations in nutrient solutions and soil cultures, and of plants collected in the field, established that Cd was absorbed and translocated by plants in a predictable manner. Plant Cd concentrations are a function of the Cd concentration of the root medium, time of collection during the growing season, and species. Grasshopper Cd concentrations, and the liver and kidney Cd concentrations of cattle, swine, and columbian ground squirrels, demonstrated Cd accumulation above plant Cd levels as a function of the animal's age. The presence of Cd in the kidney of red fox, badger, and weasel documented Cd transfer from herbivores to carnivores. A mathematical model of the ecosystem was developed depicting the system as a closed, self-contained unit. Concentrations and quantities of Cd in each ecological compartment were derived in terms of the Cd concentration at the lower trophic level, life span of the organism, and its biomass on an area basis. The half-life for the introduced Cd in the grassland ecosystem appears to be in excess of 1000 years, and shorter in cultivated lands.

Murphy, T.J. and Rzeszutko, C.P., "Polychlorinated Biphenyls in Precipitation in the Lake Michigan Basin", July 1978, De Paul University, Chicago, Illinois.

Rainfall samples were collected in Chicago, Illinois, and on Beaver Island, Michigan, and analyzed for polychlorinated biphenyls (PCBs). The precipitation weighted mean concentration of 35 samples of rain was 111 parts per trillion. This would result in the deposition of 4800 kg/yr of PCBs to the lake from precipitation. Presently available evidence on other sources of PCBs to the lake indicates that precipitation is now the major source of PCBs to the lake. The future PCB problems in the lakes will then be determined mainly by the magnitude of atmospheric inputs to the lake. The concentrations of PCBs in rainfall were found to be as high on Beaver Island as in Chicago. Results obtained from the simultaneous sampling of air and precipitation indicate that PCBs are present in the atmosphere as vapor as well as being present on particulates. This result raises doubts as to the validity of results for the dry deposition of PCBs obtained from the use of collectors covered with mineral oil or other nonpolar liquid. PCB concentrations in the parts per billion range obtained from gas samples from a vented sanitary landfill, indicate that PCB containing materials incorporated into landfills may be an important source of PCBs to the atmosphere.

Navarre, J.L., Ronneau, C. and Priest, P., "Deposition of Heavy Elements on Belgian Agricultural Soils", Water, Air, and Soil Pollution, Vol. 14, 1980, pp. 207-213.

Wet plus dry deposition processes contributed more heavy metals to Belgian agricultural soils than did chemical fertilizers. Sampling was conducted for one year in 19 purely rural sites as far removed as possible from obvious contamination sources: chimneys, roads, etc.   Metals clearly attributable to human origin were Zn, As, Se, Hg, Pb, Cd, and Sb. The sea atmosphere was responsible for the higher levels of Zn (from corroded roofing), Br, and Na found at seaside sites. Certain other metals reached higher levels near factories emitting these materials. Cd, Cr, and Ba were relatively uniform throughout the country.

Nieboer, E. et al., "Heavy Metal Content of Lichens in Relation to Distance from a Nickel Smelter in Sudbury, Ontario", Lichenologist, Vol. 5, 1972, pp. 292-304.

The Sudbury region of Ontario has large deposits of nickel, iron, and copper, and thus a number of smelting plants which produce sulfur dioxide and heavy metal pollution.   Since lichens are good indicators of SO_2 pollution levels, the pattern of heavy metal content in lichen species in the area of a copper smelter in Sudbury was correlated with distance from the smelter to ascertain whether lichens might also be good indicators of the amount of heavy metal fallout.   The lichens were analyzed qualitatively and quantitatively. All seven species of lichens contained copper, iron, zinc, nickel, manganese, and lead.   Cadmium and cobalt were detected in two species. Neither gold nor silver could be identified in lichen material with the tests used. A pollution model was developed and compared to field results. The simple dilution lichen metal content was related to the reciprocal of the distance from the pollution source. The lichens from the area could tolerate simultaneously high concentrations of several heavy metals that are known to be toxic to other plants. The mechanism of metal uptake was not clearly established. The study showed that lichens and other epiphytes are potentially the most useful indicators of heavy metal fallout around industrial plants.

Nriagu, J.O., Wong, H.K. and Coker, R.D., "Deposition and Chemistry of Pollutant Metals in Lakes Around the Smelters at Sudbury, Ontario", Environmental Science and Technology, Vol. 16, No. 9, 1982, pp. 551-560.

Since the smelters in Sudbury emitted about 2.6 tons of Ni, 2.6 tons of Cu, 0.7 tons of Pb, and 6.3 tons of Fe daily during 1977, the area around Sudbury represents a unique laboratory for studying the long-term impacts of airborne pollutants on aquatic ecosystems.   Organic matter constitutes 35-60 percent of the suspended materials in nearby lakes, but plays a minor role in the transport of metals to the sediments. Rates of metal accumulation in the sediments have been estimated typically to be 100-600, 50-300, 10-60, and 5-30 sq m/yr for Ni, Cu, Zn, and Pb, respectively.   Some of the lakes with pH values of 4.5 or less show no enrichment or accumulation of pollutant metals in their surface sediments, indicating that pollutant metals previously stored in the sediments have since been leached away.   This finding that the contaminated sediments can release substantial quantities of toxic metals to the overlying water must have interesting ramifications with regard to the limnological impacts of acid rains.

Ohta, T. et al., "Washout Effect and Diurnal Variation for Chlorinated Hydrocarbons in Ambient Air", Atmospheric Environment, Vol. 11, No. 10, 1977, pp. 985-988.

Factors reducing concentrations of several chlorinated hydrocarbons on both rainy and clear days were determined. Analyses of 1,1,1-trichloroethane ($C_2H_3Cl_3$), carbon tetrachloride (CCl_4), trichloroethylene (C_2HCl_3), and tetrachloroethylene (C_2Cl_4) in the ambient air were made every half hour using an automatic gas sample inlet system for about seven weeks. Factors influencing the concentrations of these chemicals are: amount emitted from the original source; variations in wind directions and velocities; rain scavenging or washout; and decomposition by direct or indirect photochemical reactions. The equivocal reduction of CCl_4 in daytime on clear days is compatible with its assumed low reactivity. The reduction of $C_2H_3Cl_3$ in daytime on clear days might be explained by the reaction of nitric oxide, nitrogen dioxide, and some other reactive species, such as ozone.

Oliver, B.G. et al., "Chloride and Lead in Urban Snow", Journal of Water Pollution Control Federation, Vol. 46, No. 4, Apr. 1974, pp. 766-771.

Snow and runoff samples from the winter and spring of 1972 from Ottawa, Canada, were analyzed for Cl and Pb. Chloride was analyzed using the mercuric thiocyanate and conductivity methods, and lead was analyzed by atomic absorption. Pb is a contaminant resulting from traffic and Cl from the salt used in street deicing. Chloride levels in runoff and waste water pose a burden on receiving waters but the levels are within potable water standards. Lead levels are tightly bound to snow particles and are ultimately deposited in sediments via runoff melt. Approximately 4100 kg of Pb are discharged in waste water effluent. A 30 percent reduction of Pb input into local water courses was effected by moving snow sites.

Page, A.L., Ganje, T.J. and Joshi, M.S., "Lead Quantities in Plants, Soil, and Air Near Some Major Highways in Southern California", Hilgardia, Vol. 41, No. 1, July 1971, pp. 1-31.

The lead content of 27 varieties of consumer crops and plants growing near some major southern California highways were colorimetrically ascertained. Amounts of Pb were also recorded for surface and subsurface soils and in suspended air particulates at or near the locations where the plant samples were obtained. Exposed tissues of plants grown very close to highways contained more Pb than similar tissues of plants grown some distance from the highways. This effect was most apparent at distances less than about 150 meters from the highway. Exposed tissues with smooth surfaces accumulated more lead than tissues with rough, hairy surfaces. The direction of the prevailing wind also significantly affected Pb concentrations in plants near highways; without exception, Pb in plants on the leeward side of the road exceeded that in plants on the windward side. A motor vehicle density of 35,000 vehicles/day also resulted in substantial Pb accumulations in plants. In soils and suspended particulates, lead concentrations were influenced by distance from highway and direction of prevailing winds. These results all demonstrate that the lead accumulations were caused primarily by aerial deposition and not--at least to any great extent--by absorption by the plant from Pb-contaminated soil.

Palmer, K.T. and Kucera, C.L., "Lead Contamination of Sycamore and Soil from Lead Mining and Smelting Operations in Eastern Missouri", Journal of Environmental Quality, Vol. 9, No. 1, 1980, pp. 106-111.

      Lead concentrations of both plant tissue (Platanus occidentalis L.) and soils (total and available) were determined for various sampling locations within a 10-km radius of two lead smelters and two lead mines located in eastern Missouri. Lead contents of washed foliage ranged from 1.3 to 1,120 ppm, while values in twigs, which were generally lower, ranged from 1.8 to 320 ppm. Nitric acid-extractable lead from soils ranged from 7 to 62,000 while 3 percent acetic acid-extractable lead varied from 1 to 20,400. The largest lead levels for both soils and plants occurred at the smelter sites. It appeared that both the chemical form, level and particulate size of atmospheric lead and the nature of soil lead could play an important role in the uptake of lead by plant tissues.

Patterson, C.C. et al., "Transport of Pollutant Lead to the Oceans and Within Ocean Ecosystems", NSF/INTL Decade of Ocean Exploration Pollution Survey Report, Jan. 1976, Savannah, Georgia, pp. 23-29.

      The extent of industrial lead pollution, atmospheric transport of industrial lead to the seas, dissolution of anthropogenic particle lead in sea water, and interaction of seawater lead with marine organisms are surveyed. About 40,000 tons/yr of aerosols are added by dry deposition and washout from the atmosphere to open oceans, and about 60,000 tons/yr are added by rivers and sewers. Contrary to common opinion, lead is not necessarily concentrated in higher organisms because it is transported to the ends of food chains.

Ratcliffe, J.M., "An Evaluation of the Use of Biological Indicators in an Atmospheric Lead Survey", Atmospheric Environment, Vol. 9, No. 6-7, June/July 1975, pp. 623-629.

      Atmospheric lead deposition was monitored in the vicinity of a battery factory by an integrated survey of the Pb levels in indigenous moss and grass samples and by a short-term study of weekly increments in Pb deposition at selected sites around the source using the mossbag technique. There is a highly significant correlation between the Pb content in indigenous mosses and grass. The Pb levels closely reflect the proximity to the source and the long-term wind pattern. The suspended mossbags adequately monitor the input to ground moss and grass given sufficient exposure time for slow accumulation rates. The spatial variation in mossbag Pb content after 1 mo correlates significantly with wind direction and speed and with calculated mean long-term ground level concentrations from stack emissions at each site. Only 40 percent of the temporal (weekly) variation in Pb accumulation in the mossbags can be explained by correlation with the wind direction and speed and rainfall of each site.

Ratsch, H.C., "Heavy-Metal Accumulation in Soil and Vegetation from Smelter Emissions", EPA-660/3-74-012, Aug. 1974, U.S. Environmental Protection Agency, Washington, D.C.

      Soil and plant samples were collected in the vicinity of a copper smelter and analyzed for their heavy metal content. The soil concentrations of lead, arsenic, cadmium, and mercury generally declined with increasing distance from the smelter, although concentrations at points 0.25-0.5 mi from the smelter stack were consistently lower than those at 0.5-1.0 mi due to plume rise and looping. Levels of arsenic, cadmium, lead, and mercury measured in the soil ranged from 7.3 to 457 ppm, 1.0-8.3 ppm, 9-743 ppm, and 0.2-11.0 ppm, respectively. Arsenic and cadmium were present at levels likely to be toxic. Vegetation levels of arsenic and cadmium also decreased with increasing distance from the smelter, but lead and mercury concentrations did not appear to be related to distance from the smelter. Cadmium and mercury appeared to represent a possible health hazard as constituents of leafy vegetables.

Reid, J.D. and Crabbe, R.S., "Two Models of Long-Range Drift of Forest Pesticide Aerial Spray", Atmospheric Environment, Vol. 14, No. 9, 1980, pp. 1017-1025.

Long-range drift and dry deposition of an aerial insecticide spray onto a forest are calculated from gradient transfer and Markov chain models to 80 km downwind of the spray line. Horizontal homogeneity and neutral stratification with a capping inversion are assumed. The spray droplets are seven-eights water which evaporates, and one-eighth non-volatile Fenitrothion; the initial drop-size distribution is highly polydispersed with mass mean diameter approximately 82 "mu" m.

Ruhling, A. and Tyler, G., "An Ecological Approach to the Lead Problem", Botan Notiser, Vol. 121, 1968, pp. 321-342.

In order to measure lead accumulation levels in Sweden, the concentration of lead in plants and soil was determined in transects across three large roads as well as at sites far from roads; uptake of lead by plant roots was also investigated in greenhouse experiments. Considerable accumulations were measured in plants and soil within a 50-100 m distance from the roads, although only a minor part of the lead emitted from cars settles and accumulates in the vicinity of the roads. To determine regional lead pollution, moss samples from distant roads in southern and central Scandinavia showed a distinct NE-SW gradient which decreased towards the NE. The lead concentration increased with precipitation and with decreasing distance from large population centers; in one area, much of the rain-precipitated lead originated outside of Sweden. Mosses appear to be useful indicator plants, since they naturally contain very little lead but accumulate it from the air to an unusual extent. A historical analysis of mosses collected in Skane from 1860 to 1968, showed marked increases in lead concentration in these samples, restricted to two distinct periods: the first occurred toward the end of the nineteenth century; and the second, during the last two decades. The first may be explained by the increased use of coal while the second coincides with the rapid increase in the combustion of leaded automotive fuels. Very low lead concentrations were measured in samples from northern Scandinavia.

Ruhling, A. and Tyler, G., "Sorption and Retention of Heavy Metals in the Woodland Moss Hylocomium Splendens", Oikos, Vol. 21, No. 1, 1970, pp. 92-97.

The capacity of Hylocomium splendens to absorb heavy metal ions from dilute solutions was investigated. The sorption and retention generally followed the order: copper or lead greater than nickel greater than cobalt greater than zinc or manganese. This order proved valid within a wide range of concentrations and independently whether the ions were supplied in pure or mixed solutions. The capacity of the moss tissues to sorb traces of Cu and Pb in the presence of comparatively large amounts of calcium, potassium, magnesium, and sodium was very great. This makes it probable that these ions, supplied with the precipitation, are almost quantitatively sorbed by the moss carpets. A large share of Ni, when present in precipitation, will also possibly be sorbed. A natural carpet of Hylocomium splendens showed a continuous uptake of manganese, iron, and calcium from young to old tissues, whereas the increase in the concentrations of minor heavy metals was balanced by the dry matter decrease through decomposition. The developing and mature layer below the moss carpet had no enriched heavy metals above the concentrations of the old moss tissues.

Shirahata, H., Elias, R.W. and Patterson, C.C., "Chronological Variations in Concentrations and Isotopic Compositions of Anthropogenic Atmospheric Lead in Sediments of a Remote Subalpine Pond", Geochimica et Cosmochimica Acta, Vol. 44, No. 2, 1980, pp. 149-162.

This report documents, in a manner analogous to the study of lead in Arctic snow strata, large increases in the concentrations of industrial lead in remote open-country atmospheres that have occurred since 1850. The study is based on the chronological record of precipitation and dry deposition inputs of lead deposited in pond sediments in a remote nondomesticated subalpine ecosystem which represents extensive uncultivated nonurban areas of North America. In the mountain pond studied, lead concentrations decreased 4-fold in going from the surface of sediments to layers 130 yr old. A corresponding change was noted in the Pb-206/Pb-207 ratio in the sediments, from industrial-like values of 1.18 near the surface to natural values of 1.24 at depth. Calcium, strontium and barium concentrations remained relatively constant with depth. The excess lead appeared to be of eolian anthropogenic origin. These findings prove that inputs of contaminating lead were nearly absent centuries ago in this location, but are present today in the ecosystem in amounts more than 20 times the natural inputs. As a consequence of these inputs, the lead concentrations in plants have been elevated 5-fold and in animals 50-fold beyond natural levels. Atmospheric concentrations of about 10 ng of lead/cubic meter are responsible for this increased contamination.

Siccama, T.G. and Smith, W.H., "Changes in Lead, Zinc, Copper, Dry Weight and Organic Matter Content of the Forest Floor of White Pine Stands in Central Massachusetts Over 16 Years", Environmental Science and Technology, Vol. 14, No. 1, Jan. 1980, pp. 54-56.

Lead, zinc, copper, dry weight, and organic matter content of the L, F, and H layers of the forest floor were measured in 10 white pine stands in central Massachusetts in 1962, and again in 1978, thus providing quantitative estimates of these parameters at two points in time 16 years apart. Forest floor material retained from a 1962 quantitative study of forest floor weight was compared with samples from the same plots in the same stands collected in 1978. Total lead content increased significantly. Average lead concentration increased in all layers, but the increases were not statistically significant primarily due to the dilution effect of the concurrent increase in the mass of the forest floor. The observed net increase in lead of 30 mg m^{-2} year^{-1} is approximately 80 percent of the estimated total annual input of this element via precipitation in this region during the 16-year period. There were no statistically significant changes in total zinc and copper content of the forest floor. Since the forest floor total dry weight increased significantly (42 percent), zinc and copper concentration decreased significantly. The results of this study emphasize the importance of determining both concentration and amount of tract elements in soil studies.

Soederlund, R., "Some Preliminary Views on the Atmospheric Transport of Matter to the Baltic Sea", Report No. AC-31, Mar. 1975, Stockholm University, Stockholm, Sweden.

The atmospheric fallout of various elements into the Baltic Sea was estimated. Components of the following elements were considered: cadmium, copper, lead, mercury, phosphorus, and zinc. Though the uncertainty in these estimates is relatively high, it is likely that the addition of nitrogen, phosphorus, and lead components in this way is of importance to the material balance of the Baltic Sea.

Swank, W.T. and Henderson, G.S., "Atmospheric Input of Some Cations and Anions to Forest Ecosystems in North Carolina and Tennessee", Water Resources Research, Vol. 12, No. 3, June 1976, pp. 541-546.

The atmospheric contributions of elements in precipitation and dry fallout to forest ecosystems were measured at two sites in the southern Appalachians. At both sites, relative mean annual concentrations of cations in bulk precipitation were in the order Ca greater than Na greater than K greater than Mg. At the Coweeta Hydrologic Laboratory in North Carolina, average annual inputs of Ca^{++}, Na$^+$, K$^+$, Mg^{++}, and NH$_4$-N in 1970-1973 were 4.88, 3.52, 1.62, 1.01, and 0.52 kg/ha/yr, respectively. At Walker Branch, Tennessee, the inputs of these elements during the same time period were 15.73, 3.89, 2.99, 2.94, and 2.37 kg/ha/yr. The inputs of NO$_3$-N, PO$_4$-P, and Cl$^-$ in 1972-1973 were 2.88, 0.19, and 8.53 kg/ha/yr at Coweeta. Inputs of NO$_3$-N and PO$_4$-P were 4.61 and 0.55 kg/ha at Walker Branch over the same period. One reason for differences in bulk precipitation chemistry was greater dry fallout for some cations at Walker Branch than at Coweeta. For both sites, dry fallout associated with local land use activities influenced seasonal concentrations of bulk precipitation except for Na$^+$, which appeared to be partly derived from marine sources. Total inputs of elements are considered to be minimum estimates for both forest ecosystems due to sampling and analytical methods.

Taylor, F.G., Hanna, S.R. and Parr, P.D., "Cooling Tower Drift Studies at the Paducah, Kentucky Gaseous Diffusion Plant", Cooling Tower Institute Annual Meeting, Houston, Texas, 1979, Oak Ridge National Laboratory, Oak Ridge, Tennessee.

The transfer and fate of chromium from cooling tower drift to terrestrial ecosystems were quantified at the Department of Energy's uranium enrichment facility at Paducah, Kentucky. Chromium concentrations in plant materials (fescue grass) decreased with increasing distance from the cooling tower, ranging from 251 $\pm$ 19 ppm at 15 meters to 0.52 ± 0.07 ppm at 1500 meters. The site of drift contamination, size characteristics, and elemental content of drift particles were determined using a scanning electron microscope with energy dispersive x-ray analysis capabilities. Results indicate that elemental content in drift water (mineral residue) may not be equivalent to the content in the recirculating cooling water of the tower. This hypothesis is contrary to basic assumptions in calculating drift emissions. A laboratory study simulating throughfall from 1 to 6 inches of rain suggested that there are more exchange sites associated with litter than live foliage. Leachate from each one-inch throughfall simulant removed 3 percent of the drift mass from litter compared to 7 to 9 percent from live foliage. Results suggest that differences in retention are related to chemical properties of the drift rather than physical lodging of the particle residue. To determine the potential for movement of drift-derived chromium to surface streams, soil-water samplers (wells) were placed along a distance gradient to Little Bayou Creek. Samples from two depths following rainstorms revealed the absence of vertical or horizontal movement with maximum concentrations of 0.13 ppb at 50 meters from the tower. Preliminary model estimates of drift deposition are compared to deposition-measurements. Isopleths of the predicted deposition are useful to identify areas of maximum drift transport in the environs of the gaseous diffusion plant.

Ter Haar, G.L., "The Effect of Lead Antiknocks on the Lead Content of Crops", Journal of Washington Academy of Science, Vol. 61, No. 2, 1971, pp. 114-120.

The contribution of lead naturally in the soil, lead in the air, and lead content of the soil from lead deposited from the atmosphere or added artificially to the lead content of plants is examined. Plants grown in a filtered atmosphere were compared with those of a neutral atmosphere. Most of the edible portions of the plant showed no effect from the increasing lead content in the air. Crops grown along a busy highway were analyzed. Edible portions of most compact crops, except soybeans and snap beans, showed no correlation between lead concentration and distance from the road. The inedible parts contained two to three times higher lead concentrations when grown near the road. The leafy portions of plants near busy highways contained higher concentrations of lead. An increase of lead content in the soil did not change the concentration in the plant. Rain was not a significant source of lead. Seasonal variations and stresses on the plant were studied for correlation.

Tucker, W.A. and Preston, A.L., "Procedures for Estimating Atmospheric Deposition Properties of Organic Chemicals", Water, Air, and Soil Pollution, Vol. 21, No. 1-4, Jan. 1984, pp. 247-260.

A rationale for procedures to estimate the climatological mean dry deposition velocity and precipitation scavenging ratio of organic chemicals is developed. Separate formulae are presented for vapor and sorbed fractions using their mass-weighted average to describe the deposition parameters for the total airborne concentration. Deposition of sorbed fractions is controlled solely by particle size, and methods for estimating the particle size are presented. Dry deposition velocity for the vapor fraction is derived by semiempirical analysis, and is found to depend on Henry's Law constant, molecular diffusion coefficients in air and water, and the chemical's reaction rate in water. Estimates of the dry deposition velocity for five chlorinated hydrocarbons are made and compared with observations. Estimates of the precipitation scavenging ratio are compared with observational data for 11 chlorinated or aromatic hydrocarbons.

Tyler, F. and Westman, L., "Effects of Heavy Metal Pollution on Decomposition in Forest Soils--VI--Metals and Sulfuric Acid", May 1979, National Swedish Environment Protection Board, Stockholm, Sweden.

The aims of this study are to determine to what extent soil biological processes are influenced in the coastal region of Vaesterbotten, northern Sweden, by deposition of metals and acid from a large smelter, Roennskaersverken. Elevated levels of several heavy metals in the topsoil of forested areas are measurable at least 70 to 80 km from the main source. The pollution has a measurable influence on important biological reactions in conifer forest soil over an area of at least 2500 to 4000 km^2, particularly evident northwards and southwards from Roennskaersverken. According to statistical evidence, pH and heavy metal contents of the forest moor account for a considerable part of the variability in soil respiration rate and phosphatase activity. Of the main polluting metals (Cd, Cu, Zn, Pb, As) cadmium and copper have the highest simple correlation coefficients with the soil biological variables. However, soil pH contributes roughly as much as all the metals together to account for the biological variability of the soil reactions studied, according to stepwise regression analysis. The variation in pH will partly be explained by natural differences between sites and partly by soil acidification due to deposition of acid pollutants. The mean level of pH-KCl is at least 0.2 units lower than in comparable sites of central Sweden.

Ward, N.I., Brooks, R.R. and Reeves, R.D., "Effect of Lead from Motor-Vehicle Exhausts on Trees Along a Major Thoroughfare in Palmerston North, New Zealand", Environmental Pollution, Vol. 6, No. 2, Feb. 1974, pp. 149-158.

The effect of lead from motor vehicle exhausts on trees growing along a busy thoroughfare in Palmerston North, New Zealand, was investigated. Analysis of tree samples (leaves, bark, and trunk cores) and of soils, showed that the distribution of emitted lead was influenced by the direction of the prevailing wind. Lead levels were higher on the sides of trees facing traffic. Measurements of lead concentrations in leaves, bark, and soils showed considerable accumulation in the vegetation at distances of about 5 m from the main traffic movement. Lead levels increased from ground level up to 1-2 m, then decreased.

Welch, W.R. and Dick, D.L., "Lead Concentrations in Tissues of Roadside Mice", Environmental Pollution, Vol. 8, No. 1, Jan. 1975, pp. 15-21.

Soil samples and seven different tissues from deer mice (Peromyscus maniculatus) were collected along a major highway and analyzed for lead content to determine whether lead accumulation occurs differentially in various tissues of small mammals as functions of traffic density and distance from the highway. Lead concentrations in the soil were related to proximity to the highway but not to traffic volume. Accumulation of lead in mouse liver, kidney, and bone, but not in brain, lung, stomach, and muscle, were related to both concentrations near a highway with a traffic density of 38,000 vehicles/mi ranged from 2.70 ppm for brain and muscle to 106.0 ppm for bone tissue. Thus deer mice appear to be sufficiently susceptible to existing environmental lead levels near highways to manifest different lead accumulations in various tissues.

Zajac, P.K. and Grodzinska, K., "Snow Contamination by Heavy Metals and Sulphur in Cracow Agglomeration (Southern Poland)", Water, Air, and Soil Pollution, Vol. 17, No. 3, 1982, pp. 269-280.

Samples of snow were collected on three dates in 1977 and 1978 at 27 sites in open and forest stands east of the large steel mill at Cracow, Poland. Concentrations of heavy metals, light metals, and sulfate decreased with increasing distance from the source, but snow acidity increased. There was considerable variation in pollutant concentrations in snow between sites, sample sets, type of forest, and open vs. cleared areas. Average metal concentrations (in mg per liter) were: Cd, 0.29; Pb, 3.1; Zn, 8.7; Fe, 923; Mg, 78; and Ca, 325. The mean sulfate ion concentration was 17.25 mg per liter. Snow pH varied from 10 at the source to 5 at a distance of 25 km.

Zimdahl, R.L., "Entry and Movement in Vegetation of Lead Derived from Air and Soil Sources", Paper 75-18.3, Air Pollution Control Association Annual Meeting, June 15-20, 1975, Boston, Massachusetts.

Lead uptake and translocation by plants and its subsequent effects are reviewed. Lead is taken up by many plants primarily via their roots. Large amounts of lead are deposited on plant foliage, but most remains as a topical deposit. Nevertheless, foliar uptake is also a demonstrated fact. Soil lead levels above 1000 ppm are assumed to be required to cause observable plant effects. Environmental variables, plant age, and species are important determinants of lead uptake. Increasing soil lead availability increases plant uptake; whereas increasing soil phosphorus, organic content, and pH decrease plant uptake. Mosses are excellent indicators of lead from aerial sources since they only absorb minerals from precipitation and settled dust. Based on the data at present, airborne lead cannot be ignored as a hazard to crops. A protective mechanism in plants whereby large amounts of lead taken up by plant roots are immobilized by dictyosome vesicles and deposited in the cell wall appears to exist. A similar process may be operative throughout the plant.